BIOASTRONOMY – THE NEXT STEPS

ASTROPHYSICS AND SPACE SCIENCE LIBRARY

A SERIES OF BOOKS ON THE RECENT DEVELOPMENTS
OF SPACE SCIENCE AND OF GENERAL GEOPHYSICS AND ASTROPHYSICS
PUBLISHED IN CONNECTION WITH THE JOURNAL
SPACE SCIENCE REVIEWS

VOLUME 144

PROCEEDINGS

BIOASTRONOMY - THE NEXT STEPS

PROCEEDINGS OF THE 99TH COLLOQUIUM
OF THE INTERNATIONAL ASTRONOMICAL UNION
HELD IN BALATON, HUNGARY, JUNE 22-27, 1987

Edited by

GEORGE MARX

Department of Atomic Physics,
Eötvös University, Budapest, Hungary

KLUWER ACADEMIC PUBLISHERS

DORDRECHT / BOSTON / LONDON

Library of Congress Cataloging in Publication Data

IAU Colloquium (99th : 1987 : Lake Balaton, Hungary)
   Bioastronomy : the next steps / International Astronomical Union
Colloquium 99 ; edited by George Marx.
      p.   cm. -- (Astrophysics and space science library ; v. 144)
   Includes index.
   ISBN 9027727147
   1. Life on other planets--Congresses.   2. Interstellar
communication--Congresses.   I. Marx, George.   II. International
Astronomical Union.   III. Title.   IV. Series.
QB54.I18 1987
574.999--dc19   88-3998
                                                                    CIP

ISBN 90-277-2714-7

Published by Kluwer Academic Publishers,
P.O. Box 17, 3300 AA Dordrecht, The Netherlands.

Kluwer Academic Publishers incorporates
the publishing programmes of
D. Reidel, Martinus Nijhoff, Dr W. Junk and MTP Press.

Sold and distributed in the U.S.A. and Canada
by Kluwer Academic Publishers,
101 Philip Drive, Norwell, MA 02061, U.S.A.

In all other countries, sold and distributed
by Kluwer Academic Publishers Group,
P.O. Box 322, 3300 AH Dordrecht, The Netherlands.

Printed in The Netherlands

International Astronomical Union
Commissions 51, 40

International Academy of Astronautics

Hungarian Academy of Sciences

Roland Eötvös University in Budapest
(organizer)

SCIENTIFIC PROGRAMME COMMITTEE

J.C. Tarter (USA)
R.D. Brown (Australia)
A.C. Clarke (Sri Lanka)
P. Connes (France)
F.D. Drake (USA)
H. Hirabayashi (Japan)
A. Knoll (USA)
A.R. Martin (UK)
G. Marx (Hungary)
M. Papagiannis (USA)
C. Ponnamperuma (USA)
V. Troitskii (USSR)

LOCAL ORGANIZING COMMITTEE

G. Marx (Eötvös University)
I. Almár (Konkoly Observatory)
B. Balázs (Eötvös University)
K. Barlay (Konkoly Observatory)
V. Csányi (Eötvös University)
M. Varga (Technical University)
J. Papp, secretary (Eötvös University)

CONTENTS

Session 5

IS INTELLIGENCE AN INEVITABLE EVOLUTIONARY TRAIT?

Session 6

PROSPECTS FOR DETECTING TECHNOLOGICAL CIVILIZATIONS

Session 7

WHAT IF WE SUCCEED?

WELCOME

George Marx

vice-chairman, IAU Commission 51: Bioastronomy

People were and are busy in solving their immediate everyday problems. They do not think that our world is the best one among all the possible worlds. Experiencing the difficulties and mischiefs of terrestrial existence, philosophers of each age *dreamt* about people living beyond Earth. Lucretius, Giordano Bruno, Cyrano de Bergerac, Johannes Kepler, H.G. Wells used the dreamworlds beyond to express their criticism with respect to our society. The present era of high technology made science fiction especially popular, mainly because newspapers and television made the vocabulary and technology of astronomy and astronautics known to the public. Instead of the medieval ghosts and witches simpleminded persons are willing to see U.F.O.-s and E.T-s. The young people are eager to join imaginary space missions (by books and videogames), because they anticipate their future in such ventures.

As a result of the achievements of atomic physics, molecular biology and new technology, science can treat the great problems of the origins of stars, planets, elements, molecules, life in a more exact way. We may ask about the possibilities and we may dare to give *logical* answers to these questions. The search for alien life and intelligence is not only ivory tower science, not only romantic escapism from everyday's difficulties. Nowadays the mankind faces the global threats of population explosion, environmental pollution, runaway greenhouse effect, nuclear armament. Our future looks less certain than it did in more optimistic eras like at the end of the last century. Is intelligence a transient unstable phenomenon in the Universe, a dead end, like the exaggerated muscular size of the dynosaurus was? Or has intelligence an absolute evolutionary adventage, which enables man to think rationally: not only to anticipate, but to shape future? Can we survive technology?

To understand these problems, to explore their solutions, we have to develop global models. Global models, however, can be tested only by making comparisons with other planets, atmospheres - possibly by alien biospheres and cultures. "I LOVE E.T." - is printed on the T-shirts of this Colloquium. Extraterrestrial intelligence is a relevant question for science, because scientists are concerned about terrestrial intelligence. For this reason the topics of the IAU Colloquium 99 is not only of public and scientific, but - in a wider sense - also of political interest: Bioastronomy may offer a mirror in which we may learn to know ourselves, to understand our place and role on Earth.

In the name of the host institutions - the sponsoring Hungarian Academy of Sciences and the organizing Roland Eötvös University - I offer a most cordial welcome to you, who have come from different cultures but who have the destiny to coexist on a small planet.

FERMI'S QUESTION

Sir Francis Crick, the Nobel-laureate discoverer of the DNA structure recollects in his book "The Life Itself":

The Italian physicist, Enrico Fermi, was a man with outstanding talents. It was Fermi and his Hungarian friend, Leo Szilard, who directed the design and construation of the first atomic pile. Fermi was credited with asking famous questions. There is a long preamble to Fermi's questions, rether like a shaggy dog story. It goes something like this: The universe is vast, containing myriads of stars, many of them not unlike our Sun. Our own galaxy has perhaps 100 billions of stars and there are at least 10 billion galaxies and probably more. Many of these stars are likely to have planets, a fair fraction of these planets will have liquid water on their surface and a gaseous atmosphere made up of simple compounds of carbon, nitrogen, oxygen and hydrogen. The energy pouring down from the star will cause the synthesis of numerous small organic compounds, thus turning the ocean into a thin, warm soup. These chemicals will eventually join onto each other and interact in an intricate way to produce a self-reproducing system, a primitive form of life. The simple living things will multiply, evolve by natural selection and become more complicated till eventually active, thinking creatures will emerge. Civilization, science and technology will follow and before long they will have mastered the entire environment of their planet. Then, yearning for fresh worlds to conquer, they will learn to travel to neighboring planets and then to planets of neighboring stars, choosing for their colonization those with favorable environments. Eventually they should spread all over the galaxy, exploring it as they go. These highly exceptional and talented people could hardly overlook such a beautiful place as our Earth, with its ample supply of water and organic compounds, its favorable temperature range and all its other adventages. "And so," Fermi would say, coming to his overwhelming questions, "if all this has been happening they should have arrived here by now, so w h e r e a r e t h e y ?" It was Leo Szilard, a man with an impish sense of humor, who supplied the perfect reply to Fermi's rhetoric. "They are among us," hè said, "but they speak Hungarian."

THE HISTORY OF IAA SETI COMMITTEE

Iván Almár

vice-chairman, IAA SETI Committee

ABSTRACT. The International Academy of Astronautics is fostering since 1965 activities aimed at establishing communication with extraterrestrial intelligence.

Twenty seven years ago the first scientific CETI program i.e. Project OZMA was initiated by our present chairman, dr. F. Drake. Also in 1960, another historical event took place: the International Academy of Astronautics was founded by T. von Kármán, the world famous pioneer of rocketry and astronautics. Von Kármán was born here in Hungary. He was always a strong supporter of new and bold ideas in science and technology, as well as of international cooperation. No wonder that his academy, the IAA has, from the beginning, given a forum to those imaginative people who wanted to establish the scientific basis for communication with extraterrestrial intelligence.

CETI - the concept and the word itself - was introduced into the IAA by Prof. R. Pešek in 1965, when the Board of Trustees suggested holding a 3 day symposium on the subject, either during one of the IAF congresses, or separately [1]. Prof. Pešek was charged with starting this organization. After an all-round inquiry to 50 scientists (16 of the 26 responses supported the symposium) the Board of Trustees accepted the idea and formed a special study group with the following members:

R. Pešek (chairman), C.J. Clemedson, V.L. Ginzburg, A.G. Haley, E.B. Konecci, J.S. Shklovskii, F.L. Whipple (S.M. Beresford and F.I. Ordway joined later)

This group, which held its first and only meeting during the Madrid congress of IAF on 10th October 1966, should be considered as the first international body dealing with CETI, and as the progenitor of our present SETI Committee of the IAA as well. The study group was reorganized in 1967 as Organizing Committee with the following members:

R. Pešek (chairman), C.J. Clemedson, F.D. Drake, E.B. Konecci, B. Lowell, P.M. Morrison, F.I. Ordway, C. Sagan, J.S. Shklovskii, M. Subotowicz

It held its first meeting in 1967 during the Belgrade congress of the IAF.

After some years of negotiations, the planned symposium was replaced

by a Soviet-American Conference on CETI (Byurakan, 5-11 September 1971).
The IAA CETI Organizing Committee, at its 1971 meeting, cancelled the
planned symposium and decided to organize a half-day "CETI Review Meeting" during the next IAF congress in Vienna. It was a real milestone in
the history of the committee, which has since 1971 been always involved
in the selection of papers for subsequent IAF congresses. The review
meetings held regularly each year continue to give a unique opportunity
to outline CETI-research on a truly international basis. They have generally enjoyed a large audience, and have provided one or more papers for
publication in the official jornal of the academy - Acta Astronautica.
In 1979 a special issue of Acta Astronautica was devoted to the topic of
CETI.

Only 5 papers were presented in 1972, but the number of CETI papers
soon increased to the point where it proved to be necessary to have two
half-day sessions. The first session is usually devoted to papers discussing the fundamental scientific basis for SETI, the second focusing
more on the technology and strategies. Some papers of fundamental importance have been presented, such as the description of the SETI program
of NASA, new proposals of unconventional and microwave methods, parasitic
searches for extraterrestrial intelligence, the idea of a galactic belt
of intelligent life, etc. I still remember the exciting discussion between
Prof. Shklovskii and other CETI authorities in Prague (1977) on the strategy of searching for extraterrestrial civilizations.

This year (1987) we have $2\frac{1}{2}$ sessions for CETI, one of them devoted
entirely to the formulation of an international protocol for activities
following the detection of a signal from extraterrestrial intelligence.
The number of CETI papers increased each year (Fig. 1), demonstrating
the growing interest of scientists in this rather "exotic" topic. Since
the total number of papers presented in all sessions of an IAF congress
is essentially constant, CETI has obviously gained importance.

The organizing committee was renamed "Standing Committee" in 1974
and the following membership list was accepted:
R. Pešek (chairman), J. Billingham, C.J. Clemedson, V.V. Gogosov,
A.T. Lawton, G. Marx, C. Ponnamperuma, M. Subotowicz
The Committee has grown in size since then and has been strengthened by
the addition of several members actively involved in recent observational
programs. They come primarily from Section 1 of IAA, but qualified members of other sections, as well as others who may not be members of the
academy are also included [2]. Members are appointed to 3 year terms and
may be appointed for a second 3 year term.
The present Committee consists of
R. Pešek (honorary chairman and founder)
J. Billingham (chairman)
I. Almár (co-chairman)
J. Tarter (vice chairman)
and 28 members.
It is clear that the annual opportunities for interchange and discussion
afforded by the review sessions greatly enhanced the possibilities for
international cooperation. Such initiatives have been discussed from time
to time and reported to the Academy. The Committee formulated, as one of
its aims, to coordinate national CETI programs and to incite the interest

of the United Nations in CETI. In 1982 the United Nations published "The World in Space" based on the Unispace'82 meeting; the section on SETI was written primarily by members of the Committee.

The activity of our Committee was not restricted to the organization of the review meetings. The Committee

- sponsored the regular publication of a SETI/CETI bibliography in JBIS,

- prepared the annual report on CETI for submission to IAA,

- organized or cosponsored several international conferences on SETI (like a special evening session during the Toulouse congress of COSPAR last year, or the small educational SETI meeting held in December 1985 in Sri Lanka).

The concepts of the Committee are briefly summarized in a 5 year plan. A new scientific topic that should perhaps be addressed by the Committee in the future concerns the detection of other planetary systems. A highly technical international meeting on SETI instrumentation and technology is also planned. Discussions will be held with the International Institute of Space Law on the desirability of having an internationally agreed upon format and procedure for the announcement of the unequivocal discovery of signals from extraterrestrial civilizations, as well as on the need to identify critical regions of the microwave spectrum as being protected for SETI purposes in the future.

Since 1986 the IAA Committee has changed its name from CETI to SETI in order to express its growing interest in every possible kind of search after extraterrestrial civilizations. The Committee would like to act as a focal point for the initiation of cooperative ventures in SETI science and technology as a long term activity well into the future.

References

1. R. Pešek, Acta Astronautica 6 No.1-2 pp. 3-9 1979
2. Minutes of CETI Committee Meetings and Annual Reports

Number of papers presented

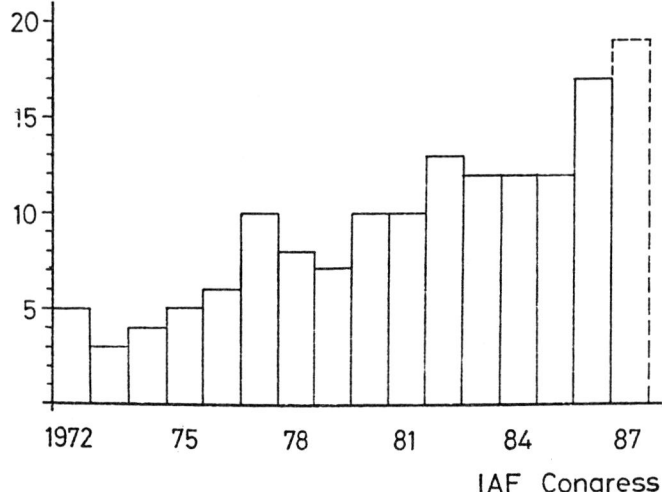

IAF Congress

COLLOQUIUM 99. IAU. BALATONFURED. HUNGARY.
JUNE 22-27. 1987

CLOSING REMARKS

Frank D. Drake

president, IAU Commission 51: Bioastronomy

After five days of meetings, we see once again that there is probably no more fascinating subject than the one we address here – Bioastronomy, the search for extraterrestrial life. The potential rewards for success in our search are beyond calculation. At the same time, we grope in the dark for ideas as to how best to pursue our search. As always, our human creativity has come forward to produce provocative, stimulating, and perhaps, correct ideas as to how we should pursue the searches of the future.

Also, as always, one of the greatest joys at a meeting like this is to meet with colleagues, to learn of new activities and new opportunities in distant lands. In the excitement and good will that his produces, we see a rehearsal for the excitement and progress which will accompany the detection of extraterrestrial life.

As always at meetings of Commission 51, we have seen human creativity and imagination at its best. There were new ideas here in abundance, and they will serve to stimulate us over the years which will pass before our next major meeting.

At the same time, we have seen substantial real progress in our search. On the one hand, particularly, in the search for extrasolar planets; on the other, in the search for radio signals for other civilizations.

Perhaps the most remarkable development at this Colloquium were the reports of many new projects aimed at detecting the planets of other stars. Not one, but several groups reported truly remarkable improvements in instrumentation which bring the detection of other planetary systems within reach. This is a development we only dreamed of a few years ago; now, not just one but several groups have demonstrated the accuracy necessary to accomplish this formidable task. Already we are glimpsing evidence of the existence of planets around the nearest stars. If indeed those planets are there, it means that planetary systems are abundant throughout the galaxy, and presumably throughout the universe. We wish our colleagues well in this challenging endeavor.

At the same time, others of us here have developed some very sophisticated observing equipment for the radio searches now underway, or soon to be underway. These searches for extraterrestrial intelligent radio signals will dwarf all previous searches put together, and will give us a real chance of finding other civilizations. We are gratified to see that, at last, both governmental and private funding organizations are beginning to provide funds for Bioastronomy at a level which is substantial, yet modest in comparison to the importance of the subject and its potential to benefit all the people of our planet. We are gratified to see that the most powerful technology of our era is being applied to one of its most challenging problems. In this, we see the talents of our species applied in a way which is to be applauded.

We thank our hosts from Hungary. They have made our stay here both comfortable and delightful. They have provided the finest facilities. They have provided the atmosphere in which we could conduct our work most productively. We have seen the best of both the old and new, here at Lake Balaton. We hope this tradition continues; if so, our next meeting will also be one to remember with very special pleasure.

Session 1

ASTROPHYSICAL CLUES
TO THE HABITABILITY

We know for certain that planet Earth is habitable. SETI raises immediately the question: are there other habitable environments in our Galaxy? If the answer is yes, where are they?

A. Katsuki emphasized that liquid water is a low-energy and low-entropy material needed as a fundamental substance for entropy elimination by circulation. The situation on Earth is just optimum. Gerald Soffen discussed Mars: Viking data indicated that Mars had had an aqueous phase, some impact craters showing a rosette form suggest the presence of water near its surface. Although the existence of life was undetected by the Viking landers, future Mars-missions are needed, aimed at the exploration of the aqueous history of the red planet. Steven Squyres revealed that observations as well as heat flow models raise the possibility that there is a deep liquid water ocean underlying a 10-30 km thick ice crust in the Jovian satellite Europa. There might be some geothermal activity on the ocean floor.

Water is one of the conditions for life, excess radiation is a danger. Ultraviolet radiation may be responsible for the lack of organic material on Mars. According to Béla Balázs intelligent life in our Galaxy is presumably limited to a narrow belt. The lifetime of civilizations is restricted by catastrophic events experienced during each crossing of a galactic arm. Leonid Marochnik came to the same conclusion: the co-rotation zone must be an exclusive place in a galaxy. In spiral arms the natural radioactivity background (due to frequent supernove explosions) is high enough to prevent any civilization. Several participants argued, however, that supernovae were not restricted to spiral arms and that life on Earth seemingly got accustomed to higher radiation in some exceptional regions on the Earth.

A number of authors previously suggested that the center of our Glaxy would be an obvious place for advanced civilizations. Antony Stark, however, emphasized that the galactic center was probably much more active millions of years ago. Finally V. Strelnitskij discussed the external (cosmological) and internal limits of the development of supercivilizations.

Ivan Almar

WHAT SHOULD WE LOOK FOR WHEN WE RETURN TO MARS?

G.A. Soffen
Goddard Space Flight Center
Greenbelt MD 20771 USA

Our moon and terrestrial planets have stimulated human interest by both the scientist and the general public. Mars, in particular, has generated voluminous romantic science fiction, and a vast body of technical literature. Thousands of articles and books, mostly in the last two decades have accumulated. This is largely due to the data returned from the dozen spacecraft sent between 1964 and 1975. Automated spacecraft data have been returned as late as 1982.

The intense public interest in Mars is in part due to the fascination with the question of possible indigenous life on the planet and of eventual human habitation; other bodies in the solar system are not as practicable or interesting. The Apollo experience tells us of the absence of life on the moon and the relative ease of human bases operating, albiet with high costs. Lunar bases will be largely an effort of macroengineering, and will require a long term commitment for sustaining scientific bases. The surface of Venus, at several hundred degrees kelvin, is too warm to sustain organic-based life. Mercury has been studied to a limited extent, and someday may come back into the public eye as we learn more about micro-environments.

Today, Mars holds the greatest public fascination for exploration within the terrestrial planets (other than the Earth). Recently, the U.S. commissioned a national effort to formulate a "bold agenda to carry America's civilian space enterprise into the 21st century". Holding numerous technical meetings and public forums, the Commission, headed by Dr. Tom Paine, published a Report last year entitled 'Pioneering the Space Frontier'. The Report recommended a "highway to space" over a period of the next twenty years, the development of the technological base for an Earth Spaceport. This to be followed by 15 years of constructing "a bridge between worlds", which will be a fully equipped, inhabited base on Mars about 2025-2030.

The Commission addressed the problems of costs, the need for national commitment and the need for international cooperation. The Report also deals with the benefits to humans, technological, scientific and societal. This heroic effort is the blueprint for the future of civilian space efforts. The details of who will do what, and how each step advances will change, but the general plan is logical and is certainly the widest distributed and the best view to date for the eventual human inhabitation of Mars.

It is appropriate now to consider what is known of Mars, our new home-to-be, what will become known within the next decade, and what is the next knowledge that we seek.

From the past data we have learned that Mars resembles the Earth in some ways, but is significantly different in other ways. Mars has an atmosphere

3

G. Marx (ed.), Bioastronomy – The Next Steps, 3–13.
© *1988 by Kluwer Academic Publishers.*

and has had an aqueous phase. Its orbit around the Sun resembles our own, requiring twice the time; rotation rate about the same as Earth. Besides the greater distance to the Sun, a key difference is the mass of Mars, about forty percent of Earth. This smaller mass allows more out-gassed atmosphere to escape, the atmosphere of Mars being only about one percent of Earth. Mars does not have a protective trapped radiation belt, it has no contemporary oceans, and there is no crustal dynamics as on the Earth.

The Surface. The surface of Mars is extensively covered with impact craters. By its close promimity to the Asteroid Belt, many of these are from meteorite bombardment. The surface also reveals a rich volcanic past. In particular, Olympus Mons, the largest known volcano, is young enough so that there are few craters on its shield. The subject of contemporary volcanism on Mars is still an open question. The surface has been heavily weathered by winds and dust storms. Valle Marineris is an extraordinary equitorial feature, thousands of km long and 6 km in depth, with dramatic patterns of geological slumping and dust deposition. The ancient flowing of liquid water on the surface is revealed by river beds and flooded plains. The presence of contemporary water on or near the surface is evidenced by three sources of data. Based on the infrared thermal measurements, we know that the polar regions are a mixture of water ice and CO_2 ice. The poles of Mars change during each Martian year, water is removed and redeposited from the surface by sublimation and condensation. Unique impact craters have been photographed that show a rosette formation suggesting the presence of a permafrost layer just below the surface. Photos taken on the surface in Mars summer and winter reveal a deposition of water on the rocks that are of volcanic origin.

The Atmosphere. The current atmosphere of Mars is much simpler than that of Earth; it is over 95% carbon dioxide with trace quantities of carbon monoxide, nitrogen, oxygen, argon, xenon, krypton, and water vapor. The curent isotopic ratios of the gases, compared with cosmic abundances, suggest that much of the nitrogen has been lost from the planet due to non thermal escape. Mars, having only forty percent of the gravity of Earth, cannot attract its atmosphere as strongly. This removal of most of its atmosphere has been significant in the different thermal and aqueous histories of Mars and Earth.

The chemistry of the surface of Earth is greatly affected by its oceans, precipitation, and complex photochemistry of the atmosphere. While we see that Mars had an early aqueous phase, this is not true in the recent geological periods. The current surface chemistry is dominated by solar radiation, simple oxidation and ion exchange. The Mars weather results in periodic dust storms that are capable of obscuring the surface. The fine material is mixed in the atmosphere, carried to all parts of the planet and is deposited; and blown about again by subsequent storms. This weathered surface is known to be as much as 20 cm deep in tested places.

A particularly important experiment was performed by the Viking missions on the surface. Small surface samples were analyzed for volatile compounds by a sensitive gas chromatograph/mass spectrometer and several growth and metabolism experiments were performed. The samples were taken from two sites several thousand km apart, and showed no organic material present. The sensitivity of the instrument was in the range of parts per billion detectivity for most organic compounds. This is important since most scientists anticipated that the surface of Mars should have received organic material from the incoming carbonaceous chondrites of the nearby astroid belt. The larger of these would penetrate to the surface, break up, and the fragments being weathered, the organic compounds in the dust would be carried by the winds, deposited, and detected in any samples taken from the surface. The complete absence of organic compounds (to the level of sensitivity) has led many scientists to conclude that the incoming organic material is destroyed by an oxidizing feature of the Martian surface. This is suggested by the results from one of the biological experiments. Small amounts of water vapor added to the surface samples resulted in the release of oxygen. The experimenters concluded that the Martian samples contained a small amount of iron peroxide, likely formed as the result of unfiltered ultraviolet solar irradiation reaching the surface. This oxidizing capacity, which possibly destroys the incoming organic material, is also important in preventing any current de novo organogenesis!

Recent laboratory results have demonstrated the relative ease in the formation of organic compounds from simple carbon sources using various kinds of energy. Many experiments have been done to help reconstruct the history of the Earth's early organic and subsequent biological periods. The early geological history of Mars is still unknown. Presence of life on Mars was undetected by three biological experiments performed on samples by the Viking landers! The absence of detectable organic compounds and the results of the three biological experiments have led many biologists to conclude that there is no present life on Mars. Other biologists do not regard the evidence as strong enough to permit this conclusion. They regard the answer still as ambiguous. A few biologists regard the Viking results as "positive" and believe that some of the data, coupled with photos from the surface, supports the hypothesis that there is life on Mars! The question of life on Mars, or any other planet in the solar system, is regarded by biologists as extremely important because we know only of the occurrence of one "origin of life" - that of terrestrial life.

Our knowledge of the chemistry of terrestrial life leads us to believe that most of the known terrestrial life is from one set of events that took place over four billion years ago that led to the "Origin". Recent discovery of life at great ocean depth has raised the question of a separate set of events that could have led to a separate Origin, but the data is still not conclusive. At this time, we know only of a common DNA-RNA protein-dominated life that appears to have a common heritage. Biologists and biological chemists mostly argue about which pathway led to the specific origin, and what happened on the early Earth; less about pathways on other planets.

Current Planned Missions: Phobos and Mars Orbiter. The U.S.S.R. plans to launch a spacecraft to Mars in 1988. The spacecraft will rendezvous with Phobos, the larger of the two moons of Mars and land a probe on the surface. Onboard will be instruments for measuring chemical composition, surface imagery, thermal mapping, physical properties, and radiometric characteristics. The spacecraft will examine large scale structures and measure possible seismic activity. While near Mars, the scientific objectives of the Phobos mission are:

Surface: Chemical composition; Mineral composition; Imagery of the planet; Thermal map; Radiometric characteristics.

Atmosphere: Chemical composition; Density profile; Temperature profile; Dust density; Dynamics; Evolution.

Ionosphere: Density profile; Dynamics.

Magnetosphere: Parameters of the Martian magnetic dipole, moment and orientation, plasma: Ion and energy composition; Three-dimension function Plasma distribution; Characteristic of plasma waves; Structure of magnetosphere and boundaries.

In addition, this mission will investigate the Sun and interplanetary medium. The scientific objectives are:

Corona and Upper Chromosphere: Large-scale structure; Active regions; Bright spots, Coronal holes.

Solar Flares: Precursors and transient sources; Explosion stage; Localization of the acceleration stage.

Solar Oscillations: Helioseismology; Spectrum of pressure and gravity modes.

Solar Activity Forecasting: Direct observations of solar activity; Spectral distribution of solar constant.

Solar Wind: Energy, ion and charge composition; Velocity distribution function.

Interplanetary Shocks: Diagnostics of shocks, Structure of interplanetary shocks; Shock near Mars.

Cosmic and Solar Gamma-Bursts: Localization; Spectra.

The current plan includes a synchronous orbit with Phobos in which the spacecraft will be lowered to within 50 m of the surface for 10-20 minutes. During this time the elemental and isotopic composition of Phobos' surface will be studied by active methods. "One of them is based on measurements of the soil substances evaporated and ionized by a laser beam focused following the laser ranger outputs. Then freely scattered ions are subject to mass spectrometric analysis in a reflectron with a retarding field. The ion time-of-flight from the study area to the spacecraft is recorded. The data are to be processed aboard the spacecraft".

The objective of another active experiment is to study the Phobos' surface soil composition (a layer of about 10A) and to determine elements implanted by the solar wind. The methodology includes the injection of an ion beam from the spacecraft and measurements of mass spectra of secondary ions knocked out from the surface layer. Mass spectra of secondary ions will be measured with a foil reflectron which ensures parallel recordings of ions with a quqdrupole mass spectrometer. The experiment is also planned to study topography, surface structure and electrophysical characteristics of Phobos' soil with a method of radio sounding from the spacecraft drifting at a low height over the Phobos surface. This is a very impressive mission and will contribute greatly to our knowledge of Mars.

Ths U.S. plans to launch the Mars Orbiter in 1992. This mission is largely devoted to geoscience and the climatology of the planet. The spacecraft will carry an array of scientific instruments that will be operated for at least one Martian year (687 earth days). The entire surface and atmosphere will be mapped. The scientific objectives of the mission are "the qualitative and quantitative determination of the elemental and mineralogical composition of the surface; measurement of the global surface topography, gravity field, and magnetic field; and the development of a synoptic data base...for determination of climatological conditions, including the role of volatiles in climate". The mission will be launched in September of 1992 onboard the shuttle and placed into low Earth orbit. After adequate separation, the Orbiter will orient itself to achieve proper injection initial conditions, and the upper stage is ignited lasting about 144 sec. After the motor burnout, the spacecraft commands separation from the upper stage and performs a small propulsive maneuver. The upper stage is also used for a post-separation collision/contamination avoidance maneuver.

At Mars, the spacecraft will be inserted into an elliptical orbit. Insertion will take place with an approach over the north pole; periapsis near the pole. After some time the orbit will be circularized for the mapping orbit which is sun-synchronized. The eccentricity of Mars' orbit causes the angle of the Sun out of the orbit plane to vary between -38^0 and -14^0. A "frozen orbit" geometry is used to reduce the orbital radius variations due to the rough gravity field. An altitude of 360 km has been chosen, based on spacecraft ground track and planetary protection considerations. At the end of the mission the spacecraft will be manuevered form the mapping orbit to a higher altitude quarentine orbit.

There are three "facility" instruments operated by teams of scientists and five instruments with principal investigators. The three "facility" instruments are:

Gamma Ray Spectrometer: To determine the elemental composition of the surface with a spacial resolution of a few hundred km, hydrogen depth in top tens of cm, atmospheric column density and arrival time and spectra of gamma-ray bursts.

Radio Science: To measure the atmosphere, profile of refractive index, number density, temperature, and pressure in both hemispheres, daily;

Seasonal variation of stratification; Thermal reponse to dust loading; Thermal structure of boundary layer; Height and peak density of ionosphere; Small scale structures.

Infrared Mapping Spectrometer: To produce mosaics of the surface to map mineralogical and chemical units especially for distribution of volatiles and structure of the regolith.

The five other instruments are:

Camera: Synoptic views of the Martian atmosphere and surface looking for atmospheric and surface changes, and possibly with very selective high resolution (1.4 m/pixel).

Magnetometer: Measure Mars magnetic field; Crustal remanent field; Solar wind/plasma interaction; Ionosphere.

Infrared radiometer: Thermal structure of atmosphere; Atmospheric dust loading; Distribution of water vapor; Spacial distribution of condensates; Seasonal pressure changes; Polar radiance.

Radar Altimeter: Topographic height; Slopes; Surface brightness temperature; Surface properties.

Thermal Emission Spectrometer: Composition of surface minerals, rocks and ices; Atmospheric dust; Water-ice and CO_2 condensate clouds; Atmospheric temperature, pressure, water vapor and ozone profile variations.

What is the next step?

The results of the U.S.S.R. and U.S. spacecraft, which will be making measurements and performing observations during the early part of the next decade, are likely to lead to a more detailed understanding of the Martian climatology, more insight into the chemistry of the surface and its evolution, interaction of Mars with the incoming solar radiation, its magnetic field, details about the Martian moons and some specific details of certain regions of interest on the surface. As with all scientific investigations the knowledge we gain will continue to pose more questions than answers. It is interesting and useful to speculate on some of the major lines of investigation that lie ahead.

What are the details and early geological processes of impact cratering, volcanism and tectonics that resulted in the Martian surface features?

The rates of erosion and weathering on Earth are rapid compared to other geological processes, so that small scale relief on the continents is completely dominated by the force of running water. By contrast, the absence of contemporary liquid water on Mars presents an opportunity to observe the accumulation of geological events and their subsequent changes on a planet whose surface features may be several billion old. The earliest impact cratering resembles the moon. To that has been added the results of extensive volcanism and the changes in elevation caused by internal forces. Superimposed on this is the continuous weathering by atmospheric conditions. Presently, we know only the gross features and must extrapolate to the smaller scale. The extensive crustal activity of the Earth constantly reworking the surface leaves little evidence of nonaqueous events of our geological past. On Mars the evidence is mostly intact. When the geological events are reassembled into a coherent history, we will be able to understand the physics of each of the forces that affect planetary formation and evolution for planets of this nature.

This geological work will extend for decades as the pieces of the puzzle are identified and placed in the proper perspective. We already have photos of features at the km. size and pictures of rocks and fine material with cm. resolution. What is needed are detailed studies of some of the general and special features to better understand some of the secondary processes and those events that obscure or obliterate the more subtle changes. This will require photographic information at all scales and in some cases down to the mm range and possibly into the microscopic. We will also require geochronological measurements to determine the sequence of events. For gross measurements this may be done with an automated instrument, but for the detailed work samples must be brought back into laboratories where exact conditions of analyses can be maintained. Geochemical data will be used to corroborate the chemistry and dynamics.

The results of terrestrial volcanism are altered by the forces of plate tectonics which rework and bury the outpouring lava. On Mars, the long-term events are preserved and geologists are offered an extraordinary opportunity for study of an uninterrupted process. Volcanism is a dominant theme of Mars. This must bring the internal volatiles to the surface each time there is new activity. Considering the youth of some of the volcanos, there may be evidence of this material locked in the subsurface material or frozen out in the ices of the polar regions.

Some regions of Mars have densely cratered areas that resemble the moon and are believed to be of the order of 3-4 billion years old. Other younger areas have very few craters. The plains close to the large Tharsis shields have very few craters and appear to be less than a hundred million years old. Most of these plains are in the northern hemisphere. Is there a real hemispheric difference? The character of the plains changes dramatically between 30^0 and 45^0 latitude in both hemispheres. At the lower latitudes the topography is sharp and features are crisp. The flow features are similar to those seen on the moon. At higher latitudes the plains are different; the topography is complex and do not relate to any known kinds of terrestrial or lunar features. These may be of volcanic origin or possibly modified by ice.

How has liquid water affected the planet?

The channels of Mars have been among the most controversial issues of all the literature. Most scientists now believe that the channels have been formed by liquid flowing water but this matter is still not completely resolved in the minds of some skeptics. Because of the instability of water at current Martian pressures (about 4 mb) and temperatures (about -40 to +10 degrees kelvin) water is believed to continuously sublime and freeze out---but does not accumulate in pools or lakes. Evidence of past flooding is strong suggesting that Mars has experienced a period of temporate climate during which liquid water flowed freely across the surface. The duration of these conditions must have been relatively brief, because a complex drainage system was never formed (unless erosion rates were extraordinarily low). Other alternatives to water erosion in channel formation have been considered, but haven't been sustained. These include erosion by lava, liquid, hydrocarbons, wind, ice, and faulting; none have gained sufficient credibility as the explanation. Most of those studying the problem still believe that at one time there was running water active on Mars.

The formation of these channels require climatic conditions different from tody. An interesting mechanism proposed for the massive release of water onto the surface is a glacial burst. These results,from melting an ice sheet at the base by volcanic heating. The water accululates under the ice and is periodically released as massive floods. Another idea is the subsurface heating of hydrated minerals. Water form deep-baked zones may migrate and accumulate just below the frozen surface and periodically break out to form flood features. Other ideas include aquifers where water could percolate to lower areas and eventually break through in a catastropic flood. There may be artesian springs that are cut by ice.

The transient water history of Mars is important to understand because of its role in chemistry and what it implies. In an aqueous medium many salts are soluble and many reactions will take place. Complex material like clay and very complex organic chemical reactions could take place. Presumably at the time of flooding there was no accumulation of the oxidizing iron peroxides seen today. The organic chemical reactions should have been possible. The extent and duration of the flooding leads to a massive search for evidence of the ancient chemistry.

What is the current volatile (CO_2 and water ice) dynamics and chemistry?

Carbon dioxide and water are the main components of exchange between the surface and the atmosphere. Both are important in the dynamic and chemical processes governing short term changes on the surface and the climate of the planet. Besides the vast quantitites of water at both poles most of the planetary water is frozen in the permafrost layer just beneath the surface. Rates of removal and deposition of CO_2 and water are important in constructing a model of the energy balance of the planet. Diurnal variations of water vapor differ at different latitudes and locations. The column density appears to be influenced by scattering from dust. The physics of this is still poorly understood. Local variations in the vertical abundance occurs where there are known topographic features characterized by abrupt elevation changes. This has been explained by the short radiative relaxation time of the Martian atmosphere and the expected low wind velocities. This relation between topography and water vapor content of the atmosphere is important in determining the extent of photochemistry and the patchy nature of the atmosphere.

Carbon monoxide and oxides of nitrogen have been detected in the atmosphere. This could lead to the notion that organic compounds might be found somewhere on the surface. These compounds in the presence of water vapor are highly reactive. Since no trace of organic material was detected at either of the landing sites, the explanation may be that the chemistry of the atmosphere is biased against organic formation, the rate of destruction is more rapid than the rate of formation, or there are small amounts being produced but only sequestered in "protective" regions. One such region might be the polar caps.

These frozen regions acting as cold traps might actually be a preserve of the long past history of the Martian organic chemistry. Just as our own polar ice cores are the sources of records of the Earth's ancient climate, the polar region of Mars should be a repository of the trace materials of the atmosphere. The poles are known to be layered and hence act as a sedimentary record. Chemical and age analysis of cores will be among the richest records of past events and the ongoing chemical processes. To perform this essential analysis, it is necessary to operate on the samples directly. No known remote sensing technique can accomplish this. There is great technical difficulty in obtaining this data since landing and operating at the pole is much more difficult than at an equitorial site. Obtaining terrestrial samples at significant depth has only been possible in the last few decades; to do so on Mars would appear to be very difficult! New ideas could change this.

What is the chemistry of the iron-rich clays?

The discovery of oxidizing material within the Martian soil was a major discovery of the Viking landers. The release of oxygen when in contact with water vapor has led to the conclusion that there are iron peroxides on the surface presumably formed by reactions energized by the solar radiation. The determination of inorganic elemental composition has led to the suggestion that the oxidized iron is part of a clay-like material resembling terrestrial smectite. Scientists have reconstructed the Viking experimental data using simulated assembled soils in the laboratory. These data support the notion that Mars has a complex inorganic material capable of hydration and catalytic activity. Understanding the surface chemistry will be one of the essential next steps. It is unlikely that Mars has a uniform chemistry. Even though the dust storms tend to homogenize the surface material, there will be imhomogenites as a result of different microenvironments, different minerals, different levels of hydration, different solar exposure and different pressures. It is interesting to note that the atmospheric pressure at the surface of Mars is near the "triple point" of water. This means that at the lowest altitude (where the pressure is in excess of 4.6 mb) water goes through a brief liquid state as it passes from freezing to boiling. Depending on the unique microgeography this could be transitory or last for a considerable period. As a thin film frost disappearing on a rock, the liquid state would be very brief. But nature is not made only of thin films. Vesicles formed in pitted rocks (of which there are numerous on the volcanic rocks of Mars) might be warmed from below (especially where the rocks are dark or have low albedo) and liquifaction might take place where the vesicle opening is plugged with ice. A small natural vessel of liquid water with dissolved CO, NO and various minerals could become a reaction flask that would entertain an organic chemist for a lifetime. While none of this has been detected, there are no results that precludes these ideas.

Where are all the organic compounds?

If the surface is litered with peroxides, the new material would be destroyed upon release to the wind exposed surface. Nevertheless, indigenous organic material discovered on Mars would dramatically change the general view of Martian prebiotic synthesis!

Another mystery associated with this oversimplified reconstruction is the general contradiction that the perioxides released in the Viking biology experiment should have been destroyed by the water vapor that was detected in the Mars atmosphere. Why was the peroxide discharged by water vapor in the experiment and not in its native state. The generally accepted explanation is in the small amounts of water vapor which is much less than the saturation in this experimental set-up. However, the Mars surface is exposed for very long times and might be expected to be free of the peroxide for parts of the year (maybe during the dust storms when less solar radiation reaches the surface. Indeed, we know that the Viking landing sites are areas of high water vapor since frost formation was photographed during the northern Martian winter by the Viking lander. This mystery needs cleaning up. Is the peroxide layer surficial? Would digging deeper uncover material of a different nature? How deep? What is below?

The Biological Issue!

Is there a consensus among biologists? The Viking biological experiments were performed by a very talented team of experts who worked for ten years designing the experiments. Carrying them out and interpeting the results (and with the world's biologists looking over their shoulders) the interpretations of the seven biologists range from absolutely negative to slightly positive, to at least one who feels that a biological explanation of the results is as plausable as a chemical explanation. Dr. Howowitz, in his recent book "Utopia and Back", is adamant. "Viking found no life on Mars, and just as important, it found why there can be no life. Mars lacks that extraordinary feature that dominates the environment of our own planet, oceans of liquid water in full view of the Sun; indeed it is devoid of any liquid water whatsoever". But, we know there was transient water over vast areas of the planet presumably when the surface pressure was higher (before nitrogen escaped) and it is not hard to imagine small recesses of liquid water. To an organism the size of a bacterium a droplet is the size of a lake - cupful of water is the size of an ocean. Scaling becomes very important.

Dr. Levin, whose experiments showed release of CO_2 from soils to which nutrients were added believes that biological activity is **just as feasible** an explanation for the oxidation of the media used as is a chemical cause. Most biologists feel that the results of this first set of metabolic experiments are indecisive. They believe that no life was detected, but that we cannot state for certainy that we have exhausted the possibility. A critical data point was the absence of organic in the soil. It is difficult to imagine a biological ecology with no organic residue. But what if there were small regions which had some organic material? Would that have changed our views?

Biologists are sufficiently used to exotic solutions to biological organisms which have evolved to deal with extremely hostile environments. Living organisms are found under extreme conditions, cope with meager energy sources, and are able to repair massive damages. Biologists are cautious about sweeping statements that rule out remarkably complex chemistry or machinery. We still cannot build a computer as complex as the brain, synthesize microstructures as small as a bacterium or pack information as tight as on a DNA molecule. To dismiss biology once and for all on Mars with our meager data may be permature. What is not clear is what experiments may be performed by automated robots before samples are returned to the earth or laboratories are set up on the Martian surface. Perhaps a microscope or a more extensive search for organic materials, new and sensitive techniques.

Other important questions.

There are numerous other scientific areas of inquiry that will be pursued. The planetary interactions with the Sun's radiation and the solar wind, internal structure detected by active seismology and chemical analysers, study of the physics of the dust storms, penetration down to the permafrost, to name a few.

Finally, what knowledge is needed to establish a scientific base? We need to determine the correct location, which hemisphere, at what elevation and what latitude. What geological features will determine this? We need to know local weather conditions and be able to forecast the dust storms. We will need to understand the engineering of life support systems, closed ecological systems, and deal with the human factors and physiological needs of extended stays.

The sequence of missions is clear and has been documented. After the new orbiters, there will be a return to the Martian surface by automated robots. These will be used to move over long distances. Samples will be returned to terrestrial, or perhaps Space Station laboratories, and finally human exploration of the Martian surface and its moons will begin in the first two decades of the next century.

WATER: AN ABSOLUTE REQUIREMENT FOR LIFE

A. Katsuki

Department of Physics, Shinshu University

Matsumoto 390 Japan

ABSTRACT. The characteristic of life and the situation of the Earth are
analyzed from the viewpoint of entropy. Living being is an open system,
which takes low-entropy materials from and puts out high-entropy ones
into the environment; for living beings to be able to maintain their
lives, the environment itself has to be kept in a low-entropy state. As
the environment of the living beings, the Earth has a mechanism of entropy elimination, the <u>circulation of water.</u> Living beings have two
kinds of low-entropy materials, clean liquid water of low energy, and
carbohydrate of high energy. <u>For all living beings, the "low-energy and
low-entropy" material is liquid water.</u> Water is absolutely important
for life not only as the place where the elementary bio-process proceed,
but also as the low-energy and low-entropy material indispensable for
entropy elimination from the concerned living system.

1. LIFE, ITS ENVIRONMENT AND ENTROPY

The shortest way to understand extraterrestrial life, even if it
were somewhere in the Galaxy, is thinking about the life on the Earth ;
the key concept for understanding life is entropy.

From the viewpoint of entropy, the phenomena presented by living
beings, for example maintenance and growth of the individual, propagation and so on, seem to be unaccountable, because they seem to be contrary to the law of increasing entropy. A living being, though it is
very special, is one of the general modes of existence of materials, and
must be subject to the laws governing these general modes; a living being, then, must not be an exception to the general law of increasing entropy. If this is the case, why does a living being appear to be an exception to this law ?

Schrödinger characterized a living being by saying of it that "it
feeds upon negative entropy".[1] This statement is not in fact an answer
to the above question, but is a question much the same as the one above,
while suggesting that the key concept to understanding life is entropy.

Again, why does a living being seem to be an exception to the law
of increasing entropy ? It is because of the fact that a living being
is an open system, which takes low-entropy materials from the environ-

15

G. Marx (ed.), Bioastronomy – The Next Steps, 15–19.

ment and puts out high-entropy ones into the environment; this is the real process of Schrödinger's idea of feeding upon negative entropy.

For living beings to be able to maintain their lives, the environment has to be kept in a low-entropy state; if not, they could not take low-entropy materials from the environment. For the environment to remain in a low-entropy state, the environment itself has to have a mechanism to eliminate entropy. For living beings existing on a planet, the environment is the planet itself. Let us call a planet on which living beings exist a "living" planet.

In order for a planet to be a "living" one, it must have a mechanism to dissipate entropy. Generally, radiation from the centred star as a heat source with a high temperature and space as a heat absorber with a low temperature offer the possibility to eliminate entropy from the concerned planet. In the case of the Earth, these are the Sun's radiation and the space surrounding Earth. But this is only a possibility, and a special mechanism is needed to realize it. Indeed, as far as the Sun's radiation is concerned, the situation is essentially the same among the Earth, Mars, Venus and the Moon. What is the special mechanism which distinguishes the Earth from the others and makes possible to eliminate entropy from the Earth ? It is the circulation of water.

2. EARTH AND WATER CIRCULATION, THE MECHNISM OF ENTROPY ELIMINATION

The mechanism of entropy elimination of the Earth due to water circulation is as follows : Absorbing heat Q at the ground (about 300K) water becomes vapour, then goes up to the upper atmosphere, emitting the heat there (250K) as infrared radiation to space, becomes again liquid (or solid) water, then returns as rain (or snow) to the ground. By this process, a net amount of entropy $Q/250 - Q/300$ is eliminated from the Earth.[2] The Earth satisfies miraculously the very severe conditions necessary for entropy elimination through water circulation to be realized.

One of the necessary conditions concerns the temperatures of the upper atmosphere and the ground. It is very important for entropy elimination due to water circulation that the temperature of the ground T_G should be sufficiently higher than the temperature of the upper atmosphere T_U and should be such a temperature that water exists as liquid and easily vapourizes. The physical properties (albedo and so on) and the astronomical situation in the solar system of the Earth and the balance of heat have determined the temperature T_U to be about 250 K ; the existence of atmosphere has made T_G sufficiently higher than T_U due to greenhous effect, so that water can exist as liquid on the ground.

The other necessary condition concerns keeping water on the Earth in a liquid state. Mean velocity of H_2O molecules of vapour at 300 K, 645 m/s, is much lower than the first astronautical velocity, 7.9 km/s. This means that Earth is sufficiently heavy for water to be kept on it. On the other hand, molecules with this velocity are able to go up to a height of 21 km; this is enough to reach the upper surface of the atmosphere. This means that Earth is sufficiently light to permit the rising of vapour to a height where the temperature T_U is sufficiently lower than T_G. The fact that ice is lighter than water has great importance for the circulation of water to proceed. When water has been solidified due to lowering of temperature, ice is easily able to return to a liquid state by receiving heat from the Sun, as it floats on the surface of water. If ice were heavier than water, then when it froze, it would sink to the bottom and would not be melted by the Sun's radiation, so early in the history of the Earth all the water would have frozen as cold rock and only a thin layer of the surface would have melted when warmed by the Sun. There would not be abundant liquid water on the Earth and there would be no water circulation.

The high latent heat of vapourization of water makes the elimination of entropy by water circulation effective.

By generalizing the situation of the Earth, we can summarize the necessary conditions for a planet on which life exists as follows: it has to have a mechanism to eliminate entropy from it; there must be:

(1) Centred star as a heat source with an appropriate temperature and appropriate size at an appropriate distance, to make the "living" planet able to take in heat at a higher temperature and disperse it at a lower temperature into space.

(2) An appropriate "fundamental substance" to embody the mechanism above.

(3) A circulation system with the "fundamental substance" as working substance.

For the Earth, (1), (2) and (3) are the Sun, water and water circulation.

If it is most effective for the elimination of entropy from the planet to use the difference in entropy between gaseous and liquid states of the "fundamental substance", then it has to have a high vapourization heat, to be liquid under the temperature and pressure at the ground level of the planet and to be easily vapourizable under the prevailing conditions.

In order for the vapourized "fundamental substance" to be able to reach the upper atmosphere, an atmosphere heavier than the vapourized "fundamental substance" should exist on the surface of the planet; and the atmosphere should be essentially transparent for the radiation from the centred star in order to make the temperature at the ground of the planet sufficiently higher than that in the upper atmosphere where emission of heat(=elimination of entropy) takes place.

The planet has to be sufficiently heavy in order to keep the vapourized "fundamental substance" on it; it has to be sufficiently light in order for the vapourized "fundamental substance" to be able to reach the upper atmosphere, where the temperature is suitably lower than that at the ground.

When the "fundamental substance" is solidified due to lowering of temperature, in order for it to return easily to a liquid state by receiving heat from the centred star, the solid "fundamental substance" should float on the liquid one, i.e., the density of the solid "fundamental substance" should be lower than that of the liquid one.

For most celestial bodies, these conditions seem generally to be too severe to be realized. The situation of the Earth and the unique physical properties of water are, miraculously, just right for water circulation.

3. LIFE AND THE TWO KINDS OF LOW-ENTROPY MATERIALS

To live, a living being must keep itself in a low-entropy state or construct its own organs of various materials. For example, living beings synthesize proteins from amino-acids. In this process, roughly, the assemblage of concerned nitrogen atoms decreases in entropy because of reduction in range of distribution — initially they are distributed over a wider range, as in amino-acids, and finally they are concentrated in a narrower region, as in proteins. On the other hand, any process must involve an increase of entropy in general, so there should be an associated process to compensate for the decrease in entropy of the assemblage of nitrogen atoms. This process should be one in which certain low-entropy materials become high entropy and leave the system. What are the low-entropy materials that compensate for the entropy reduction of the living system in biochemical reactions and enable living beings to live?

Living beings take in two kinds of low-entropy materials: clean liquid water with low energy and carbohydrates with high energy; to maintain themselves in a low-entropy state, living beings take in water and carbohydrates with low entropy and discharge waste matter with high entropy (vapour, "soiled" water, CO_2, etc.) and heat.

Clean liquid water is able to function as low-entropy material because it enters a high-entropy state by dissolving waste products or by absorbing heat and goes out of the living body into the environment as soiled water or as vapour.

Carbohydrates are able to function as low-entropy material because they enter a high-entropy state by oxidation and by generating heat and go out of the living body into the environment as CO_2 and as heat carried off by vapour. Oxygen is needed for this process, i.e., to make carbohydrates function as low-entropy material.

The thermal properties of liquid water and carbohydrates contrast with each other. When liquid water functions as low-entropy material, it absorbs heat and becomes vapour, so it may be called "low-energy and low-entropy" material, while when carbohydrates function as low-entropy material, they generate heat by oxidation, so they may be called "high-energy and low-entropy" material. The heat generated by the oxidation of carbohydrates is essentially absorbed by water in organs.

Some living beings do not need oxygen to live. In such cases, the "high-energy and low-entropy" materials are carbon compounds which function as low-entropy materials, for example, by fermentation needing no oxygen (not by respiration needing oxygen). For all living beings, the "low-energy and low-entropy" material is liquid water.

Also, in regard to the progress of the reaction from a low-entropy to a high-entropy state, water and carbohydrates contrast with each other. The vapourization of water actually acts against the vapourization (by a fall in the temperature of the water due to loss of latent heat and by increasing vapour pressure due to vapourization), i.e., the vapourization of water has a negative feedback mechanism.

Oxidation of carbohydrates generates heat, and the elevation in temperature accelerates the oxidation, i.e., the oxidation of carbohydrates has a positive feedback mechanism. Rapid growth or rapid prop-

agation of living beings is possible because the oxidation of carbohydrates has a positive feedback mechanism. What prevents an excessive
progression of oxidation is the vapourization of water.

The delicate process of life has been made possible by the fact
that living beings use two kinds of thermally antithetical low-entropy
materials.

Water circulation regenerates clean liquid water and photosynthesis
regenerates carbohydrate. For photosynthesis to proceed, several hundred moles of water, which do not appear in the chemical equation, are
needed per the molar chemical equation $6CO_2+6H_2O \rightarrow C_6H_{12}O_6+6O_2$. The glu-
cose produced is a material with high energy and low entropy. The Sun's
light fixed as chemical energy has made it high energy, while the water,
vapourized and escaping through stomata, has made it low entropy. The
role of water is very important for carbohydrate to be low entropy in
photosynthesis.

4. CONCLUSION

Water, especially liquid water, is absolutely important for life
not only as the place where the elementary bio-process proceed, but also
as the low-energy and low-entropy material indispensable for entropy
elimination from the concerned living system.

When we discuss extraterrestrial life, we must consciously ask what
does function as the low-energy and low-entropy material (corresponding
to water for the life on the Earth), what does as the high-energy and
low-entropy material (corresponding to carbohydrates) and what is the
mechanism to eliminate entropy from the planet on which the life exists.

Concerning the conditions of existence of living beings, we must
not view the conditions (temperature, pressure, humidity, chemical components and so on) of the immediate environment alone, but in the global
framework of the hierarchical multiple structure of entropy elimination
through which water circulation runs.[3]

References

1) E. Schrödinger : What is Life ?——The Physical Aspect of the Living
Cell, Cambridge Univ. Press, 1944.
2) The essential idea of entropy elimination from the Earth through
water circulation was first put forward by Tsuchida, in embryonic form
in 1976 and in more extended form in 1978. A. Tsuchida : 'Limitation
of Nuclear Fusion and Physics of Resources' ('核融合の限界と資源物理
学') Proc. Phys. Soc. Japan (日本物理学会誌) 31 938 (1976) (in Japanese); A. Tsuchida : 'An Attempt at a Physics of Resources' ('資源物理
学の試み') Kagaku (科学—Science) 48 76, 176, 303 (1978) (in Japanese); A. Tsuchida : Introduction to Physics of Resources (資源物理学
入門) Nippon Hoso Shuppan Kyokai, Tokyo 1982 (in Japanese).
3) More detailed discussion, see : A. Katsuki : 'The Earth, Living
Beings and Entropy' J. Fac. Sci. Shinshu Univ. 21 71 (1986). Reprint
is available for request to the author.

EUROPA: THE PROSPECTS FOR AN OCEAN

R.T. Reynolds, C.P. McKay, J.F. Kasting

NASA Ames Research Center
Moffett Field CA 94035, USA

S.W. Squires

Radiophysics and Space Research, Cornell University
Ithaca, NY 14853, USA

ABSTRACT. Tidal dissipation in the satellites of a giant planet may provide sufficient heating to maintain a liquid water ocean below a thin ice layer. In our own solar system, Europa, one of the Galilean satellites of Jupiter, may have such an ocean. Both theoretical calculations and certain observations support its existence, although proof is lacking. The putative ocean would probably have temperatures, pressures, and chemistry conducive to biologic activity. However, the environment would be severely energy limited. Possible energy sources include transient transmission of sunlight through fractures in the ice and hydrothermal activity on the ocean floor. While temporary conditions could exist that are within the range of adaptation of certain terrestrial organisms, origin of life under such conditions seems unlikely. In other solar systems, however, larger satellites with more significant heat flow could provide environments that are stable over an order of aeons and in which life could perhaps evolve.

1. INTRODUCTION

As far as we know, Earth is the only planet in our Solar System that supports life. It is natural therefore that our understanding of life as a planetary phenomenon is based upon Earth-like planets. In particular, estimates of the probability that life exists on planets orbiting other stars have been based upon the concept of habitable zones for earth-like planets /1,2/. Within this classical habitable zone the only source of heating for the planetary surface is sunlight. This restricts the zone to the inner region of a solar system.

However, with the spacecraft exploration of the outer reaches of our own solar system it has become evident that some of the satellites of the giant planets can have appreciable generation of internal heat due to tidal dissipation of orbital energy. Most dramatically, Io, the Galilean moon closest to Jupiter, has been observed to have extensive and violent

G. Marx (ed.), Bioastronomy – The Next Steps, 21–28.
© 1988 by Kluwer Academic Publishers.

tidally-driven volcanic activity. Tidal heating on Europa may well maintain a liquid ocean below a relatively thin solid ice crust. There is possible evidence for former or ongoing tidal heating on Enceladus in the Saturn system and Uranus' Ariel and Miranda as well.

In considering the satellites of outer planets as a site for biologic activity, it is important to recognize also that they can be quite large in size. Ganymede and Titan are larger than the planet Mercury, and Callisto is not much smaller. Titan, the largest natural satellite of Saturn, has a 1.5 bar atmosphere composed primarily of N_2, with some CH_4. Over nine organic molecules have been detected in Titan's atmosphere.

The combination of these characteristics could result in a habitable satellite in the outer part of a solar system. A thin ice crust above a liquid ocean could, if fracturing occurred, produce a small transient region with both physical conditions and sufficient sunlight to support photosynthetic life. Geothermal energy could also make a significant contribution. While an environment in which life could actually evolve is unlikely in our solar system, other solar systems could have considerably more optimum conditions. Furthermore, if a large natural satellite could form with an atmosphere and undergo sufficient tidal heating over its history that its surface would be maintained above freezing, life could conceivably evolve there.

In this paper, we examine the prospects for aqueous environments on satellites in the outer solar system. We consider specifically the case of Europa, and discuss the possibility that it has a tidally heated ocean beneath the observable layer of ice. We then briefly consider some possible habitats that might arise in other solar systems.

2. TIDAL HEATING

Tidal forces arise in a satellite because the gravitational field of the planet varies across the body of the satellite. Regions on the satellite which are nearer the planet feel a stronger attractive force than those that are further from the planet. The effect of these tidal forces is to distort the shape of the satellite. In general, the tidal distortions will move across the satellite, causing it to flex and heat by the dissipation of non-elastic strain energy. Conservation of angular momentum and energy in such a process result in the eccentricity of the orbit being damped to approach a circular orbit, and the rotation period being forced toward that of a synchronous orbit. A synchronous orbit is one in which the satellite rotates with the same period as its orbital period about the body, such that the same face (and the tidal bulge) always face the planet. This tidal heating process is self-limiting and the heating will cease when the circular orbit and synchronous rotation conditions are reached. For many of the satellites of the outer solar system, however, mutual interactions among the satellites act to maintain non-circular orbits (although they do achieve synchronous rotation). The prime example of this phenomena is the so called "Laplace resonance" between Io, Europa, and Ganymede which maintains relatively high values for the eccentricities. Physically, the energy dissipated by the tides comes from the orbital energy of the satellite and the rotational energy of the planet. For a more complete discussion, the reader is referred to Cassen *et al.* /3/.

As a result of the comparatively close spacing of natural satellites about the giant planets, as seen in the Jupiter, Saturn, and Uranus systems, and the fact that the inner satellites evolve to larger orbits faster than the outer satellites, thereby overtaking them, resonances often occur among these objects and play a major role in the dynamical evolution of these systems. Thus significant eccentricities and strong tidal heating are expected to occur in such systems.

For a satellite of uniform density ρ and uniform rigidity μ, in an orbit with an eccentricity much less than unity, $e \ll 1$, the total rate of tidal dissipation of energy $\dot{E}_T$ is expressed by /3,4/

$$\dot{E}_T = \frac{21}{2} k_2 \frac{G M_p^2 R_s^5 e^2 n}{a^6 Q} \tag{1}$$

where G is the gravitational constant, M_p is the mass of the planet, n is the mean angular velocity of the satellite in its orbit, a is the semi-major axis of the satellite's orbit, R_s is the radius of the satellite, Q is the dissipation factor, and k_2 is the potential Love number of the second degree /3,4/ given by

$$k_2 = \frac{3/2}{1 + (19\mu/2\rho g R_s)} \simeq \frac{3\rho g R_s}{19\mu} \tag{2}$$

where g is the satellite's surface gravity. Noting that the mean angular velocity in a Keplerian orbit is given by $n = G M_\rho a^{-3}$, and expressing the tidal dissipation in terms of a surface heat flow, H_T, we have,

$$H_T = \frac{\dot{E}_r}{4\pi R_s^2} \simeq \frac{21}{38} G^{\frac{5}{2}} \times \frac{\rho^2 R_s^5}{\mu Q} \times e^2 \left(\frac{M_p}{a^3}\right)^{\frac{5}{2}} \tag{3}$$

The right hand side of Eq. 3 has been grouped into three sets of terms. The first term is a constant, the second term contains variables that depend on the radius of the planet and its internal composition, and the third term contains the variables that depend on the orbit of the satellite about the planet and the mass of the planet.

The tidal heat flow is strongly dependent on the size of the satellite. The increase in radius of a satellite from that of Io to that of Ganymede or Titan results in an increase in tidal heat flow by a factor of over six, other things remaining equal. If Io had an eccentricity as high as Europa's it would increase the heating by a factor of four. Both μ and Q are poorly known and highly variable. The mean heat flow for Io has been determined observationally to be $\simeq 1500$ erg cm^{-2} s^{-1}. For a review of the observations as well as a discussion of heating in a non-homogeneous body, see Cassen et al./3/.

As discussed above, satellites in orbit about giant planets can be expected to have nonzero eccentricities. What may vary the most from one system to another is the mass of the planet and the orbital distance of the satellites. The innermost satellites of a given planet are the most likely candidates for tidal heating. This is consistent with observations of the saturnian and uranian systems, as there is some evidence for a tidally heated ocean on Enceladus /5/ and Miranda and Ariel may also have experienced extensive tidal heating /6,7/. Signficantly higher heat flows might be possible for large satellites around giant planets in the range of $1 - 10$ Jupiter masses.

3. AN OCEAN ON EUROPA?

The images of Europa taken during the Voyager flybys of the Jovian system show a very bright surface transected by a network of long linear features of lower albedo and brownish color. The features are seen down to the resolution of the images (~ 4 km/lp) and are as large as tens of kilometers across with maximum lengths comparable to the radius of Europa (~ 1500 km). Their geometry indicates that they are probably fractures caused by extensional stresses in Europa's crust. Other lineaments include low ridges of unknown origin. Overall, however, the surface of Europa is remarkably level. Only about three small impact craters have been positively identified, and it is doubtful that topography anywhere on the satellite exceeds a few hundred meters.

Even the most basic pre-Voyager information about Europa appeared enigmatic. The surface is composed primarily of H_2O frost, yet the satellite's density clearly indicates that it is dominantly a silicate body. Simple but powerful cosmochemical arguments suggest that all of the Galilean satellites should be composed primarily of silicates and varying amounts of H_2O /8/. Water frost was first positively identified as a major component of Europa's surface by Pilcher et al. /9/, based on the presence of strong infrared H_2O absorption features. This conclusion has been confirmed by later observations /10,11,12/ which have led to the conclusion that more than 95% of the spectroscopically detectable material on the surface of Europa is H_2O. Despite the nearly pure water ice surface composition, the density of Europa is known to be 2.97 g cm^{-3}. A density this large indicates a composition dominated by silicates, with only a relatively small admixture of H_2O. If the silicate component of Europa is largely deydrated and has a density like that of Io (3.57 g cm^{-3}), then Europa is composed of 7 to 8% H_2O by mass. If, however, the silicates are hydrated, then the density is consistent with a composition including little or no free H_2O.

Three models have been proposed for the internal structure of Europa that are consistent with the satellite's surface composition and density. These are the *thin ice*, *thick ice*, and *ice/ocean* models. The thin ice model was proposed by Ransford et al. /13/. They suggested that the silicates in Europa's interior are largely hydrated, and that the surface H_2O ice is a very thin (a few km) layer lying over hydrated silicates. In the thick ice model /14/, enough internal heat is produced and retained to dehydrate the silicates, driving the H_2O to the surface to form a layer of solid ice on the order of 100 km thick. In the ice/ocean model /15/, still more heat is produced and retained, melting much of the ice. The internal structure of Europa then would consist of a dehydrated silicate interior, an ocean of liquid water on the order of 100 km thick, and a thin ($\leq$ 10 km) ice layer at the surface.

All of these models are consistent with Europa's density and surface composition, and choices among them must be made on the basis of surface morphology and models of internal evolution. The thin ice model appears to have the most severe problems. Squyres et al. /15/ have pointed out the importance of tidal heating in the silicate portion of Europa, and have argued that the combined tidal and radiogenic heat flow is sufficient to deydrate the bulk of the silicates in Europa's interior. More importantly, Thomas and Schubert /16/ have shown that the thin ice model should lead to long-term retention of large impact craters, in strong contrast with Europa's nearly crater-free surface. Choosing between the thick ice and ice/ocean models, however, is more difficult. The critical problem concerns the rate of heat production by radionuclide decay and tidal dissipation, compared to the rate of heat removal by thermal conduction and solid state convection. Conduction alone cannot remove heat rapidly enough from Europa to keep most of the ice in a thick outer layer from melting. On the other hand, if solid state convection occurs, it can remove heat rapidly enough to prevent melting and maintain a solid ice layer. Solid state convection in an outer ice layer will occur only if the layer exceeds some critical thickness, which is about 30 km on Europa. The outcome of a particular evolutionary model therefore hinges on the highly uncertain calculation of the tidal heating rate, and whether the heat flow ever became low enough to allow the onset of convection. The calculations of Squyres et al. /15/ suggested that tidal heating rates exceed those that would allow convection and freezing by more than a factor of two. More detailed consideration of the rheology of an ice shell by Ross and Schubert /17/ has suggested that tidal heating in such a shell would be smaller than that calculated by *Squyres et al.*, and preservation of a liquid layer correspondingly more difficult. On the other hand, Ojakangas and Stevenson /18/ have considered tidal dissipation in an ice layer of inhomogeneous thickness, and have suggested that inhomogeneities could lead to enhanced local heating and melting. Based on theoretical arguments alone, then, the presence of a liquid ocean appears plausible but not proven.

If an ocean does exist, through-going fractures could occur in the thin ice layer covering it. This possibility is certainly supported by the appearance of the features on Europa's surfaces. Fracturing could be caused by tidal stresses or from movement of the shell with respect to the tidal bulge of Europa. Liquid water exposed to the surface by fracturing would rapidly boil, creating a cloud of vapor that would condense over the surface of the satellite. If the surface conductivity was reduced below that of solid ice by the presence of an insulating layer such as porous frost, the equilibrium thickness of the ice shell could be greatly reduced.

Several lines of observational evidence support the hypothesis of liquid water below a thin ice shell on Europa and active resurfacing of the surface, possibly by release of water through cracks in the ice /15/. First, the lack of craters is difficult to explain in any other way. Second, the photometric function of Europa's surface indicates that there is a textural difference between that surface and impact ejecta deposits of similar albedo on Ganymede and Callisto /19/. The photometric properties may be more consistent with a surface layer of condensed frost. Third, the SO_2 concentration observed on Europa's surface can be interpreted as a result of uniform addition of S ions to a surface on which a much larger amount of H_2O is continually deposited /15,20/. Deposition rates of H_2O of ~ 0.1 μm yr^{-1} are inferred /15/.

In a previous paper /21/, we considered the input of biologically useful energy into an ocean on Europa. The largest potential energy source would be thermal energy generated by radionuclide decay and tidal dissipation in the satellite's silicate interior. On the Earth, the concentration of thermal energy in local high temperature regions at submarine vents has led to large scale hydrothermal activity /22/. Reactions between sea water and hot basalt produce solutions that ascend to the sea floor where they sustain oases of life. At the base of the food chain in these oases are chemosynthetic bacteria that derive their entire energy supply from low temperature oxidation of the geothermally reduced sulfur compounds in the vents /23/.

Analogous hydrothermal regions might exist on Europa. The calculated mean heat flux from Europa's interior is less than a third of the Earth's, and about 50% more than that of the Moon. The Earth is presently highly active volcanically, while the Moon is inactive. A substantial fraction of the Earth's heat flow takes place along mid-ocean ridges, and a similar concentration of Europa's heat flow in spatially limited regions would be required to produce submarine volcanism. However, present data and modelling techniques are wholly insufficient to permit a quantitative assessment of the possibility that such submarine volcanic activity is occurring on Europa.

It is also possible that cracks in an ice cover could admit small amounts of light into an ocean on Europa. In a simple calculation of refreezing of ice after a fracturing event, Reynolds et al. /21/ determined that amount of photosynthetically useful radiation that could penetrate the ice as a function of time after the event. They found that under certain conditions it could take over 1000 days for light levels to drop below the illumination levels at which algae are capable of photosynthesizing in perennially ice-covered aqueous environments in Antarctica.

In order to continue to exist after such an illumination event, organisms in such an ocean would have to survive until the next such event. One Earth analog community considered by Reynolds et al. /21/ was that of sea-ice diatoms. Survivorship of winter adapted cells kept in complete darkness at 0°C for 6 months was reported by Sullivan and Palmisano /24/ to be between 0.1% and 10%. This gives values of for the time constant for loss of biomass of 26 and 80 days, respectively. More recent experiments with the sea-ice organisms indicate that the survivorship is ten times higher, ranging from 1 – 100% under these same conditions /21/. Apparently the ability to survive lengthy dark periods is not confined to polar organisms. Anita /25/ found three temperate benthic diatoms capable of autotrophic growth after 12-36 months of darkness at 2°C. However, the time between fracturing events on Europa, if indeed they occur at all, is unknown.

4. OTHER SOLAR SYSTEMS

The prospects for life on Europa appear slim. All we can say at this point is that there are certain Earth organisms with adaptations that might allow them to survive and grow if introduced into an Europan ocean. However, other solar systems may exist with satellites oribiting giant planets in which tidal heating may be considerable. In addition, these satellites may possess atmospheres. In this section we consider the effect of the combination of high tidal heating and an appreciable atmosphere on the surface temperature of a satellite.

The possibilities of large amounts of volatiles being retained in large satellites forming near giant planets are dependent upon details of satellite formation processes whose general outlines are only poorly understood. In lieu of a theoretical understanding of such systems we will calculate a few illustrative examples based upon atmospheres of satellites in our own solar system. Elaborate radiative-convective equilibrium codes have been developed at NASA Ames for the study of the thermal structure of planetary atmospheres /26,27,28,29/. We have used these codes to calculate the surface temperature on a tidally heated satellite with a thick atmosphere in order to illustrate the possible importance of tidal heating in maintaining oceans and possible habitable conditions on a satellite in another solar system.

As a first example, we consider the case of a satellite similar to Titan. We have calculated the surface temperature of an object with a Titan-like atmosphere for the solar flux at Jupiter's orbit and a range of tidal heating rates. In these calculations we have assumed an atmosphere composed of 5% CH_4, 0.2% H_2 and 94.8% N_2, with a total pressure of 1.5 bars (conditions similar to the present Titan atmosphere). We have also included the photochemical haze, assuming its concentration in all cases to be the same as that in the present Titan atmosphere. The model includes pressure-induced absorption as the source of gas opacity in the infrared, Rayleigh scattering in the visible, absorption and scattering by haze, methane absorption in the visible, and convective adjustment in the lower layers. The model does not include absorption by water vapor, which could become important for temperatures near 273 K. Hence at temperatures approaching 273 K the model may underestimate the surface temperature. It is reasonable to expect that the N_2 would remain in the atmosphere even in the presence of an ocean, since an ocean of liquid water on Titan 100 km deep in equilibrium with a 1.5 bar N_2 atmosphere at 273 K will have only have ~20 mb of N_2 in solution.

As one would expect, our results show that surface temperature increases as tidal heat flow is increased. For a heat flow 100 times that experienced by Io (as could be produced in Io by, for example, increasing its orbital eccentricity to 0.04), the surface temperature in our model is 268 K. If CO_2 is included in the atmosphere, the temperature rises to ~ 300 K. We find that an ocean of depth of a few tens of km could survive over geological times on such a satellite.

5. CONCLUSIONS

Based on our present understanding of the processes in the outer solar system we suggest that tidally heated oceans on satellites of giant planets are possible and that one may exist on Europa. Depending on the size and location of the satellite and the mass of the planet, such aqueous environments may be conducive to the formation and sustenance of life over geologic time. It is interesting to note that the intensity of sunlight on the surface of the

Earth far exceeds the level required by plants for net photosynthetic production. In fact, the threshold for production reported for Antarctica lake plankton by Parker *et al.* /30/ corresponds to the solar flux at the orbit of Neptune.

Conventional wisdom has held that there is a "zone of habitability" that exists around a given star, the limits of which are fixed primarily by the radiative energy flux from the star. We suggest that there may, under certain plausible circumstances, be zones of habitability around giant planets. The limits of these zones would be set, according to Eq. (3), by the mass of the planet and the radii and orbital eccentricities of the satellites surrounding it. The discovery of giant planets around other stars more massive than Jupiter would favor the existence of tidally heated habitable zones. An example such as this of previously unexpected potential sites for life supporting environments serves to reemphasize the importance of searches for other planetary systems.

REFERENCES

1. Huang, S., Occurrence of life in the universe, *American Scientist* **47**, 397-402 (1959).
2. Hart, M.H., Habitable zones about main sequence stars, *Icarus* **37**, 519-528 (1979).
3. Cassen, P.M., S.J. Peale, and R.T. Reynolds, Structure and thermal evolution of the Galilean satellites, *The satellites of Jupiter* (D. Morrison, Ed.) U. of Arizona Press, 93-128 (1982).
4. Peale, S.J. and P.M. Cassen, Contribution of tidal dissipation to lunar thermal history, *Icarus* **36**, 245-269 (1978).
5. Squyres, S.W., R.T. Reynolds, P.M. Cassen, and S.J. Peale, The evolution of Enceladus, *Icarus* **53**, 319-331 (1983b).
6. Squyres, S.W., R.T. Reynolds, and J.J. Lissauer, The enigma of the Uranian satellites orbital eccentricity, *Icarus* **61**, 218-223 (1985).
7. Smith, B.A., L.A. Soderblom, R. Beebe, D. Bliss, J.M. Boyce, A. Brahic, G.A. Briggs, R.H. Brown, S.A. Collins, A.F. Cook II, S.K. Croft, J.N. Cuzzi, G.E. Danielson, M.E. Davies, T.E. Dowling, D. Godfrey, C.J. Hansen, C. Harris, G.E. Hunt, A.P. Ingersoll, T.V. Johnson, R.J. Krauss, H. Masursky, D. Morrison, T. Owen, J.B. Plescia, J.B. Pollack, C.C. Porco, K. Rages, C. Sagan, E.M. Shoemaker, L.A. Sromovsky, C. Stoker, R.G. Strom, V.E. Suomi, S.P. Synnott, R.J. Terrile, P. Thomas, W.R. Thompson, and J. Veverka, Voyager 2 in the Uranian System: Imaging Science Results, *Science* **233**, 43-64 (1986).
8. Lewis, J.S., Low temperature condensation from the solar nebula, *Icarus* **16**, 241-252 (1972).
9. Pilcher, C.B., Ridgway, S.T., and McCord, T.B., Galilean satellites: Identification of water frost, *Science* **178**, 1087-1089 (1972).
10. Pollack, J.B., Witteborn, F.C., Erickson, E.F., Strecker, D.W., Baldwin, B.J., and Bunch, T.E., Near-infrared spectra of the Galilean satellites: Observations and compositional implications, *Icarus* **36**, 271-303 (1978).
11. Clark, R.N., Ganymede, Europa, Callisto, and Saturn's rings: Compositional analysis from reflectance spectroscopy. *Icarus* **44**, 388-409 (1980).
12. Clark, R.N., and McCord, T.B., The Galilean satellites: New near-infrared reflectance measurements (0.65-2.5 μm) and a 0.325-5 μm summary, *Icarus* **41**, 323-339 (1980).
13. Ransford, G.A., Finnerty, A.A., and Collerson, K.D., Europa's petrological thermal history, *Nature* **289**, 21-24 (1981).
14. Cassen, P.M., S.J. Peale, and R.T. Reynolds, Tidal dissipation in Europa: A correction, *Geophys. Res. Lett.* **7**, 987-988 (1980).
15. Squyres, S.W., R.T. Reynolds, P.M. Cassen, and S.J. Peale, Liquid water and active resurfacing on Europa, *Nature* **301**, 225-226 (1983a).

16. Thomas, P.J., and Schubert, G., Crater relaxation as a probe of Europa's interior, *J. Geophys. Res.* **91**, D453-D459 (1986).
17. Ross, M.N., and Schubert, G., Tidal heating in an internal ocean model of Europa, *Nature* **325**, 133-134 (1987).
18. Ojakangas, G.W., and Stevenson, D.J., in preparation.
19. Buratti, B., and J. Veverka, Voyager photometry of Europa, *Icarus* **55**, 93-110 (1983).
20. Eviatar, A., A. Bar-Nun, and M. Podolak, Europan surface phenomena, *Icarus* **61**, 185-191 (1985).
21. Reynolds, R.T., S.W. Squyres, D.S. Colburn, and C.P. McKay, On the habitability of Europa, *Icarus* **56**, 246-254 (1983).
22. Corliss, J.B., Dymond, J., Gordon, L.I., Edmond, J.M., von Herzen, R.P., Ballard, R.D., Green, K., Williams, D., Bainbridge, A., Crane, K., and van Andel, T.H., Submarine thermal springs on the Galapagos Rift, *Science* **203**, 1073-1083 (1979).
23. Jannasch, H.W., and Wirsen, C.O., Chemosynthetic primary production at East Pacific sea floor spreading centers, *BioScience* **29**, 592-598 (1979).
24. Sullivan, C.W., and A.C. Palmisano, Sea-ice microbial communities in McMurdo Sound, *Antarctic J. U.S.* **16**(5), 126-127 (1981).
25. Anita, N.J., Effects of temperature on the darkness survival of marine microplanktonic algae, *Microb. Ecol.* **3**, 41-54 (1976).
26. Kasting, J.F., J.B. Pollack, and T.P. Ackerman, Response of Earth's atmosphere to increases in solar flux and implications for loss of water from Venus, *Icarus* **57**, 335-355 (1984a).
27. Kasting, J.F., J.B. Pollack, and D. Crisp, Effects of high CO_2 levels on surface temperature and atmospheric oxidation state of the early Earth, *J. Atmos. Chem.* **1**, 403-428 (1984b).
28. Kasting, J.F., and T.P. Ackerman, Climatic consequences of very high CO_2 levels in Earth's early atmosphere, *Science* submitted (1986).
29. McKay, C.P. and J.B. Pollack, Radiactive-convective model of Titan's atmosphere, *Bull. Amer. Astro. Soc.*, **17**, 739-740 (1986).
30. Parker, B.C., G.M. Simmons, Jr., K.G. Seaburg, D.D. Cathey, and F.T.C. Allnutt, Comparative ecology of plankton communities in seven Antarctic oasis lakes, *J. Plank. Res.* **4**, 271-286 (1982).

IS URANUS THE MOST PROMISING PLANET FOR SETI?

N.G. Bochkarev
Sternberg State Astronomical Institute
Moscow, USSR

ABSTRACT. The layer with T=300 K and density 0.07 g cm in Uranus (and maybe Neptune) atmospheres is the most suitable extraterrestrial locations in the Solar system for the origin and sustentation of life. The high density of CH_4, NH_3, NH_2HS, H_2O etc, the presence of water drops and electrical discharges facilitate the emergence of life.

The space missions aimed at exploring the Moon and the terrestrial planets have not found any traces of life on them, thereby throwing doubt upon the prospects of finding extraterrestrial life within solar system. At the same time, C. Sagan and E. Salpeter have long argued in favour of feasibility of life on giant planets (see Shklovsky, 1965, Goldsmith and Owen, 1980).

The aim of the present note is to draw attention to the fact that the latest data (in particular, those of Voyager-2) are indicative of Uranus (and probably Neptune) as more suitable candidates for life germinate and be sustained compared with Jupiter and Saturn. The trouble C. Sagan's hypothesis faces in case of Jupiter and Saturn is that the atmospheres of two planets are of solar abundance, i.e. consist basically of hydrogen and helium with minor admixtures of heavier elements, necessary for organic molecules to be synthesized. The higher average density of Uranus and Neptune and the Uranus contraction degree are indicative of a strong deficit of hydrogen and probably, helium in the lower atmospheres of the planets (Hubard 1984) which seem to consist mainly of methane, ammonia, ammonium hydrosulphide (NH HS), and water. Thus the atmospheres of Uranus and, probably, Neptune have high abundaces of heavy elements which constitute foundation of life. Besides, they form just the compounds which have been shown experimentally (see, e.g., Oparin, 1966, Calvin, 1969) to give rise readily to carbon-based prebiological compounds.

In contrast to the very cold surfaces of the satellites of giant planets, the atmospheres of the planets

G. Marx (ed.), Bioastronomy – The Next Steps, 29–30.
© *1988 by Kluwer Academic Publishers.*

proper contain a layer of 300 K temperature facilitating development of life. The expected pressure of the layer of Uranus and Neptune is 10^8 dyne.cm^{-2} i.e. 10 times as high as in Jupiter and Saturn (see, e.g. Hubbard 1984). With the gas molecular weight of 16-18, the density of such an atmospheric layer is 0.07 g cm^{-3}, which is as small as an order below the density of the ocean on the Earth. At the same time, the giant-planet atmospheres must contain clouds of drop-liquid phase of water, possibly with admixtures of other aerosole. Therefore, the chemical processes there may proceed almost as intensively as in the Earth's oceans.

The Uranus atmosphere seems to contain also the high-energy sources which are necessary for formation complex compounds. E.g., the electric discharges induced by electrization of the layers of the intensively circulating planetary atmosphere may prove to be such sources. The occurence of the discharges is indicated by the U-emission bursts detected on Voyager-2 (see, e.g. Ingersoll, 1987). Such discharges, together with their associate bursts, are also common in atmospheres of other planets, including Jupiter and Saturn.

Contrary to Mars, comets, and asteroids, the physical conditions in the Uranus atmosphere are extremely stable because of thermal inertness and rapid circulation of the atmosphere. The Voyager-2 observations have shown that the temperature of even the outer atmosphere cover both poles of the planet is the same, although one of the poles has not been sunlit for 20 years. The stability of the atmospheric conditions must undoubtely facilitate life to originate. On the other hand, the extreme stability of the atmospheric parameters does not make it necessary for life to fit continually to changing conditions, thereby affecting adversely the evolution rate of the evolving forms of life.

Thus Uranus and, probably, Neptune should be regarded as most suitable bodies in the solar system for the carbon-based primordial forms of life to originate and be sustained.

References
Calvin, M., 1969, Chemical Evolution
        The Clerendon Press, Oxford
Goldsmith, D. Owen, T., 1980, The Search for Life in the
        Universe. Benjamin/Gummings, Melo Park.
Hubbard, W., 1984, Planetary Interiors
        Van Nortrand Reinhold Co.
Ingersoll, A.P. 1987, Scientific American, v.256, No.1.
Oprin A.I., 1966, Origin and Initial Development of Life
        Nauka, Moscow
Shklovky, I.S., 1965, Universe, Life, Intelligence
        Nauka, Moscow.

WHAT DO WE KNOW ABOUT THE NUCLEUS
OF COMET HALLEY?

K. Szegő

Central Research Institute for Physics
P.O.Box 49 Budapest 114, 1525, Hungary

ABSTRACT. In this paper an overview is given about the most
interesting physical and chemical properties of the nucleus
of Comet Halley learned from the VEGA space mission.

In March 1986 five spacecraft, two Soviet: VEGA 1 and 2,
two Japanese: SUISEI and SAKIGATE and ESA's GIOTTO
encountered Comet Halley to reveal its secrets. The
scientific objectives of the missions were manyfold: to
image first time in history a cometary nucleus, to study
its composition, to study the basic physical and chemical
processes on its surface and in the cometary environment
including the solar wind – comet interaction.
  The interest in comets and in the primitive bodies of
the solar system stems from the belief that these bodies
preserved the pristine material of the solar system since
geological evolution did not modify them. This however does
not exclude other factors. Radioactive heating due to e.g.
Al27 decay (Wallis 1986) might have heated up the internal
part which cooled down later on. Surface irradiation might
have changed the uppermost layer which also eroded during
consequent activity around perihelion. So, when the results
of the Halley-encounters will be interpreted and
generalized to other comets certain cautiosness is
appropriate. The more so, since Halley is definitely an
exceptional comet. Out of the 121 comets registered as
periodic till 1982 (Marsden 1982) only 11 were observed
more than 10 times. Though P/Halley is only the second from
the top with its 30 apparition, this has been the most
spectacular and active one in the family of comets.
  Our knowledge about comets, their chemistry and
physics was beautifully summarized before the advent of the
cometary missions by Mendis et al. (1985). Though in many
respects it has been proved to be true, the missions
revealed many surprises; and it is evident that earlier

G. Marx (ed.), Bioastronomy – The Next Steps, 31–37.
© 1988 by Kluwer Academic Publishers.

ideas should be modified. Some of the surprises concerns
the interaction of the cometary plasma with the solar wind
(Sagdeev et al. 1987a). These questions will not be
discussed in the present paper but I feel compelled to
mention at least some of the new features: cometary ion
acceleration due to the intensive magneto-hydrodynamical
turbulences, new type of bow-shocks around comets in
comparison to other known celestial bodies, surprising
boundary structure in the plasma surrounding the cometary
nucleus.

But surprises were registered when the nucleus itself
was investigated which is already our subject here. The
first was its size. Earlier the size (5-6 km in diameter)
was estimated by brightness measurements, assuming 0.1
albedo which was a reasonable number deducted from
Whipple's dirty snowball model. The VEGA images however
revealed that it is bigger, it is a consolidated body of
irregular shape with major dimensions of 16x7x8 km, its
volume is about 500 km3, its surface is 350 km2. The albedo
is lower, about 0.04 (Sagdeev et al. 1986a) which means
that it is black as charcoal; so it is one of the darkest
object of our solar system. The crude three dimensional
shape has also been reconstructed from the data (Sagdeev et
al. 1986b). This is exhibited in Fig. 1, whereas Fig. 2
shows the best image of the nucleus taken by VEGA 2 on 9
March 1986 at 7:20 UT at a distance of 8030 km from the
nucleus. The shape and size reconstruction was possible
since the VEGA s/c imaged the nucleus both on 6 and 9
March, exposing about 66 images where the nucleus is
clearly resolved. The relative view angle between the
images changed considerably, about 160^0 for the VEGA-1 and
about 60^0 for the VEGA-2 images. A detailed summary of the
imaging experiment aboard the VEGA s/c was published by
Sagdeev and Szegö (1987). The results were confirmed by the
HMC data obtained during the GIOTTO encounter with the
comet on 14 March 1986 (Keller et al. 1986). In the case of
GIOTTO the relative view angle between the image set
changed less than 15^0, so it provided only a two
dimensional view but with better resolution than the VEGA
image set.

One of the new observational evidence has changed a
paradigme. Earlier it was thought that the density of the
nucleus is about 1 g/cm3, reasoning that the most abundant
material in it is water. The pioneering work of Rickman
(1986) revealed that the nucleus might be considerably less
dense. The density derived from non-gravitational forces
turned to be about 0.1 - 0.3 g/cm3. This was reanalysed by
Sagdeev et al. (1987b) obtaining an average density about
0.6 g/cm3 which is still less than anticipated. So the
nucleus material seems to be less dense. Greenberg (1986)
suggested a fluffy structure of silicate and other grain

components where the voids are filled with volatiles. This accomodates well with the observations, and yields low surface albedo. Whipple (1987) called this type of nucleus as a dirty snowdrift to distinct it from the previous snowball model. We conclude that the material is more fluffy, less dense and probably more friable than it was thought.

Earlier the surface was considered to be homogeneouos in accordance with the snowball model. In radial direction the core-mantle structure was accepted. After the missions the author of the present paper is convinced that there are two types of cometary surfaces, at least in the case of P/Halley. The first is a thicker, hot, inactive crust, impervious to volatiles; the second is a fresh cometary material rich in volatiles. This is the basic source of the jets.

The complete set of equations describing the thermodynamics of the core-mantle model and estimating the gas and dust outflow was first solved in the friable sponge model of Horanyi et al (1984). From this model if follows that the inactive places should be hot, about 400 K. Such surface temperature was actually measured by IKS aboard VEGA (Emerich et al., 1986); from their measurements they inferred a hot object of 5 km radius; but they are able to find solution with different vertical and longitudinal sizes, or even with disintegrated hot spots. The heat must come from the cometary surface. In principle other factors may influence the measurements, e.g. the dust temperature in the atmosphere etc. Whereas it is true that the grains leaving the surface are heated up during the first few ten meters according to coupled dust-gas hydrodynamical equations (Gombosi et al., 1983), later on they are cooled down if only the hydrodynamics is considered. However the sun heats the grains too. To estimate this temperature we may use the same heat balance as for the surface. As there is no outflowing heat, the grains should be slightly warmer than the inactive surface. But the temperature of small, absorbing grains depends on their size too, since if the size is much smaller than the wavelength, the size determines their emissivity . As showed by Hanner (1984) 0.1 micron grains can be 300-400 K hotter then big ones. However, because of the low optical thickness of the dust, (Sagdeev et al, 1985) this contribution should not be important. Hence we assume that IKS really measured the inactive layer temperature and size.

It is known from the literature that there are possibilities for building up inactive layer. This was studied by Moore et al. (1983) experimentally. They exposed cometary type ice mixtures to conditions persisting in the Oort cloud, including the simulation of cosmic ray radiation. They concluded that 1 % of the material was converted into a nonvolatile residue. After the first few apparition this can turn to be e.g. the inactive crust. The procedure was discussed by Johnson et al. (1986) also.

It was Ip (1985) who concluded first based on GIOTTO HMC data that the active surface is only about 15 % of the total. The 3-dimensional analysis of the most prominent jet features seen on Vega-2 images also resulted that the total surface of the most active jet sources is small. From the images it is also evident that the general activity is low on the dark hemisphere. This proves the low heat conductivity of the surface which is consistent with the fluffy material referred above. Whereas it is true that the most active dust jets are well localized, the source of the other brightness features on the VEGA images is not known. These can not be just part of the coma since what we call as coma in ground based observation is not visible on the VEGA images. So there is an activity unlocalized on the sunward surface which may consist of small jets or may be different and their source extension is unknown.

To analyze jet sources the VEGA-2 imaging data are unique. During the VEGA-2 encounter several large size images were exposed from a distance closer than 30.000 km from the nucleus. On these images extended jets and other brightness features are distinguishable (Smith et al. 1986). That was used to reconstruct the spatial orientation of the most prominent jets. Recently the brightness distribution of the jets were also analysed (Szego et al., 1987). The preliminary analysis suggest that grain disintegration takes place in the vicinity of the nucleus. As the disintegration can not be caused by collisions, it was assumed that the grains leaving the surface are still rich in volatiles. This does not contradict to the observation of PUMA aboard VEGA (L.M. Mukhin, private comm.). As the grains are heated up, the volatiles evaporate and this process leads to the disintegration. If it is so, we have to conclude that at least the jet sources are not covered by dust, there the naked pristine ice is exposed to the solar radiation. A rough estimate of the source surface, based on (Smith et al., 1986) for the VEGA-2 fly-by, can be about 5% of the surface. The rest of the surface is partly inactive, partly covered by thin dust (Szegö, 1987), the ratio is about half and half. A quickly changing thin dust layer is very convenient to explain the strong variability of the activity. It is an important but yet unanswered question whether the jet sources are fixed relative to the nucleus or they may change position. Temporal changes in the activity has been reported (Kaneda et al., 1986).

The rotation of this irregularly shaped nucleus is complicated. As it is summarized e.g. by Sagdeev and Szegö (1987), both ground based and spaceborn observations converged to two different characteristic periods, one of 2.2 d, the other of 7.4 d long. We only recently were able to present an acceptable solution to this question (Sagdeev et al., 1987c). The nucleus being a very irregular object rotates as an asymmetric top. In this case the rotation axis, the position if which is fixed in space, is not attached rigidly to the body, the major inertial axes of the nucleus are wandering about the rotation axis. However, if the angle between the direction of the rotation axis on one hand and the direction of the largest inertial axis on the other is not too big, the set of equations describing the motion of the general top can be simplified. The motion in this case is characterized by two periods related by the inertial momenta of the body. So the rotation of the nucleus is as follows: it rotates about the axis with a period of 2.2 d and while it rotates, it 'nods' periodically in the rotation axis direction with a period of 7.4 d. The amplitude of the nodding is about $12^0 - 14^0$. The orientation of the rotation axis in space can also be deduced from the data. This solution assumes that the torques due to jet emissions are negligible which is consistent with the observed stability of these periods (Festou et al., 1987). We are now trying to predict what brightness-curve characterizes this motion and whether it coincides with the one observed by ground based astronomers.

In what follows the chemical composition of the nucleus is brifly described. The atomic composition was measured by the devices PUMA aboard VEGA and PIA aboard GIOTTO. Molecular compositions were measured by spectrometers on VEGA and from ground. I confine myself only to PUMA results based on two papers of Sagdeev et al. (1897d and 1987e).

The PUMA experiment was a difficult one. It measured the atomic components of dust particles by evaporating them on impact on a silver target. The impact velocity was equal to the relative velocity of the spacecraft to the comet, i.e. about 80 km/s. As such high velocity is unattainable in laboratory the device can not be calibrated on ground. The impact evaporation procedure, the formation of a plasma cloud on impact is also not fully understood theoretically. So in this case the data reduction is a very tedious job; the newly published results differs from quick-look data and will be enriched in future.

The most important conclusions are the following. Based on the isotope composition of dust, the nucleus is of solar system origin. Light atoms are abundant in the dust; the most natural explanation is the presence of organic materials, similar to compounds found in carbonaceous condrites (Kerosene, aminoacids). The presence of complicated organics can not be proved at this stage. The

composition of several dust particles can be interpreted by
the presence of silicates or in some cases triolit.
     The dust particles can be grouped into different
classes. Dust belonging to the first contains only H, C, N
and O. From a set of 266 particles 17 belonged to this
class. The atoms probably came from organic compounds of
high boiling point (mineral oil derivatives, aminoacids,
higher alcohols), since these signals were generated by
very small and consequently about 400-500 K hot dust. In
the second, most populated group the particles consist of
light elements and metals. This group may contain organics,
pure carbon, metal carbids are also present. Generally they
can be called quasi-chondrits. About 80% of all particles
belong to the first and second groups. The third class is
characterized by the presence of metals and hydrogen, or
only metals. Olivine, periclase, cristobalit, piroxen etc.
can be the mother particle. For further details see the
papers referred above. There were also particles out of
these classes. Their spectra need further investigations.
In three cases the analyses can not exclude the presence of
pure ice particles.

REFERENCES

Emerich et al., ESA SP-250, Vol.II. 381, (1986);
Festou et al., Astron. & Astrophys., to be published 1987).
Gombosi et al., Proceedings of the Internnational
          Conference of Cometary Exploration, Budapest,
          Vol.II. 99,1982.
Hanner, M.S., Icarus 64, 51, 1986.
Horanyi et al., The Astrophys. J. 278, 449, 1984.
Ip, W.H., Adv., Space Res., 5, 233, 1985.
Johnson et al., ESA SP-250, Vol.II. 269, 1986.
Greenberg et al., ESA SP-250, Vol.II. 255, 1986.
Kaneda et al., ESA SP-250, Vol.I. 397, 1986.
Keller et al., Nature 321, 320, 1986.
Marsden, B., Catalog. of comet. orbits, Cambridge, 1982.
Mendis et al., Fundametals of Cosmic Phys., 10, 1, 1985.
Moore et al., Icarus 54, 388, 1983.
Mukhin, L.M., private comm.
Rickman, H., ESA SP-249, 149, 1986.
Sagdeev et al., Adv. Space Res., 5, 12, 95, 1985.
Sagdeev et al., ESA SP-250, Vol.II. 295, 1986a.
Sagdeev et al., ESA SP-250, Vol.II. 307, 1986b).
Sagdeev and Szegö, KFKI preprint, KFKI-1987-35/c, 1987.
Sagdeev et al., KFKI preprint, KFKI-1987-50/c, 1987a.
Sagdeev et al., to be published, 1987b.
Sagdeev et al., KFKI preprint, KFKI-1987-75/c, 1987c.
Sagdeev et al., to be published, (1897d and 1987e).
Smith et al., ESA SP-250, Vol.II, 327, 1986.
Szegö, K., Astrophys. & Space Sci., in print, 1987.

Szegö K., et al, to be published, 1987.
Wallis, M.K., ESA SP-249, 63, 1986.
Whipple, F.L., Sky and Telescope, March, 242, 1987.

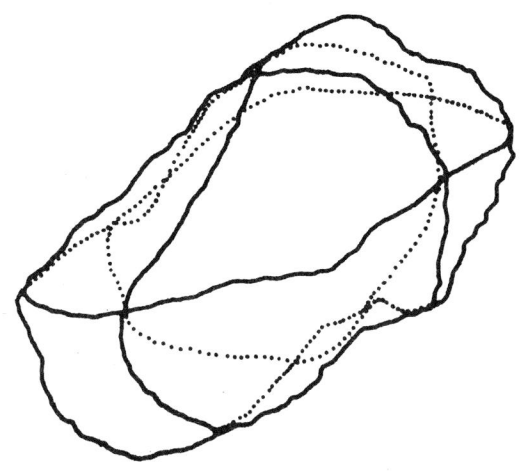

Fig. 1

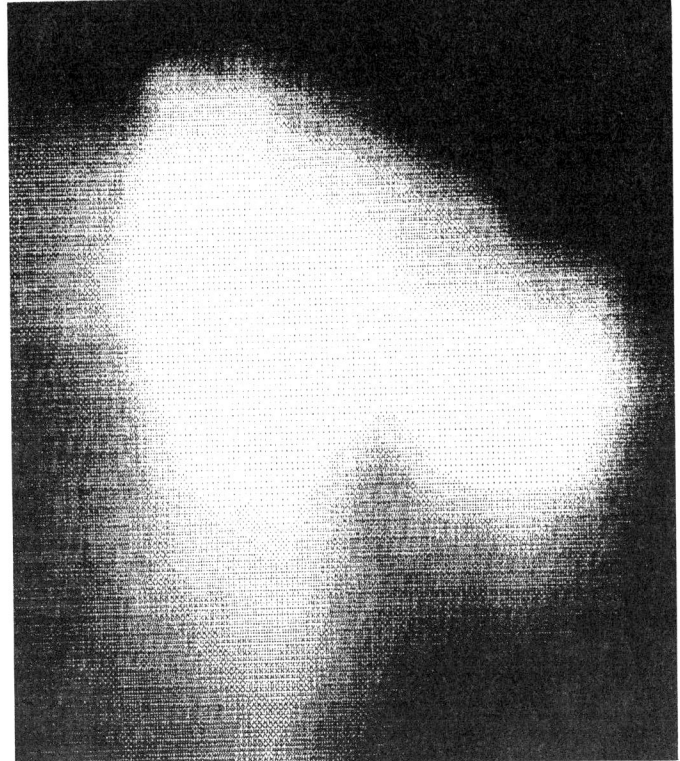

Fig. 2

COMETS AND LIFE

A. Hajduk

Astronomical Intsitute, Slovak Academy of Sciences
Bratislava 84228 Czechoslovakia

ABSTRACT. The study of the orbital evolution of long-period comets, observing their debris, is proposed to bring the question of catastrophic collisions of Earth with large interplanetary bodies (connected possibly with mass extinctions) from the stage of probability estimates to the much precise determinations of such events.

After the paper of Alvarez et al. (1980), explaining the abnormal enrichment in iridium in Cretaceous-Tertiary boundary as caused by the extraterrestrial event (collision with an asteroid), many articles occurred, suggesting periodicity of such events, causing mass extinctions. But a critical review of these results by Hallam (1984) has shown that the periodicity is controversial and they give 26 Myr, 28.4 Myr or 30 Myr periods, or in a longer scale 65 Myr period, with deviations of 6-8 Myr. However, the strong periodicity is related to such hypotheses as of Nemesis; but, in general, for mass extinctions we do not need any periodicity.

From the distribution of orbits of larger interplanetary bodies and from the mass distribution - over the large scale of masses, we can estimate the probability of catastrophes of different orders. Kresák (1978a) has determined the probability of collision of Earth with 1-10 km size body to 60 Myr (which corresponds to the Alvarez explanation of iridium layer). Going back in time the frequency of such events should be higher, fitting the cratering statistics. Hence, this secondary effect of comets, not bringing organic material, but changing strongly the living conditions, is indisputable. Whether these were followed by extinctions or not, is the question beyond astronomy, but other explanations seem to be more hypothetical. The other question is, whether such events are necessary conditions for the biological evolution or not, the question connected with the Anthropic principle.

In connection with Halley comet research a new possibility occurred to postpone the question of catastrophic collisions from a rough probability state to a more precise estimates of past or also future events. This comes from the relations between the evolution

39

G. Marx (ed.), Bioastronomy – The Next Steps, 39–40.

of comet orbit and of orbits of their debris, and between the mass loss of the comet at each perihelion passage, and the mass of the stream. In the case of Halley from these relations the time of the capture of this comet by Jupiter was estimated to 175 000 years from its previous long-period orbit (Hajduk, 1985). We meet particles from Halley twice a year; fortunately they are relatively low density objects, which during the slow diffusion process towards the Earth's orbit in 10^4-10^5 years are eroded mostly to dust and grains. However, the Tunguska event, caused probably by a fraction from comet Encke (Kresák, 1978b) is typical for splitting comets and such local events cannot be excluded from other comets too.

The research in association of meteor showers with long-period comets are at the beginning. A very interesting case was the comet 1983 V - Sugano-Saigusa-Fujikawa with a semimajor axis of 2 500 AU and period of 125 000 years (Everhart and Marsden, 1987). The comet approached the Earth on 12 June 1983 at the distance of 0.063 AU (after the perihelion passage on 1 May 1983). Two days later the Earth passed at the distance of 0.048 AU from the comet's orbit, and within one hour of the predicted moment a substantial increase of large meteors was observed by the radar equipment of the Czechoslovak Academy of Sciences at the Ondřejov Observatory (Šimek and Pecina, 1984). The ejection velocities for a large-meteoroid size particles are small and these particles were not able to reach the Earth, when released at the present perihelion passage. Moreover, the particles were recorded on the following year at the same position of Earth. This means that the comet was not the first time near Sun and that the cometary debris survive along such long-period orbits forming streams, similarly as in case of short-period comets. This gives us the opportunity to estimate the evolution of the comet's orbit in very long terms of 10^6-10^7 years. Studying the evolution of the orbits of long-period comets, observing their debris, also via space probes in the vicinity of their orbits may bring us much closer to the estimates of possible collision times with large bodies and enabling nongeological timing of mass extinctions. This gives also new reasons for the studies of comet-meteor shower associations.

References

Alvarez, L.V., Alvarez, W., Asaro, F., Michel, H.: 1980, Science 208, 1095.
Everhart, E. and Marsden, B.G.: 1987, Astron. J. 93, 753.
Hajduk, A.: 1985, in Dynamics of Comets, Their Origin and Evolution, D. Reidel Publ. Co., Dordrecht, 399.
Hallam, A.: 1984, Nature 308, 686.
Kresák, Ľ.: 1978a, Bull. Astron. Inst. Czechosl. 29, 114.
Kresák, Ľ.: 1978b, Bull. Astron. Inst. Czechosl. 29, 129.
Šimek, M., Pecina, P.: 1984, Bull. Astron. Inst. Czechosl. 35, 375.

DISTRIBUTION OF PLANETARY SYSTEMS

T.V. Ruzmaikina
Smidt Institute of the Physics of Earth, Academy of Sciences
Moscow 123810 USSR

ABSTRACT. Planets are formed in the gas-dust protoplanetary disc surrounding the solar type star at early stage of its evolution. The star and disc originate as a result of collapse of a cloud which angular momentum J is small enough to form a star-like core of minimal mass in the center of the cloud. However, the angular momentum must exceed the maximal angular momentum admissible for the formation of a single star. This puts the following limits on the cloud's angular momentum:

$$10^{51} \lesssim J \lesssim 2\,10^{52} \text{ g cm}^2 \text{ s}^{-1}.$$

The ability of planetary systems depends on the number of protostellar clouds having the angular momenta within this interval. The rotation of dark cores in the clouds comes from the interstellar turbulence. The turbulence is intermittent over the scales less than 0.1pc. The study of observational data on the rotation of the dark cores in molecular clouds shows that a fraction from 10^{-4} up to $\gtrsim 0.1$ of all clouds have the appropriate angular momenta. Thus, a fraction of stars having the planetary systems can amount to $\sim$ o.1 of all stars of solar mass in the Galaxy.

The energy of a rotating graviting system with the fixed total angular momentum and mass is minimal when a bulk of the mass is concentrated in the center while the most of the angular momentum is in a distant outer part containing a negligibly small mass (Lynden-Bell and Pringle, 1974). Both the present and primitive solar systems can be refered to such energetically profitable systems. In fact, more than 95% of the total angular momentum of the solar system is concentrated in the orbital motion of planets which total mass is 1/700 $M_\odot$ only. The reconstraction of the solar nebula by augmenting the planets with H and He to

41

restore a solar composition results in a mass of the solar
nebula of 0.01 to 0.1 $M_\odot$ and a total angular momentum of
$3 \cdot 10^{51}$ to $2 \cdot 10^{52}$ g cm2 s^{-1} (Hoyle,1960; Weidenschilling,
1977) while the Sun had an angular momentum not greater
than $(1 \div 3) \cdot 10^{51}$ g $cm^2 s^{-1}$ (in oder to be stable against
destractive non-axisymmetric pertubations pointed out by
Ostraiker and Bodenheimer (1973)). Hence, it is energetically
more profitable to form a star of solar type with the planetary system than to create, say, a binary stellar system.
One could expect therefore that the stars with planetary
systems are widely spreaded.
However, the situation is not so simple. The solar type star
is forming in a dense interstellar molecular cloud as a
result of the cloud core collapse. The distribution of the
angular momentum for the star with planetary system is quite
different from the distribution of the angular momentum in
the cloud. Thus, an effective outward transfer of the angular momentum is needed to concentrate it in a small fraction of mass, i.e. into the protoplanetary disc.
    The efficiency of the transfer changes during the
collapse. It is negligible at an early stage of the collapse. In the lapse of contraction of the rotating cloud a
ratio β of the rotational energy to the absolute value
of gravitational enrgy increases. If the cloud rotated fast
enough initially it can reach a critical value $\beta = 0.1-0.3$.
After that a central part of the cloud becomes unstable and
breaks up. The process seems finally terminates in the formation of binary or multiple stellar system (Regev and
Shaviv, 1981).
    Observations show that about a half of all stars are
the members of the binary and multiple systems (Abt,1978).
The mean angular momentum for the systems is much higher
than the angular momentum of the solar system.
    In the opposite case of slowly rotating cloud a star-
like core is formed in the center of the cloud before β
can reach the critical value. It follows from the numerical
simulations of collapse that an initial mass M and the
central density ρ of the core are close to 10^{-2} $M_\odot$ and.
10^{-2} g cm^{-3} ., correspondently (Larson,1969). The core
evolves on the time scale of the accretion which changes
from 10^2 to 10^5 yrs as the core mass grows from $10^{-2} M_\odot$ to
1 $M_\odot$. This time scale is longer than the free fall time.
Besides, the core contains a magnetic field of $10^2 - 10^3$ G
which is a product of an intensification of the interstellar magnetic field (Ruzmaikina,1985). An outward transfer of
the angular momentum associated with a torque produced by
the magnetic field prevents the core breaking and results
in a formation of the protoplanetary disc.
    A mechanism for the formation of a star with a low

mass disc was found first by Ruzmaikina (1980) later by
Cassen and Moosman (1981) and studied in Ruzmaikina (1982),
Cassen and Summers (1983), Ruzmaikina and Maeva (1986).

The disc originates from the equatorial layer of the
core and spreads simultaneously with the growth of the core
mass due to the accretion. The disc's spreading results
from the outward transfer of the angular momentum first by
magnetic field and then by the turbulent stresses. One can
believe that the disc should be turbulized due to shear
flows caused by accretion of the cloud gas onto the disc.
If the solar nebula is turbulized it can expand due to the
turbulent diffusion. The radius R_d of the disc increases
according to the diffusion law

$$R_d \sim (\nu t)^{1/2} ,$$

where ν is a turbulent viscosity. It follows from this
equation that a rather weak subsonic turbulence with
$10^{-2} c_s h$ (h/R $\sim 10^{-1}$) is needed for the increasing of
the disc radius up to $10 - 10^2$ AU in $10^5 - 10^6$ yrs which
is a time of accre tion for the central star. To the end
of the accretion the disc has a mass of $10^{-2} - 10^{-1} M_\odot$ and
contains a significant part of the angular momentum. Only
few percents of the kinetic energy of accreting gas is needed to support the turbulence in the disc.

Evidences on the presence of disc around some pre-
main-sequence stars give a new support to the idea of a
joint formation of the star and disc(Beckwith et al 1984).

A further evolution of the disc inevitably results in
the formation of planets. The following sequence of events
is proposed : a settling of solid particles to the central
plane, the formation of the dust swarms and planetisimals,
the accumulation of planets and satelites accompanied by
a removal of the gas from the region of the terrestrial
planets and its accretion onto the cores of the giant planets (Safronov, 1969; Safronov and Vityazev, 1985).

The possibility of the formation of the core during
the initial stage of the collapse restricts the value of
angular momentum of the cloud. It can be formed in an initially uniform and uniformly rotating cloud of mass $M \approx 1 M_\odot$
with the angular momentum

$$J_o \lesssim 2 \, 10^{52} \, (M/M_\odot)^{5/3} \text{ g cm}^2 \text{ s}^{-1}$$

and a correspondent initial value of $\beta \lesssim 4 \, 10^{-6} - 10^{-5}$
(Ruzmaikina,1981). Note, that a fragment of one solar mass
has such a value of angular momentum if at onset of collapse
(when the density is 10^5 cm^{-3} and the temperature is 10 K)

it corotates with the Galaxy $\Omega \sim 10^{-15} s^{-1}$.

On the other hand the cloud evolves to a single star with disc only if its angular momentum exceeds the maximal angular momentum admissible for a single star, e.i.

$$J \gtrsim 10^{51} (M/M_\odot)^{5/3} g \ cm^2 s^{-1}.$$

Thus, a star of solar mass with the planetary system is, likely, formed from the cloud which angular momentum is within interval

$$10^{51} < J \lesssim 2 \ 10^{52} g \ cm^2 s^{-1}. \qquad (1)$$

Kracheva et al (1978) revealed a gap between single and unevolved binary stars in the distribution over the angular momenta. The gap is closed to the interval (1). So the stars with disc fill the gap between single and binary stars in distribution on angular momentum. The solar nebula angular momentum is inside the interval.

The ability of planetary systems depends on the number of protostellar clouds which have the angular momentum within the interval (1). The absence of a preferable orientation both of rotational axes of single stars and orbits of binaries meens that a rotation of protostellar clouds is caused by the interstellar turbulence. Larson (1981) pointed out that in the scale range $0.5 \lesssim L \lesssim 60$ pc the turbulent velocity is

$$v_t(km \ s^{-1}) = 1.1 \ L^{0.38} (pc).$$

Subsequent studies for an extended set of molecular clouds have confirmed this scaling law, but the scatter in the data is appreciable(Myers,1983). It seems resonable to suppose that the scattering reflects the intermittent character of turbulence (Ruzmaikina 1986).

An intermittency means that turbulent vortices are distributed nonuniformly in time and space, i.e. the small regions of intense rotational motions are separated with the extended quiet regions. Using as tipical for the intermittent turbulence the log-normal distribution for the angular momenta one can estimate a fraction of clouds with $J \gtrsim 10^{52}$ g $cm^2 s^{-1}$. One can obtain that the fraction of clouds which have angular momenta $J \leq \zeta \langle J \rangle$ is equal

$$I(\zeta) = \frac{1}{2}\left\{ 1 + erf (\ln \zeta + \frac{D^2}{2})/ 2 D\right\} \qquad (2)$$

where $D^2 = \langle (\ln J - \langle \ln J \rangle)^2 \rangle$.

Unfortunately, there are a few measurements of the rotation of dense cloud cores with dimentions less than 0.1 pc and the mass of about 1 $M_\odot$. Using the data for 17 cores (Myers and Benson,1983)we have found that $J = 7.8 \ 10^{53}$ g $cm^2 s^{-1}$

and D≳0.93 and hence I more than 10^{-4} of the cores have the angular momentum appropriate for the formation of a single star with planetary system. In **fact**, a fraction of the co- res is,possibly, much larger than this value.(It was not revealed rotation of 10 from 17 cores. Only upper limit on rotational velocity was obtained. We suggest that angular velocity of a core is equal 1/2 of limit value, i.e. we take $\Omega = 10^{-14} s^{-1}$ if $\Omega < 2\ 10^{-14} s^{-1}$ was measured.) It is enough for one core to be rotating remarkably slowly than suggestered to obtain I ~ 10^{-1}.
Thus,a distribution planetary systems reflects a distribution of protostellar clouds,i.e. dense cores in molecular clouds, on the angular momenta. Observational data available permit that a considerable fraction of the clouds rotate slowly enough to evolve into the single stars with planetary systems. It follows from (2) that a number of cores whose angular be- long to interval (1) is greater than a number of cores with $J \gtrsim 10^{51} g\ cm^2 s^{-1}$. Therefore most of single stars may have a planetary systems. The observed frequency of single stars is 45% among F3-G2 main-sequence stars(Abt,1978). So, we can ho- pe to find the planetary systems in the vicinity of the Sun.
    A recent observation of the envelope of solid particles biger 10 μm around the nearest relatively young (3 10^8 years old) star Vega gives s strong support to this possibility (Au- mann et al ,1984).
    Taking into account the searches for the planetary sys- tems Zakhozai and Ruzmaikina (1986) present catalogue of stars brighter 5^m which worth to be studied in first turn.The stars for catalogue were selected from the solar vicinity of radius 10 parsecs. The following principles for the star se- lection were used:
1) the single stars, the most like the Sun by the spectral type and the class of luminosity
2) the stars which are suspected to have satelites of low mass
3) the stars wth IR excess
4) all stars at the distances shorter 3 parsecs.
It was included also 4 wide bynaries with distances between components greater than 30 **AU** because they can have plane- tary systems around components. Here the stars are enumerated, i.e. name and spectral class of earch star is given:
1) ζ Tuc (F8), τ Cet (G8 Vp), $\varkappa$ Cet (G5 V), 82 Eri (G5 V), π^3Ori F 6 V), α Men (G5 V), UMa (G5 V), β Com(G0 V), 61 Vir, δ Pav (F6 V), α Cen A(G2 V), β **Hyi**;
2) η Cas A (G0 V), η Csa B (dMO V), ε Eri (K2 Ve), G 175-34 (dM4), Laland 21185 (M2 V), Cen A (G2 V), Cen B (KO V), Proxima (M5e), DM +68°946 (M4 V), Barnard (M5 V), 70 Oph A(KO Ve), 70 Oph A (KO Ve), 61 CygA(K5 e), Krüger 60 A(M4);
3) DM +61°366 (K5 V), ε Eri (K2 V), δ Eri (KO IVe), α CMa A (A IV), α CMa B (DA5), αCMi A(F5 V), α Cmi B (DF), α Lyr(A0 V); α Aql (A5 V), α PsA (A3 V);

4) L726-8 (dM6e), UV Cet (dM4e), α CMa A(AO V), α CMa B(DA5), Wolf 359 (dM6e),Laland (M2 Ve), α CenA(G2 V), α Cen B (KO V), Proxima C (dM5e), Barnard (M5), V 1216 Sgr (dM4e);
5) (wide binaries) η Cas A (GO V) & ηCas (d MO V), γ Lep A (F 6 V) & γ Lep B(K2 V), ξ Boo A(G8 Ve) & ξ BooB(K5 Ve), 41 Ara A(G8 V)& 41 Ara B (MO V).

One can hope that the realization of planed projects of search for planetary systems will result in their discovery. It will give an additional knowledge for solving the problem of the origin and evolution of planetary systems, and will increase enthusiasm in search for extraterrestrial civilizations.

REFERENCES

Abt,H.A.1978.'The binary frequency along the main sequence! In Protostars and Planets,ed.T.Gehrels (Tucson: Univ.of Arisona Press),pp.323-340.

Aumann,H.H.,Gillett,F.C.,Beichman,C.A.,de Jong,T.,Houck,J.R., Low,F.,Neugebauer,G.,Walker,R.G.,and Wesselius,P.1984!Discovery of a shellaround Alpha Lyrae!Astrophys.J.278:L23-27.

Beckwith,S.,Zuckerman,B.,Skrutskie,M.F., andDyck,H.M. 1984. 'Discovery of solar system-size halos around young stars. Astrophys.J.278:793-800.

Hoyle, F.1960.'On the origin of the solar system'.Quart.J. Roy.Astron.Soc. 1:28-55.

Cassen,P.,and Moosman,A.1981!On the formation of protostellar disks!Icarus 48:353-376.

Cassen,P.M.,and Summers,A.L. 1983.'Models of the formation of the solar nebula! Icarus 53:26-40.

Krajcheva,Z.T.,Popova,E.T.,Tutukov,A.V. and Jungelson,L.E. 1978.'Some propeties of specroscopic binary stars'. Astron. J.(SSSR) 55,1176-1189.

Larson,R.B.1969.'Numerical calculations of the dynamics of a collapsing protostar.' Mon.Not.Roy.Astron.Soc.145:271-295.

Larson,R.B.1981!Turbulence and star formation in molecular clouds.'Mon.Not.Roy.Astron.Soc. 194:809-826.

Lynden-Bell,D.,and Pringle,J.E. 1974.'The evolution of viscous disks and the origin of nebular variables.' Mon.Not. Roy.Astron.Soc. 168:603-637.

Myers,P.C.1983.'Dense cores in dark clouds.III. Subsonic turbulence. Astrophys.J. 270:105-11'.

Myers,P.,and Benson,P.J. 1983. Dense cores in dark clouds.II. NH_3 observations and star formation. Astrophys.J.239:309-320.

Ostriker,J.P.,and Bodenheimer,P. 1973.'On the oscillations and stability of rapidly rotating stellar models.III. Zero - viscosity polytropic sequences! Astrophys.J. 180:171-180.

Regev,O.,and Shaviv,G.1981.'Formation of protostars in collapsing,rotating,turbulent clouds! Astrophys.J. 245:934-959.

Ruzmaikina,T.V. 1980.'On the role of the magnetic field and turbulence in the evolution of the presolar nebula.' 23rd COSPAR Meeting, Budapest,June 1980. Adv.Space Res. 1981,1:49-53.

Ruzmaikina,T.V. 1981. Angular momentum of protostars giving the birth to preplanetary discs.' Pisma Astron.J.(USSR) 7:188-192.

Ruzmaikina,T.V. 1982. In Diskussion forum'Ursprung des Sonnensystems',ed.H.Völk. Mitt.Astron. Ges. 57:49-54.

Ruzmaikina,T.V. 1985. 'Magnetic field in the collapsing presolar nebula.'Astron.Vestnik(USSR) 18:101-112.

Ruzmaikina,T.V. 1986.Origin of the angular momentum of the presolar Nebula.' Astron. Zirkulyar 1439:1-3.

Ruzmaikina,T.V.,and Maeva S.V. 1986.'Investigation of the formation of the solar nebula.' Astron.Vestnik (USSR) 19:212-227.

Safronov,V.S. 1969. Evolution of the protoplanetary cloud and the formation of the Earth and planets (Moscow: Nauka Press).

Safronov,V.S.,and Vitjazev 1985.'Origin of the solar system.' In Astrophyssics and Space Physics:(Soviet Scientific Reviews),ed. R.A.Sunjaev.

Weidenschilling,S.J. 1977.'The distribution of mass in the planetary system and solar nebula.'Astrophys.Space Sci. 51:153-158.

Zachozhay V.A., and Ruzmaikina T.V. 1986.'Stars to search for the Planetary Systems! Astron.Vestnik (USSR),19,No.2.

BELT OF LIFE IN THE GALAXY

L.S. Marochnik and L.M. Mukhin
Space Research Institute, Academy of Sciences
84/32 Profsojuznaya Str., Moscow 117810 USSR

ABSTRACT.   The hypothesis is substantiated, according to
which technological civilizations of our type can originate
in the Galaxy in a 'belt of life', a narrow circular zone
with the galactic orbit of the Sun inside. This zone is characterized by specific conditions due to the corotation
circle, an exceptional region of the Galaxy, passing through it. In the belt of life birth and death of civilizations should apparently occur in spiral arms because of supernova explosions. In this case, the lifetime of a civilization is the time of its motion along Galactic orbit between
the adjacent spiral arms. In $3.3 \ 10^9$ years our civilization
will enter the Perseus arm. It can be expected that in the
Galaxy the ratio of the probable number of civilizations at
a higher than ours - technological level to that at a lower
technological level is of the order of 0,7.

1.   The aim of this paper is to estimate a possible number
of technological civilizations in the Galaxy based on astronomical concepts that seem to point to the specific conditions within a narrow ring region of the Galaxy, which encloses the galactic orbit of the solar system (Marochnik
1981).
     The question whether any other civilizations exist in
our Galaxy, some other galaxies or in the Universe in general (or any other forms of life, intelligent or merely organic) remains open. This is so first because only example of

---

x) This paper was published as Preprint of Space Research
Institute of USSR Academy of Sciences N 761, 1983.

49

G. Marx (ed.), Bioastronomy – The Next Steps, 49–59.

the existence of the civilization and life in general we
know about is our own; hence, this example is unique and
consequently not representative. Second, so little is known
about the initial conditions of the origin of the solar system and the life in it so bad that it is not clear if processes of the formation of systems similar to the solar system occur somewhere in our or other galaxies. Theoretically
a great number of hypothesis (including alternative may be
assumed: there are grounds to suppose that we are alone in
the Galaxy (Shklovskii 1976a) opposite is possible either
(Kardashev 1981). If the last point of view is true, the
existence of civilizations of other types (Kardashev 1965)
and, probably, of various forms of life (Molten 1973) can
be assumed.

The aim of this paper, as has been mentioned, is more
narrow - to estimate a possible number of technological civilizations of our type in the Galaxy proceeding from the
concept of the exclusive position of the solar system and
other objects in it close to the Sun's galactocentric orbit.
As the solar system in the Galaxy originated and has evolved in specific conditions (see below) which take place into the entire narrow ring zone, including the Sun's galactic
orbit, it can be assumed that this ring is 'belt of life'
in the Galaxy, i.e. a place where systems can originate,
similar to the solar system, with planets and life.

There is an antropic cosmological principle based on
the hypothesis that the Universe has been evolving so as
to produce the human being (Dicke 1961; Carter 1974). This
paper formulates another principle - a galactic anthropic
principle (GAP). We suggest a hypothesis according to which
life forms and civilizations of our type can originate only
in galactic 'belts of life'. Possibly, it is not so. However, the fact that the solar system in the Galaxy is under
exceptional conditions makes this hypothesis attractive.

2.   The ring zone including the orbit of the solar system
in the Galaxy is exceptional due to the following circumstance. As is known, spiral arms of our and other galaxies
are density waves propagating through the stellar population
of galactic disks. The angular velocity of the Galaxy rotation Ω determined by its rotation curve is the decreasing
function of the galactocentric distance R, i.e. $\Omega = \Omega(R)$.
In the region close to the Sun's galactic orbit $\Omega_\odot = \Omega(R_\odot) =$
$= 25$ km/s kpc   and $R_\odot = 10$ kpc (according to the Schmidt
model (1965). To date there are some other models of the Galaxy (see review in Marochnik and Suchkov's book "The Galaxy" (1984) and refs. therein). However we do not change parameters $\Omega_\odot$ and $R_\odot$ used by Marochnik and Mukhin (1983)

because they differ only slightly from those used in new models. In this case, for example, Ω (R = 5 kpc) = 45,3 km/s kpc and Ω(R = 15 kpc) = 14 km/s kpc. In contrast to the Galaxy which rotates differentially ($\Omega = \Omega(R)$), density waves (spiral arms) rotate with a constant angular phase velocity Ω_p = const. The value Ω_p is a free parameter of the spiral-structure wave theory and is determined from observations. According to this theory, two types of density waves can be excited in galaxies - short-wave and long-wave modes forming a spiral structure. Along with other differences, it is essential that various values of Ω_p correspond to them. According to Lin et al. (1969) the spiral structure of the Galaxy is caused by shortwaves mode and for them Ω_p = 11-13 km/s kpc. According to Marochnik et al. (1972) the Galaxy's spiral structure is caused by long waves for which Ω_p = 23-24 km/s kpc, which point of view is correct can be solved only by observations.

The recent observation data confirm more confidently the second point of view (in more detail, see (Marochnik and Suchkov 1981). Apparently the most accurate self-consistent estimation of parameters of the Galaxy spiral structure from observational data gives the value Ω_p = 23.6 + + 3.6 km/s kpc Mishurov et al. (1979).

As $\Omega = \Omega(R)$ and Ω_p = const in each spiral Galaxy there is a corotation circle with the radius R = R_o where the condition $\Omega(R_c) = \Omega_p$ is satisfied. Within the corotation circle density waves are rotating in synchronism with the Galaxy rotation. Therefore the corotation zone (a narrow region around the corotation circle) in each Galaxy is an exclusive and unique place. As it follows from the figures given above, the solar system Galaxy is situated in the corotation zone of the Galaxy, i.e. it exists in specific conditions like other objects in this zone. According to Marochnik (1983), such a specific position of the Sun in the Galaxy makes it possible to solve an old problem of the co-existence of time scales in the solar-system cosmogony determined by the radioactivity of various nuclides. Besides, Marochnik et al. (1983) calculated a large-scale flow of the interstellar gas across the spiral arms of the Galaxy as was shown by them (see also Marochnik 1983), the conditions of the star-formation and the evolution of clouds of the interstellar gas inside and outside the corotation zone are essentially different (see Fig. 1). These circumstances together with the fact of the existence of the solar system and of life in it lead us to the GAP hypothesis.

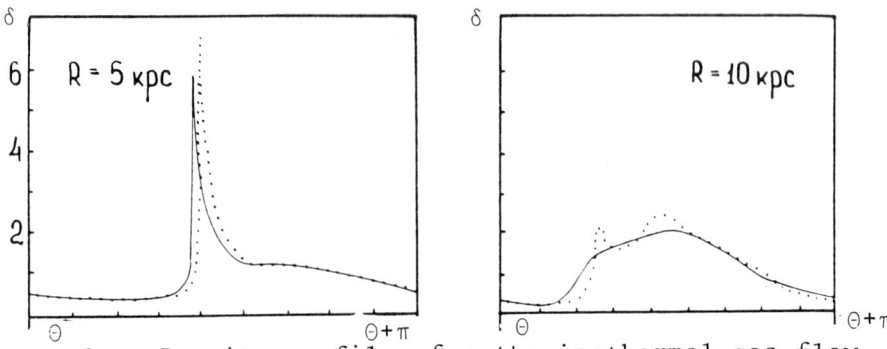

Figure 1.   Density profiles for the isothermal gas flow
in the density wave calculated for the two distances from
the galactic center, R = 5 kpc and R = 10 kpc (in the
neighbourhood of corotation). Two cases are presented:with
the gas self-gravitation (dotted lines) and without it (solid lines). The undisturbed surface density of the gas is
taken to be equal to that of the HI: μ_o = 3 $M_\odot$ pc^{-2} (Ma-

rochnik et al. 1983).

3.   When the interstellar gas rotating together with the
galactic disk flows into the spiral arm, on its internal
edge the galactic shock waves form where the gas compression
and, as a consequence, star formation occur (see review in
Marochnik and Suchkov (1984) and refs. therein). Small massive stars (like the Sun) and type II massive supernovac can
form here.
    The Sun in the Galaxy is situated between spiral arms
of Perseus and ogittarius and not within any of them. It
means that $\Omega_p \neq \Omega_\odot$ as stars are forming in arms though
the difference between Ω_p and $\Omega_\odot$ should be slight. Using the values of $\Omega_\odot$ and Ω_p given above we find for:

$$\Delta\Omega = \Omega_\odot - \Omega_p = 1.4 \pm 3.6 \text{ km/s kpc} \tag{1}$$

It follows that $\Delta\Omega/\Omega_\odot \sim \Delta\Omega/\Omega_p \sim 0.1$, i.e.   T, the period
of the Sun rotation relative to density waves (spiral arms)
is by the order of magnitude higher than the rotation period of the Galaxy in the co-rotation zone, equl to $T_o \approx$
$\approx 2.10^8$ years. So, the time scale ΔT is given in billions
of years. It is the order of the solar system 'life time'
estimated as $T_{ss} \approx 4.6 \ 10^9$ years. In other words, the
'lifetime' of the solar system is of the order of time which

it spends between the arms. Hence a scenario of Marochnik
(1983) according to which the collapse of the presolar cloud began when it was crossing the Sagittarius arm under the
effect of supernova explosions. Then cloud has been evolving and the solar system forming in the interarm space, i.e.
under quiet conditions (far from outbursting supernovae).
It is just in the interarm space where the solar system and
the life in it have reached their present stage. Now, the
Sun is moving in the direction of the Perseus arm. As will
be shown below, while entering this arm our civilization
may perish being subjected to the radiation of exploding
supernovae there. Apparently other civilizations formed in
the corotation zone may suffer the same fate. Therefore,
τ_{tot} is the total life-time of the civilization of our
type: according to GAP this is the time during which the
corresponding star with its planetary system (with life in
it) is moving from arm to arm. In this case civilizations
that have not reached our evolution level evidently have
$\tau_- = \tau_t < T_{ss}$ and $\tau_+ = \tau_t > T_{ss}$ for those that have. The time
left to our civilization till its probable death is $\tau_d =$
$= \tau_{tot} - T_{ss}$.

4.   Let us estimate the distances at which the supernova
nearest to the Sun may explode. The problem how the nearby
supernova explosion may affect the 'Earth biology' was first
considered by Krasovsky and Shklovskii (1957). The state-of-
-the-art of the problem is given by I.S. Shklovskii in his
book (1976). Let us consider the effect of the nearby supernova explosion on the 'Earth biology' from the view-points
of this paper.
      As a rule, spiral arms of galaxies are well approximat-
ed by a logarithmic spiral. Then the length of an arc section of the spiral arm is

$$L = \frac{\sqrt{1 + k^2}}{k} (R_2 - R_1),$$

where R_1, R_2 are distances from the Galaxy center up to
the ends of the spiral arc section in polar coordinates,
$K = 1/2 \, tgi$, i is the spiral pitch-angle. As $i \simeq 7\text{-}8^o$
$k \simeq 0.06$. From here with $R_2 = 14$ kpc $\gg R_1$ we find that
$2L \simeq 450$ kpc. The number of explosions SNII (of their total
number) at the distance R from the Sun is equal to the
ratio of volumes (Shklovskii, 1976b):

$$P = V/v, \tag{1}$$

where $\quad V = \frac{4}{3}\pi R^3$, $\qquad\qquad v = 2L\cdot\pi r^2$

In this case it is assumed that the spiral arm has a circular cross-section whose radius is r. On the other hand, the number of the explosion SNII at the distance R from the Sun when the latter is passing through the spiral arm $(\Delta T_\odot)$ is

$$P_1 = \frac{\tau_{SNII}}{\Delta T_\odot} \cdot P, \qquad\qquad (2)$$

where $\quad \tau_{SNII} = v^{-1}$, $\quad \Delta T_\odot = r/(\Omega_\odot - \Omega_p) R_\odot \sin i$; comparing (1) with (2) we find $R = |1.5\cdot L\cdot r\cdot \tau_{SNII}\cdot R_\odot(\Omega_\odot - \Omega_p)\sin i|^{1/3}$; with $r \simeq 150$ pc $\quad R_\odot = 10$ kpc $\Omega_p = 24.6$ km/s kpc (see below), we find $R \simeq 10$ pc.

As Shklovskii (1976b) showed the main effect of the nearby supernova explosion is the increase of the intensity of cosmic ray by about two orders of magnitude in the region with the radius ~ 10 pc surrounding the supernova. The natural radioactivity background of the Earth due to cosmic rays leads to a radiation dose of the order of $D_\odot \simeq$ $\simeq 0.04$ rem/year (Kozlov 1977). A hundred-fold increase of the intensity because of the nearby supernova explosion $(R \sim 10$ pc) leads to a dose of the order of $D \sim 4$ rem/ year. According to Shklovskii (1976) the solar system is moving through the radionebula that formed around the exploded supernova in about 10^4 years.

At the same time it is known (see refs. in Marochnik and Mukhin 1983) that the total average probability of a homo sapiens' death because of the radiation explosure normalized to 1 rem-dose

$$\rho = \rho_1 + \rho_2 = 1.40\ 10^{-4}\ \text{rem}^{-1}, \qquad\qquad (3)$$

where ρ is the total risk of death, $\rho_i = 10^{-4}$ rem^{-1} is the risk of death from cancer, $\rho_2 = 0.4\ 10^{-4}$ rem^{-1} is the risk of death from lethal mutations.

Hence, the rough estimate shows that with a dose of $D \simeq 4$ rem/year 0.056% of the world population should die every year. So, the entire population can die in 10^4 years if the mortality is exceeded by the re-production - by the natural increase of the population due to the birth rate.

In modern conditions the annual increase in the population is about 2.3% (Dubinin 1976) that essentially exceeds the risk of death. However, in earlier epochs the increase in the population was considerably lower. The Table summarizes the data from Dubinin (1976).

As follows from the Table, at the early stages of our civilization evolution the explosion of a nearby supernova should have been fatal for a human population as the natural increase of the birth-rate of population was lower than the death rate (3). On the other hand, in future the increase in the population will reduce sharply since hardly can our planet sustain more than 10 billion people (Dubinin 1976).

Therefore, in the future the probability of death due to radiation may be considerably higher than a possible increase of population.

Certainly, at present it is difficult to predict our future because, on the one hand, our civilization may kill itself in the course of a global war, and, on the other hand, it may found effective means for its protection from long radiation.

Hence, we can assume the lifetime (τ_{tot}) for a civilization identical to ours as a time for its star being between the Galaxy' spiral arms.

A phase angle characterizing the Sun's position between the Galaxy' spiral arms is $\chi_o = 151 \pm 9^o$ (Mishurov et al. 1979). Respectively, polar angles θ_1 and θ_2 (see Fig.2) are

$$\theta_1 = \frac{\chi_o}{2} \simeq 75^o, \qquad \theta_2 = \pi - \frac{\chi_o}{2} \simeq 105 \qquad (4)$$

From (4) we obtain

$$\tau_d = \frac{\theta_1}{\theta_2} \, T_{ss} \qquad (5)$$

Thus, with $T_{ss} \simeq 4.6 \times 10^9$ years we find the lifetime value left to our civilization (if some possible global cataclysms are not taken into account), i.e. $\tau_d \approx 3.3 \times 10^9$ years.

Using GAP let us estimate the upper limit of the total number of the our-type civilizations in the Galaxy. Assume that in the co-rotation zone the number of stars with plane-

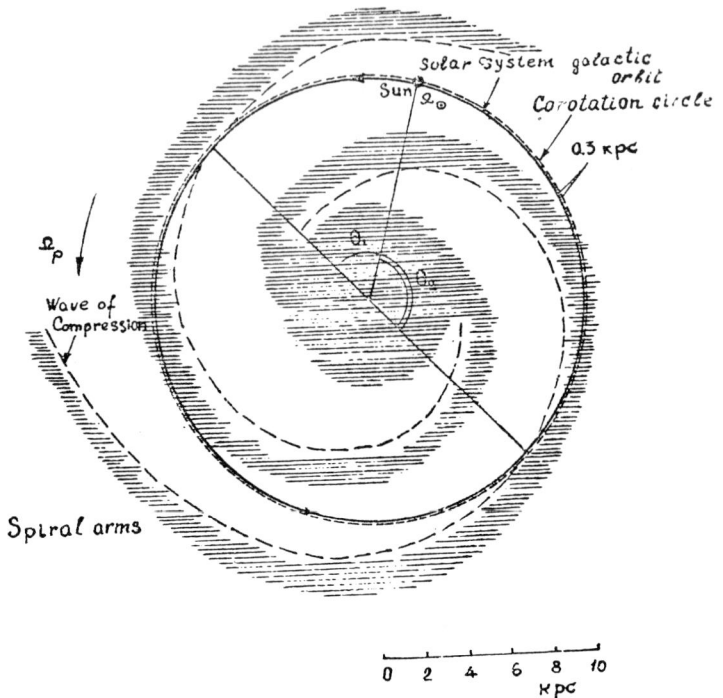

Figure 2. Schematic for spiral arms of the Galaxy and the present position of the Solar system moving along the galactic orbit.

tary systems is of the order of the number of sun-type stars (G_2-dwarfs). In the region around the Sun (a radius of 25pc), G_2-dwarf number is about 2 to 3% (Wolley et al. 1970; Starikova 1960).

The co-rotation ring thickness can be estimated from simple considerations. On the one hand, at present according to Marochnik (1983) the Sun is located from the co-rotation circle approximately by 300 pc. On the other hand, the Sun is moving over the epicycle relative to the galactocentric circular orbit on the scale $\Delta x = \dfrac{V_\odot}{\varkappa_\odot} \simeq 600$ pc where $V_\odot \simeq 20$ km/s is the Sun's peculiar velocity in the Galaxy and

$$\varkappa_\odot = 2\Omega_\odot \left(1 + \frac{r}{2\Omega_\odot} \frac{\partial \Omega_\odot}{\partial r}\right)^{1/2} \simeq 32 \text{ km/s.kpc}$$

Table

Epoch	Time during which the population doubles	Corresponding annual increase in the population %
From Ancient to Middle Paleolithic	170.000	0.0004
15,000 years since the New Paleolithic	10.000	0.007
1700 years since the beginning of our era	400	0.17
Since 1830	100	0.7
Ourdays	30	2.3

is the epicycle frequency at the solar neighbourhood. Thus, it is clear that the co-rotation ring zone thickness $\Delta R = R_c - R_\odot$ is of the order of $\Delta R \simeq 300\text{-}600$ pc (it is really narrow, as with the Galaxy's radius $R_G \simeq 15$ kpc $\Delta R/R_G \simeq 0.02\text{-}0.04$. Extrapolating for the whole co-rotation zone we shall find that the number of G_2-dwarfs is of the order of

$$n_{G_2}^{corot} \simeq 2\pi R_0 \Delta R \cdot \sigma_{G_2} \simeq 7 \cdot 10^7, \qquad (6)$$

where the surface density is $\sigma_{G_2} \simeq 0.03\sigma_{tot}$, $\sigma_{tot} \simeq 114$ M_0/pc^2 according to the Schmidt model (1965).

The number of civilizations in the Galaxy at the τ-level of their technological development can be obtained from the Drake formula (1961):

$$n = NP_1 \cdot P_2 \cdot P_3 \cdot P_4 \frac{\tau}{T_G}, \qquad (7)$$

where N is the number of stars in the Galaxy; P_1, P_2, P_3, P_4 are the probabilities that near the star there are a planetary system, a life, an intelligence, and a technology, repsectively; τ is the time scale of the technological era, T_G is the Galaxy's age.

According to GAP (7) can be rewirtten as follows:

$$n = n_{G_2}^{corot} \, P_1 \cdot P_2 \cdot P_3 \cdot P_4 \, \frac{\tau}{\tau_{tot}} \,, \qquad (8)$$

where τ is the time from the moment when the given civilization escapes (together with its star) the spiral arm, the place of its birth. We now find the upper limit n_τ if $P_1 = P_2 = P_3 = P_4 = 1$. It means that in the co-rotation zone the number of stars with planetary systems and the intelligent life on them is of the order of $n_{G_2}^{corot}$. In its essence this hypothesis follows the GAP concept. As $\tau_{tot} = \tau_d + T_{ss} = 7.9.10^9$ years according to (8) and (6) the possible number of the civilizations being at our technological level in the corotation zone is by the order of magnitude

$$n_{T_{ss}} \simeq n_{G_2}^{corot} \, \frac{T_{ss}}{\tau_{tot}} \simeq 4.10^7 \qquad (9)$$

From (5) we can find the ratio of the possible number of upper-level civilizations to the number of lower-level civilizations:

$$\frac{n_\tau^+}{n_\tau^-} = \frac{\tau_d}{T_{ss}} = \frac{\theta_1}{\theta_2} \simeq 0.7 \qquad (10)$$

Estimates (9) and (10) give the upper limit of the possible number of technological civilizations within the frame of the GAP hypothesis. If we are not alone indeed the orbit of the solar system can be figuratively called a 'road of life' similar to the name for the co-rotation zone - a 'belt of life' - in the Galaxy. It is evident that the lower limit of the number of technological civilizations is $n_\tau = n_{T_{ss}} = 1$ that speaks about the uniqueness of our civilization. In this case our civilization existence can be an accidental event. Using the symbolics used our famous fantasts A. & B. Strugatskie we say that this situation reminds the "picnic on the road side" of the Galaxy (the solar system is almost on the Galaxy's edge) made up by the Nature, the accidental event had led to developing our civilization.

All concepts given in this paper are equally referred to other spiral galaxies.

Acknowledgements

We should like to express our deep gratitude to Professor I.S. Shklovskii for essential remarks and to Professor A.M. Poverenny for the fruitful discussion.

REFERENCES

Allen K.U. 1973 "Astrophysical Quantities", Univ. of London, The Athlone Press.
Carter B. 1974, in Proc. IAU Symp. No. 63, ed. M. Longair, Reidel Publ. Comp. Boston, USA.
Dicke R.H. 1961 Nature, 192, 440.
Drake F. 1961. Discussion at Space Science Board Nat. Acad. Sci. Conf. on Extraterrestrial Intelligent Life, Green Bank, USA.
Dubinin N.P. 1976. Obscaya genetika, Moscow, Nauka (Russian).
Kardashev N.S. 1965, in the book (Russian): "Vnezemnye civilizatsii", Publ. by Armenian Acad. of Sci., Erevan, p. 37.
Kardashev N.S. 1981, in the book (Russian): "Problema poiska vnezemnykh civilizatsii", Moscow, Nauka, p. 29.
Schmidt M., 1965, in "Stars and Stellar Systems", v. IV, p. 513.
Krasovskii V.J. and Shklovskii J.S. 1957, "Doklady" Acad. Sci. USSR 116, 197.
Kozlov, V.F. 1977. Spravochnik po radiatsionnoi bezopastnosti (Russian), Moscow, Atomoizdat.
Lin C.C., Ynan C. and Shu F. 1969. Astrophys. J. 155.
Marochnik L.S. 1981 Acad. Sci. USSR "Doklady", 261, 571.
Marochnik L.S. 1983, Ap. Space Sci., 89, 61.
Marochnik L.S., Berman V.G., Mishurov Yu.N., and Suchkov A.A. 1983, Astrophys. Space Sci., 89, 177.
Marochnik L.S., Mishurov Yu.N., and Suchkov A.A. 1972. Astrophys. Space Sci. 19, 285.
Marochnik L.S. and Mukhin L.M. 1983. Preprint of Space Research Institute of USSR Acad. Sci. N. 761.
Marochnik L.S. and Suchkov A.A. 1981. Astrophys. Space Sci. 79, 337.
Marochnik, L.S., and Suchkov, A.A. 1984. "The Galaxy", Moscow, Nauka (Russian).
Mischurov Yu.N., Pavlovskaya E.D. and Suchkov A.A. 1979. Astron. J. (Russian), 56, 268.
Molten P.M. 1973, Space flight, 15, 139.
Shklovskii J.S. 1976. Voprosy philosophii, No. 9, 80.
Shklovskii J.S. 1976b. "Supernovae", Moscow, Nauka.
Starikova G.A. (1960). Astron. J. (Russian), 37, 476.
Wolley R., Epps E.A., Penston M. and Rocock S.B. 1970, Catalogue of Stars within 25 parsecs of the Sun, Roy Obs. Ann., 5.

THE GALACTIC BELT OF INTELLIGENT LIFE

B. Balázs
Department of Astronomy, Eötvös University
Kun Béla tér 2, Budapest, Hungary 1083

ABSTRACT. Current estimates of the likelihood, galactic distribution
and accessibility of extra-terrestrial civilizations generally contain
three shortcomings: They treat our Galaxy as a homogeneous, isotropic
and steady-state system and not as an object of specific geometric and
kinematic properties with reasonably well understood morphology and path
of evolution. If we assume that the case of mankind is about average and
accept the idea that the longevity of a civilization might be limited
with high probability by catastrophic events threatening during the
crossing of galactic arms, intelligent life is presumably concentrated
on a belt in the Galaxy which is a narrow annulus including the co-
rotation circle and the galactic orbit of our Sun. If the galactic belt
of intelligent life is a reality at least the first and last factors in
the "Drake Equation" must be reevaluated.(The number of suitable stars
in the belt is only of the order of 10^8 and the average longevity of a
civilisation needs to be judged in comparison with the time which its
system spends between two neighbouring spiral arms.) Supposing that intelligent life will develop on the same time-scale, by the same rules
wherever the proper surroundings and the needed time are given, it is
possible to locate a zone of advanced civilisations where societies at
least as old as ours are primarily expected. From heliocentric point of
view the distribution of our potential extraterrestrial partners is
highly anizotropic: in a small solid angle around the line of sight
there are about 10^3 times as many of them in the tangential directions
than towards the galactic anticentre.

What is the distribution of intelligent life in the Galaxy? A simple
question, an exciting question and clearly a question whose answer is
most difficult to achieve. But it is a question that surely should be
asked and answered in order to elaborate any suitable search strategy,
and a question which is impossible to answer professionally without
taking into account the geometric and kinematic properties of our stellar
system.

    The gravitational density wave theory of Lin and his co-workers
(1969) is currently the most popular of theories which can provide an

G. Marx (ed.), Bioastronomy – The Next Steps, 61–66.

acceptable quantitative viewpoint from which it is possible to explain the large-scale galactic spiral structure in a coherent way. This theory involves the formation of stars along a rather concentrated shock front which co-rotates with the spiral pattern as a rigid body around the galactic centre.

The interstellar gas accelerated by the density-wave gravitational field forms the shock which can trigger the formation of new objects and induce star formation in general. Therefore the locus of star formation is expected to move according to the pattern velocity which -- in view of the differential rotation of the Galaxy -- generally departs from the circular velocity of the stars and interstellar clouds and thus one could expect to find systematic effects in the shifts of spiral arm tracers of different ages.

As we know, in the framework of Lin's theory the age difference, ΔT, between us and another planetary system with heliocentric galactic co-ordinates r, l, b can obviously be written in the form:

$$\Delta T = F[r, l, b, \Omega_p, \Omega(R), \mu(R), \sigma(R)], \tag{1}$$

where $\Omega(R)$ is the galactic rotation curve, $\mu(R)$ the surface density distribution in the galactic plane, $\sigma(R)$ the radial velocity dispersion of the stars and Ω_p the angular velocity of the spiral pattern. It is to be remarked that owing to the possible diffusion of stellar orbits, the epyciclic motion of the stars and the incubation spread which is time and again observed in connection with star formation, (1) actually holds only for the average age difference, $\langle \Delta T \rangle$, between us and systems in the vicinity of a galactic point with the coordinates given above.

The angular pattern speed is one of the most important model para-meters and is usually chosen so that the co-rotation radius, R_c, defined by $\Omega(R_c) = \Omega_p$, is equal to the galactocentric distance of the outher-most visible H II regions (cp. Feitzinger and Schmidt-Kaler, 1980). Ac-cording to this procedure the angular velocity of the spiral pattern must lie in the range $11 < \Omega_p < 21$ km s^{-1}/kpc if our galactocentric radius, R_o, is 10 kpc. It is important to be able to check this result in an in-dependent way and there are a number of reasons why the galactic distri-bution of the open star clusters is particularly suitable for this pur-pose (Balázs and Lyngå, 1983).

The spacial distribution of about 200 open clusters with ages bet-ween 10^6 and 10^8 years suggest that the galactocentric distance of the Sun is about 7 kpc and the angular speed of the pattern lies somewhere between 33 and 36 km s^{-1}/kpc or between 23 and 25 if one uses 10 kpc for R_o (cp. Balázs and Lingå, 1983).

It appears, therefore that Ω_p is considerably larger than was originally accepted and that not the outhermost H II regions but the stars in the solar neighbourhood lie in the zone of co-rotation. The rather small value of $\Delta \Omega_o = \Omega(R_o) - \Omega_p$ is empirically supported by an earlier work of Blaauw (1964) who found that the associations of the solar neighbourhood consist of a number of subsystems of different ages. The direction of separation seems to be nearly parallel to the galactic plane, and the linear velocity of the physical mechanism which initiated the birth of this subsystems turns out to be between 5 and 10 km/s. This is in good agreement with our result, but entirely incompatible with the rather large $\Delta\Omega_O$ proposed by Lin.

The smallness of $\Delta\Omega_O$ is further supported by the existence of old stellar groups in the solar neighbourhood. The dependence of the maximal kinematic age of stellar groups on the value of Ω_p was numerically investigated by Korchagin (1986). The change of Ω_p from 13 to 24 km s^{-1}/kpc increases the life expectancy of stellar groups from $7\,10^8$ years to roughly $4.5\,10^9$ years. So the existence of old stellar groups with an age of about 5 billion years in the vicinity of our Sun confirms the previous conclusion that $\Omega_p/\Omega_O \cong 1$.

The empirical picture outlined above is quite in accordance with the theoretical results of Marochnik and his co-workers (1972). It is easy to see that in this case the age of the solar system is comparable with the period during which the Sun remains between two spiral arms. Its present position on its galactic orbit, more exactly its radial phase ($\chi_O = 152^{\circ}$) between the Sagittarius and Perseus arms corresponds to an Ω_p value of 24.7 km s^{-1}/kpc (if $\Omega_O = 25$ in the same units) and the uncertainty due to the epicyclic motion of the Sun is about 10%. As it is known that in the vicinity of the Sun the galactic rotation curve is linear with a slope of $-2A/R_O$, where A is Oort's galactic rotation constant, the co-rotation radius comes to

$$R_c = R_O(1 + \Delta\Omega_O/2A) \ . \tag{2}$$

If $R_O = 10$ kpc and $A = 15$ km s^{-1}/kpc R_c comes to 10.1 kpc, so that the proximity of the Sun to the co-rotation circle is really striking.

It is clear from the foregoing that in the well known "Drake Equation" the Galaxy cannot be treated as a homogeneous, isotropic body, but as a system of definite geometric and kinematic properties with reasonably well understood morphology, history and path of evolution.

Now, if we -- following Shklovskii and others (Krasovskii and Shklovskii, 1957; Clark at al., 1977; Marochnik, 1983) -- accept the idea that the longevity of a civilization might be limited with a high probability by close supernova explosions, the life expectancy of advanced civilizations is the time which their system spends between two neighbouring spiral arms -- where the occurence of fatal cosmic events is rendered unlikely -- and the belt of extraterrestrial civilizations presents itself as a surprisingly narrow one.

Indeed, since the radial phase of the Sun is 152°, objects with $\Delta\Omega = (360/152)\,\Delta\Omega_O = 0.7$ km s^{-1}/kpc which left the Sagittarius arm or the Perseus one in the direction of the galactic rotation $T_{crit} = 4.6\,10^9$ years ago have by now just reached the other arm and conversely, objects of the same age and parent arm but with $\Delta\Omega = -0.7$ km s^{-1}/kpc have similarly travelled the whole way in the opposite direction between the two spirals. Denoting the galactocentric radius of these objects by R_{crit} and making use of equation (2) we get $R_c - R_{crit} = \Delta\Omega/3 = \pm0.23$ kpc for the half width of the belt where we can primarily expect advanced civilizations (Balázs, 1986a; Fig. 1).

It is known that basically as a reaction against exaggerated subservience to the "copernican principle" Brandon Carter in the early seventies introduced the so called "anthropic principle", which in its

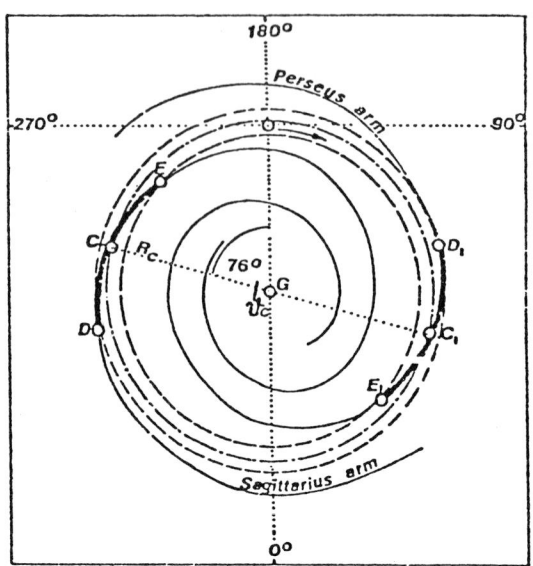

Figure 1. "The belt of intelligent life" in the Galaxy. G is the galactic centre; C, C_1, D, D_1, E, E_1 are the intersections of the spiral arms with the circle of co-rotation and with the edges of the belt.

weak form declares that "we must be prepared to take account of the fact that our location in the universe is necessarily priviliged to the extent of being compatible with our existence as observers."

Now, if we assume that the case of mankind is about average and accept the idea that highly developed planet-dwelling life is not likely to survive the catastrophic events threatening during the crossing of spiral arms, our very presence and the possibility to discuss extraterrestrial contact problems shows that we and our potential non spacefaring partners live close to the galactic circle of co-rotation.

There are thus morphological arguments favoring the concept of a "belt of advanced civilizations" based on a "kinematic ecosphere" in the Galaxy in the form of an annulus with a breadth of roughly 0.5 kpc including the co-rotation circle and the galactic orbit of our Sun. We should concentrate our efforts to contact extraterrestrial beings on this narrow belt.

In the discussion of the number (N_c) of technological civilizations in our Galaxy two general forms of the Drake Equation have been widely used (see f.i. Kreifeldt, 1971). For our purpose Cameron's formulation is the more suitable one, according to which

$$N_c = N_s f_s L_c / L_p \quad , \tag{3}$$

where N_s is the number of potentially suitable parent stars, f_s is a compound selectivity factor, L_c is the length of the communicative phase of a society and L_p is that part of the lifetime of a planet during which highly developed species can exist on it.

According to certain interpreters of the so-called Fermi Paradox f_s must be extremely small because we do not have any concrete evidence of alien activity in the solar system. It is surprising that the extra-ordinary weakness of this argumentation has gone unmentioned in most discussions.

A trans-Plutonian spherical solar system boundary encloses 260 000 AU^3 and 10^{11} km^2 planetary and asteroidal surface. Current surveys could not have detected a 1 to 10 m probe more than $0.01 - 1$ AU from Earth, so planetary space is at least 99.999% unexplored for artifacts of this size!

Less than 10 % of the Earth surface, 1 % of the Moon, 0.1 % of Mars and 10^{-7}% of Venus (total: $5\,10^7$ km^2) has been surveyed to 1 to 10 m visible resolution. This leaves 99.96 % of solar system surface area ($1.3\,10^{11} km^2$) unexamined for likely artifacts! (Freitas, 1985.) That means that we simply do not have the right to speak about a paradox!

If the galactic belt of intelligent life is a reality the first and last factors in eqn (3) must be reconsidered. As far as N_s is concerned, it is clear that in our case only stars in the belt of life can be regarded as suitable ones. Since the stellar density in the solar neighbourhood is 0.13 */pc^3 and one can take around 200 pc for the thickness of the belt, the volume of the ring comes about 6 kpc^3 and we get $N_* = 8\,10^8$ for the number of the stars contained in it. As the only stars however, that have long lasting habitable zones around them are of masses between about 0.8 to 1.2 $\mathfrak{M}_\odot$ (see f.i. Tucker, 1981) and roughly 10 % of the stars lie in this mass range, the number of potentially suitable stars in the belt of life is only about $N_s = 8\,10^7$.

One can see at the first glance (Fig. 1) that from heliocentric point of view the distribution of this stars is highly anisotropic. The an-isotropy is growing even further if we restrict the scope of our investi-gations to the "upper" half of the belt (CSC_1) and take into account that we are only at the beginning of our technical civilization and there-fore our potential partners for extraterrestrial communication must have at least as old a society as we have.

One can compute at each galactic longitude the lenght of the section of our line of sight (Δr) which falls into the advanced part of the belt of life (Balázs, 1986b). As shown in Fig. 2, the maxima lie roughly at 90° and 270° and the minimum shows up in the vicinity of the galactic anticentre. The maximum to minimum ratio is around ten, consequently the volume contained within a small solid angle ω around our line of sight in the advanced part of the belt is approximately 10^3 times larger in the tangential direction than towards the anticentre. Choosing f.i. a solid angle of 10^{-4} sr one deals with $3.7\,10^4$ stars in the first case and with 41 objects in the second one. As only around 10 % of the stars have long lasting habitable zones the number of potentially suitable planetary systems is in our cases $3.7\,10^3$ and 4, respectively. Fig. 2 is therefore of essential importance for the elaboration of a sound search strategy aiming the detection of extraterrestrial intelligence.

As regards the last factor of the Drake Equation, only L_p can be judged by astronomical methods. Basically, L_p is limited by the evolution of the parent star and by stellar kinematic factors. Limitations of the

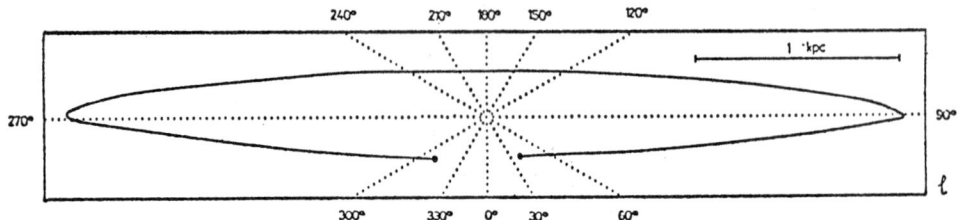

Figure 2. The length of the section of our line of sight which falls into the advanced part of the belt of life as a function of the galactic longitude. The curve is terminated at $\ell(C)$ and $\ell(C_1)$ (heavy dots).

first type were widely discussed already (cp. Zuckermann, 1985). Stars with nearly the mass of the Sun have main-sequence lifetimes (T_Z) that vary like $\mathfrak{M}^{-3}$. In the suitable mass range of 0.8 to 1.2 solar masses T_Z varies between 6 and 20 gigayears.

Regarding factors of the second type, L_p is limited by the crossing time (T_{cr}) which a planetary system spends between two spiral arms. It is easy to see that for the given galactic rotation curve and angular speed of the spiral pattern T_{cr} is inversely proportional to the distance from the co-rotation circle (Balázs, 1986b). Since the life expectancy of the suitable stars is at least 6 Gyrs and for an equally long crossing time $|R_c - R| = 0.16$ kpc, L_p is purely kinematically limited between this circle and the edges of the belt. Conversely, as the main sequence lifetime of a star with 0.8 solar masses is nearly 20 Gyrs, L_p is governed only by the evolution of the parent star for all planetary systems which are closer to the circle of co-rotation than 50 pc. In the rest of the belt, depending on the mass and co-rotation distance of the central star, both cases can occur.

REFERENCES

Balázs, B.A., and G. Lyngå: 1983, Publ.Astron.Inst.Czech.Akad.Sci. 56,37.
Balázs, B.A.: 1986a, Acta Astronautica, 13, 123.
Balázs, B.A.: 1986b, preprint IAA-86-482.
Blaauw, A.: 1964, Ann. Rev. Astron. Astrophys. 2, 213.
Clark, D.H., W.H. McCrea, and F.R. Stephenson: 1977, Nature, 265, 318.
Feitzinger, J.V., and Th. Schmidt-Kaler: 1980, Astron. Astrophys. 88, 41.
Freitas, Jr., R.A.: 1985, Icarus, 62, 518.
Korchagin, P.I.: 1986, Astron. Tsirk. No. 1462.
Krasovskii, V.I., and I.S. Shklovskii: 1957, Dokl.Akad.Nauk USSR, No.116, 197.
Kreifeldt, J.G.: 1971, Icarus, 14, 419.
Lin, C.C., C.Yuan, and F.Shu: 1969, Astrophys. J. 155, 721.
Marochnik, L.S. at al.: 1972, Astrophys. Space Sci. 19, 285.
Marochnik, L.S.: 1983, Astrophys. Space Sci. 89, 61.
Tucker, W.H.: 1981, in Life in the Universe, ed. J. Billingham, MIT Press,
                                                       Cambridge, p. 287.
Zuckermann, B.: 1985, Quart. J. Roy. Astron. Soc. 26, 56.

LANDSCAPES ON AN IMAGINARY PLANET

B.B. Mandelbrot

IBM Research Center, Yorkton Heights NY 10598 USA
Yale University, New Haven CN 06520 USA

Transcribed from video tape by George Marx
Edited by the speaker

Shapes in Nature that Elude Euclid's Geometry

It is a great honor to be invited to visit Hungary. I have always known many Hungarians very closely, and some of them have greatly influenced my life. In particular, John von Neumann was my mentor for a certain time, shortly before he left Princeton for public service in Washington. I was told he was the most Hungarian of the Hungarians, and now may I have a chance of seeing whether the story is right.

Let me first present you with a series of slides. The first is a computer generated imitation of a branching tree (Figure 1). It exemplifies very clearly the basic property of every fractal: that its parts are like the whole, except of course that they are smaller. In this tree, you see that each piece is neither more complicated nor simpler than the whole — except that as you get near the branch tips, you cross over to a very different behavior.

A straight line has the same property: it is made of parts each of which is like the whole, but smaller. Thus, the straight line is in certain sense the trivial limit example of the shapes that interest us. But of course we are interested in the many highly non trivial examples. Amazingly, we have found ways, using the computer, to generate fractals that look like clouds or mountains!

When one says that "each part is like the whole", the word "like" may have several different technical meanings, and each different meaning leads to a different theory. In order to model the world around us, it is necessary to mix different techniques. Instead of seeking theoretical purity, the scientist must seek effectiveness. We can generate pictures of mountains in such a way that, as you zoom in, newer and newer details are made visible, and that these details show the same degree of roughness like in real mountains. The effectiveness is further enhanced by adding fog in the valleys (Plate 2).

We can use the computer to construct a landscape on the Moon (or — as Lucasfilm has done — a landscape on a imaginary faraway planet). This is possible, because I have found the mathematical formula that is appropriate here. Unfortunately, I must *deny* the persistent ugly rumor that we also made the NASA pictures, for example those which show the moons of Jupiter! Fractals, however, definitely affect the budgets of science fiction films. In this way, something that used to be part of very pure mathematics, sophisticated and very mysterious, has become of interest for the millions of people who watch this kind of entertainment. I am often asked whether or not I am bothered by this development. My answer is that, *of course*, I am not bothered at all! In fact, I am very pleased. I am convinced that, in the same way as there should be no boundary between mathematics and physics, there should be no boundary between science and common experience. Any connecting bridge between the two is therefore most welcome!

G. Marx (ed.), Bioastronomy – The Next Steps, 67–76.

Computer generated fractal landscapes from the book THE
FRACTAL GEOMETRY OF NATURE by Benoit B. Mandelbrot (published by W.H.Freeman 1982).Reproduced by the permission
of the author.Copyrights by B.B.Mandelbrot and R.F.Voss.

The shape of clouds is elusive, but this shape has also been grasped by our fractal models, and imitated on the computer. Indeed, the new geometry I have the privilege of having developed and named includes shapes that have precisely the same degree of richness as the clouds.

What do I mean by "new geometry"?

Consider, on the one hand, the shapes that Euclid, Newton and so on have made so very well known, such as straight lines, triangles, planes, cubes, spheres. The beauty of the geometry of Euclid is extraordinary, yet it is limited in its variety and therefore in effectiveness. I think this is so because these shapes become increasingly simple as one gets closer and closer to them. On the other hand, mathematicians continually invent shapes that become more and more complicated as one gets closer and closer to them. On such shapes true chaos reigns; they represent complete disorder, complete lack of structure. Between these two extremes, there used to be no organized mathematics. But now there is the new field called *fractal geometry*. We may say that fractals are objects of a third kind which exist between Euclid and Chaos. The most widely available two books published about this topic are *The Beauty of Fractals* by Heinz-Otto Peitgen and Peter Richter (1986), which I strongly recommend, and of course my own book, which started the new field, *The Fractal Geometry of Nature* (1982).

There are two aspects to fractal geometry, like there are two faces to a coin: understanding natural shapes that used to completely elusive, and creating new shapes. These two faces are strongly correlated, but for didactic reasons it is best to describe them separately.

Fractal Geometry

To understand the fractals, let us trace their origin. The history of pure mathematics includes a very strange episode, that is not without parallels in psychological or political history. About 100 years ago, the mathematicians decided to separate themselves from physics. They set out to invent something new, to be called *pure mathematics,* that would no longer be intimately linked to physics, as had been the case of mathematics in the past. Quite the opposite: this new pure mathematics was expected to deal exclusively with shapes that had been created *ex nihilo* by human minds. Better: it was not to deal with the shapes themselves, but exclusively with the underlying ideas. One of the main early representatives of this trend was Georg Cantor. He did not create new shapes for the purpose of understanding the objects of Nature, but for a completely orthogonal purpose: for the pleasure of creating something completely independent from the outside world. This is why he created the Cantor set (Figure 2). Actually, the story is more complicated than that. The man who originally discovered this set was an Englishman named Henry J.S. Smith. But he did not do much out of it. It was Cantor who first saw its importance. How to represent this Cantor set or dust? Draw a straight interval, and take away the middle third of it, then take away the middle thirds from each of the two resulting pieces, and so on. By definition, the Cantor dust is self-similar, since the whole is just a three times bigger version either of its two parts. To Cantor, however, self-similarity was not of any significance; it was just a question of economy in writing down a formula. For his purposes, he might just as well have taken off a middle third, then middle fourths, then middle thirds again, etc.. Indeed, the mathematicians find greatly generalized dusts to be much more interesting as objects of study. To the contrary, when − one hundred years later − I started on the task of grasping certain elusive shapes of Nature, the notion of geometric self-similarity became very important. The reason was that I found many aspects of Nature itself to be self-similar! In the science that

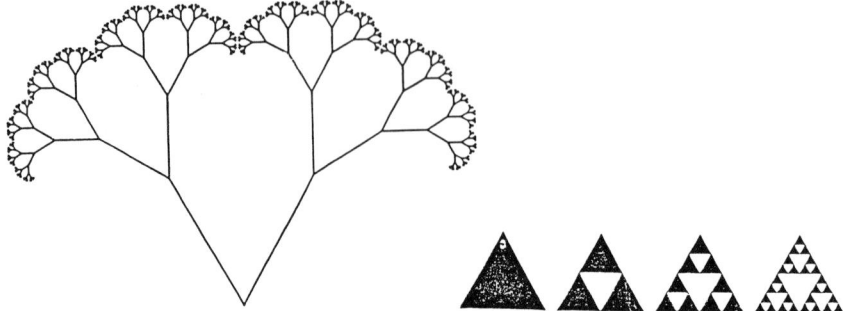

Figure 1

Figure 3

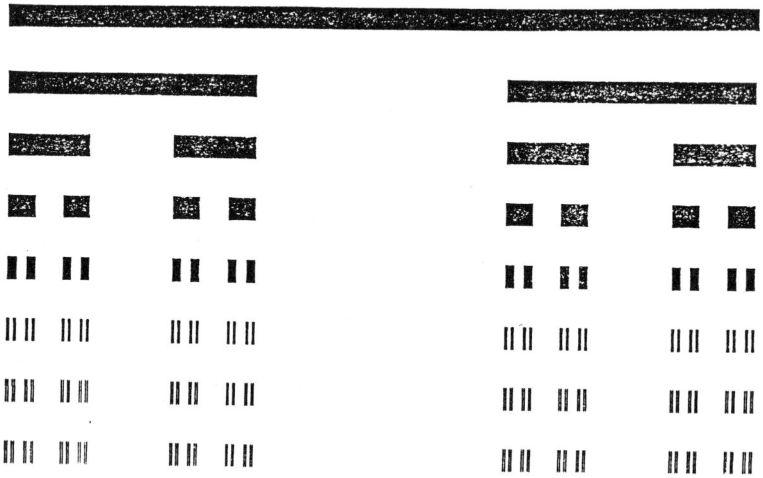

Figure 2

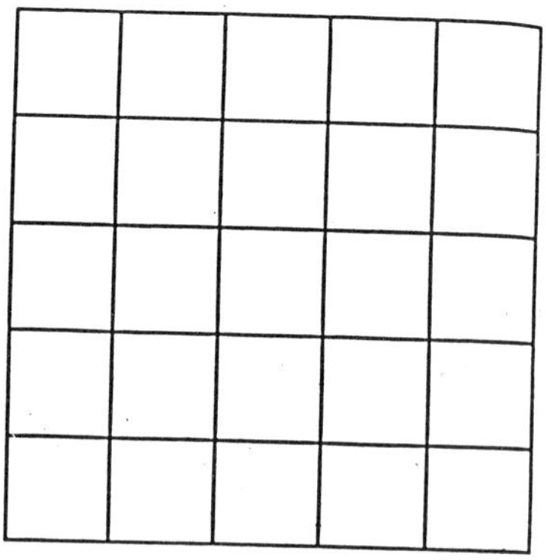

Figure 6

Figure 5

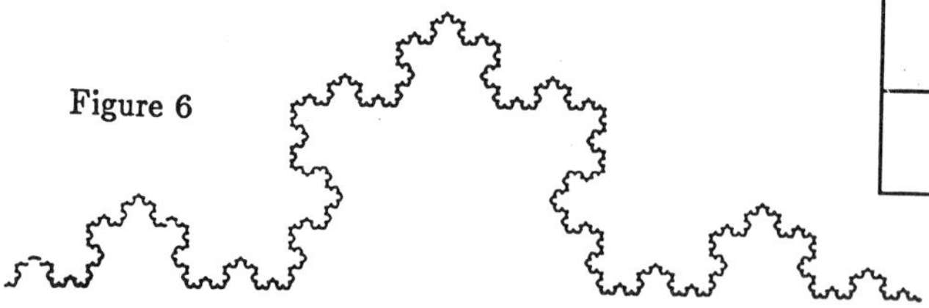

emerged, self-similarity is no longer a matter of mental economy, but a basic and surprising principle concerning Nature. This certainly has advantages when we want to imitate the shapes of Nature by computer.

A very nice self-similar shape has been introduced by Sierpinski (Figure 3). It is obtained by drawing a triangle, and then removing the central triangle with half side lengths, then in repeating this operation again and again. Why was this funny shape introduced? In order to improve upon Euclid's world of geometrical curves, in which double points are exceptional; they occur for example where the borders of three countries meet on the map. Like Cantor, Sierpinski proclaimed his imagination to be more fertile than the map! "I have the power to imagine a curve which has double points everywhere"! He removed regular triangles again and again, using an unchanging rule, but (again) using this very special case was just a matter of convenience for him.

Gustave Eiffel was a mechanical genius, but he also wrote a book about his Eiffel Tower. This book describes how he proceeded to make his construction lighter and lighter. We can make a new "Eiffel Tower" by starting from a regular tetrahedron, and by removing the center to leave four tetrahedrons, each with half side length, and by doing the removals again and again (Figure 4). In this way, we get another self-similar structure.

Henri Poincaré was a great thinker who he never made a sharp distinction between pure mathematics and Nature. He is quoted frequently as an initiator of relativity theory, but he also should be mentioned as a precursor of fractals and chaos. Perhaps one reason why he did not go farther is that the computer did not yet exist. Now, with the availability of fractal geometry and of the computers, these types of investigation have become very popular.

A first definition of fractal dimension was given by Hausdorff, but because of technical complications that definition has not proven important for physics. The simplest definition is one that applies to self-similar fractals. It is the similarity dimension defined by $D = \log N/ \log(1/r)$, i.e. $Nr^D = 1$, where r is the resolution of the geometric measurement and N is the number of resolved pieces. It is called a dimension, because in the case of Euclidean shapes it reduces to their Euclidean dimension.

If we divide a straight interval of length $1m$ into 5 pieces (Figure 5), then $r = 0.2$ and $N = 5$, therefore $D = 1$. If we divide it into $N = 100$ pieces, then $r = 0.01$; again we obtain $D = 1$. The straight line is a self-similar one-dimensional object.

If we divide a square of side length $1m$ into smaller squares of side length $r = 0.2m$, we find $N = 25$, therefore $D = 2$. The same dimension is obtained if we use a finer resolution. The square is a self-similar two-dimensional object.

If we take a cube and we cut it into small cubes of side length $r = 0.2m$, we get $N = 125$ small cubes, therefore $D = 3$. If we cut finer with $r = 0.1m$, we obtain $N = 1000$, hence $D = 3$ again.

Now, it order to introduce the concept of fractal dimension in a non trivial context, let us draw the Koch curve (Figure 6). Here, if we take $r = 1/3$, then $N = 4$, therefore $D = \log 4/ \log 3 = 1.2619$, which is not an integer. If we take $r = 1/9$, we get $N = 16$, giving the same fractional D. Because the Koch curve is self-similar, its dimension is again independent of the resolution of the geometric measurement. This quantity D can be used to characterize the Koch curve. Self-similar shapes with fractional dimension are the simplest among the shapes for which I have coined the name *fractals*.

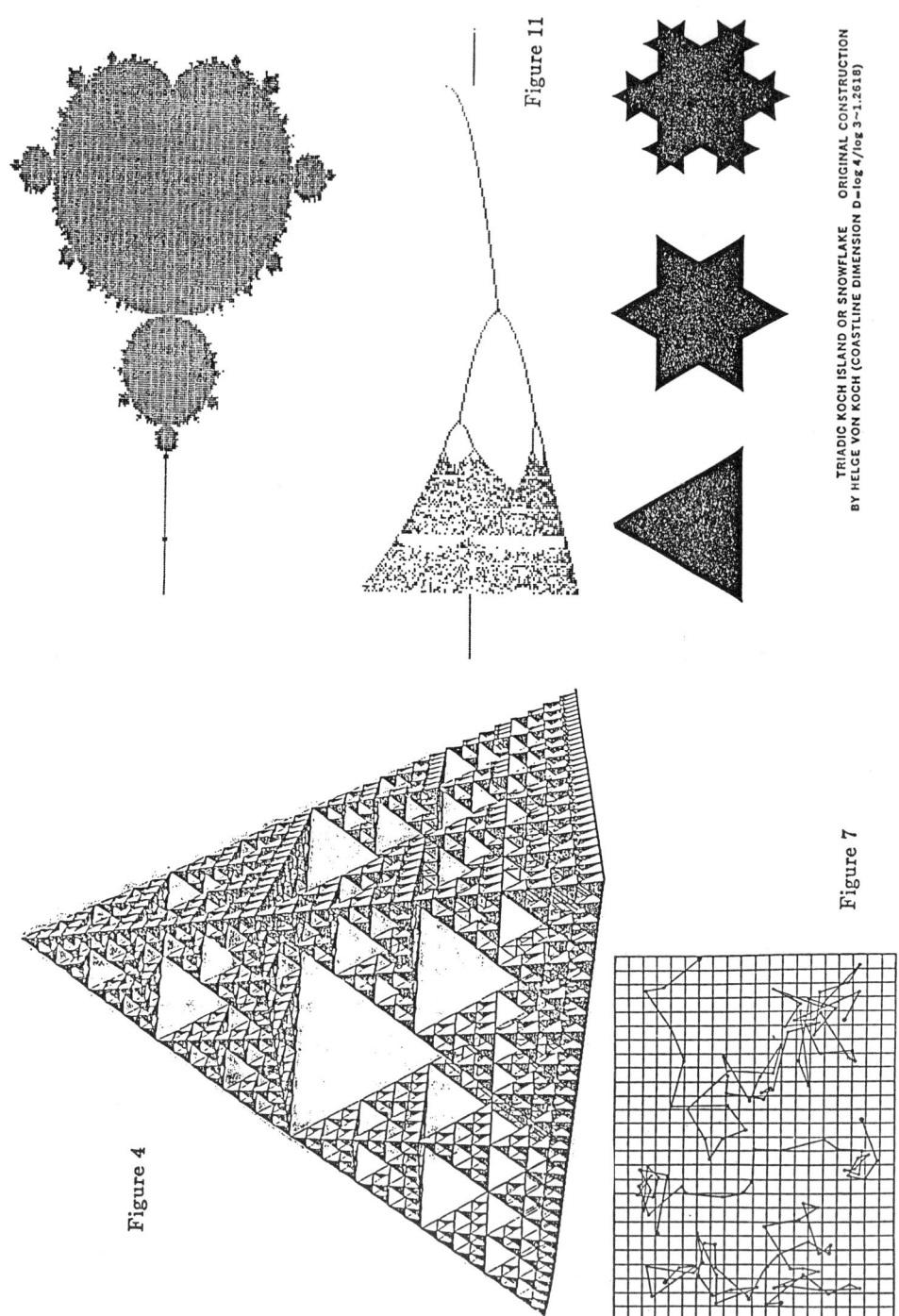

Figure 11

Figure 4

Figure 7

TRIADIC KOCH ISLAND OR SNOWFLAKE   ORIGINAL CONSTRUCTION
BY HELGE VON KOCH (COASTLINE DIMENSION D=log 4/log 3 ~1.2618)

Fractal Dimensions in Nature

The length of the coastline of England is found to depend on how we measure it, but there is order in how this length increases. Indeed, coastlines may be characterized (approximately) as having a fractal dimension $D > 1$. The reason is that the length measured to the resolution of r varies like $L = Nr \sim r^{-(D-1)}$. The empirical observation that the coastline length increases with decreasing resolution r is a consequence of $D > 1$. Smaller countries tend to use a smaller resolution r when giving the length of their border. This is why the length of a common border is often quoted to be longer in the geography books of the smaller country than in that of the bigger one. This fact in itself indicates $D > 1$.

Why are the mountains self-similar, and why does the roughness (the fractal dimension) of the surface of a mountain usually fall within a definite and narrow range of values? I don't know, and no one else knows, as yet. But when the computer draws a mountain with a fractal dimension higher than those actually measured, we immediately see that the picture fails to be irrealistic!

A young scientist has investigated pictures of clouds on the sky, and compared them at different resolutions. It turned out that all pictures indicate the same fractal dimension for the cloud boundaries. But clouds observed under different atmospheric conditions have different fractal dimensions, and at present it is not yet possible to deduce the fractal dimension of any cloud from the basic equations of physics. The fault, however, does not reside in the concept of the fractal dimension of the cloud − which is now firmly established. The fault is with the primitive state of our understanding of the basic equations. Progress of understanding meteorology is coming from this direction, but the investigations are still very phenomenological.

My models of mountains and of clouds are meant to implement the maximum extent of self-similarity "or self-alikeness" in the system, under the most economical constraints. Our forgeries do not assume anything more from Nature. Do we derive these constraints from physical equations? We do not, but it is not a failure of fractal geometry, but a failure of today's physics. Its power is not yet sufficient to predict the shape of clouds and mountains. It is a challenge for science to become able to predict the dimensions of fractal shapes in Nature. The relation of mathematics to physics is our basic point. Perrin took a picture (Figure 7) which shows the random motion of a Brownian particle, and he won the Nobel prize for it, because he was able to use it to estimate the Avogadro number. Earlier physicists had used the diffusion equation to describe the spreading of dust in the air. Now we can state that the fractal dimension of the Brownian particle trajectories is $D = 2$. Other trajectories may lie in the range from $D = 1.5$ to 1.6.

We have similarly investigated the surfaces of metal fractures, and found that the roughness is characterized by a certain well-defined fractal dimension. The value of D tells us something we have known before. Now we are in a position to investigate how fractures begin and spread in metals.

Percolation describes whether or not steam can get through the coffee grains touching each other in the coffee machine called percolator. Now, the theory of percolation has become something like a bench-mark in statistical physics. Here the physics is fully understood, and it is completely governed by geometry. Hungarians contributed much to the clarification of this topic (starting from the study of powder metallurgy).

Figure 8

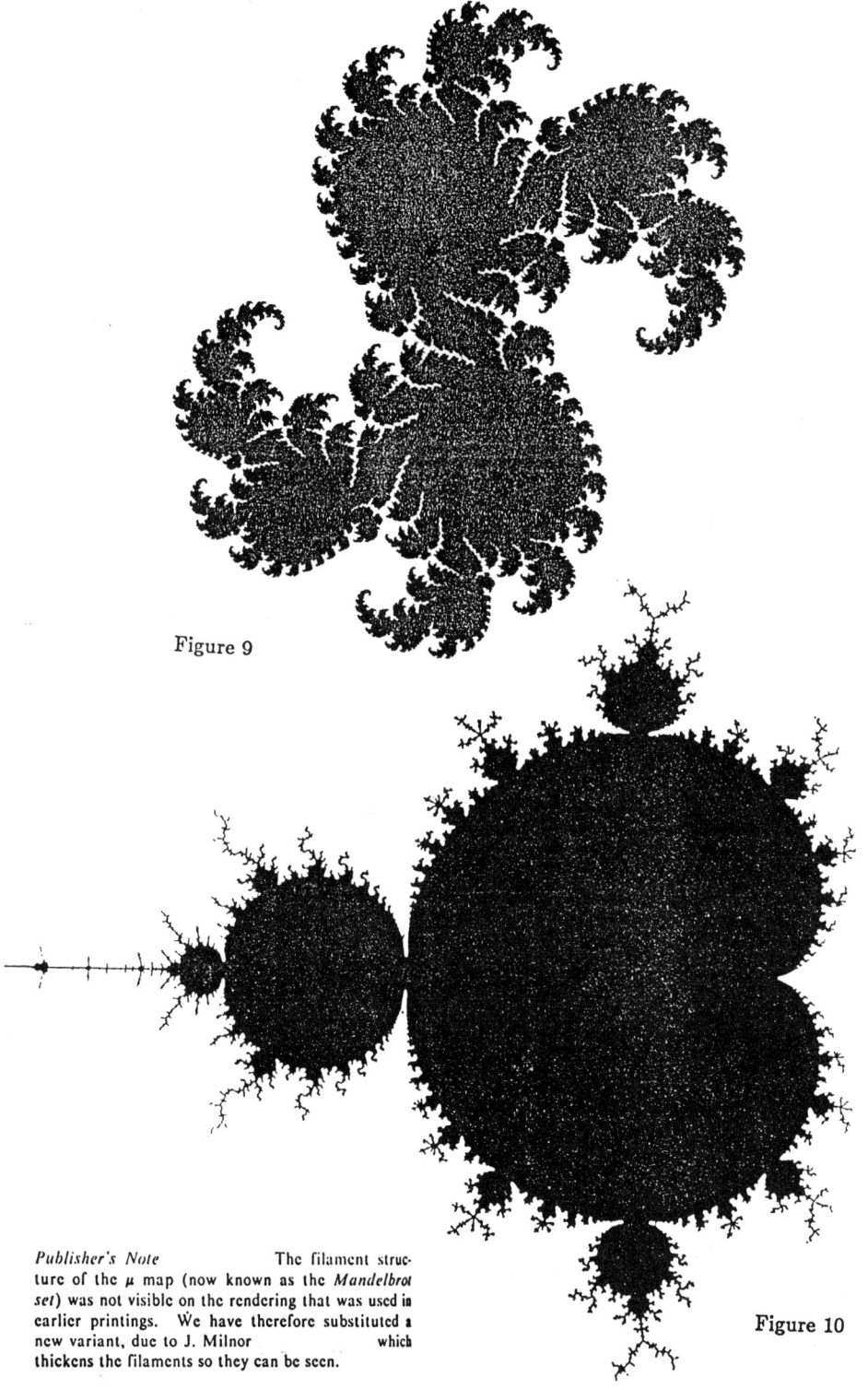

Figure 9

Figure 10

Publisher's Note   The filament structure of the μ map (now known as the *Mandelbrot
set*) was not visible on the rendering that was used in
earlier printings. We have therefore substituted a
new variant, due to J. Milnor   which
thickens the filaments so they can be seen.

Taming the Dragon

Take the following simple non-linear map of the complex plane:

$$z_{n+1} = z_n^2 + c.$$

Here c is a complex constant. Not long ago, the idea that such a quadratic nonlinearity may lead to anything interesting would have looked ridiculous. But let us iterate this map, that is, let us apply it again and again.

Starting with $z_0 = 0$, we obtain the following sequence of points, which is called the orbit of z_0: $z_1 = c$; $z_2 = c^2 + c$; $z_3 = (c^2 + c)^2 + c$, and so on. Sometimes, the dot representing the position of z_n moves away from the origin of the complex plane. If, for example, $c = 2$, the orbit of $z_0 = 0$ is made of the points 2, 6, 38, 1446, . . . If we fix c, and look for the z_0 such that the orbit *does not* diverge, we get a beautiful shape called *filled-in Julia set* (Figure 9). It is determined by just one line of formula, the computer can be programmed to draw it with no difficulty. The Eastern people immediately called this shape a dragon, and they are supposed to know how a dragon looks like. Once I had looked at many, many dragons, the following question came up: to identify in the complex plane of the constant c, the values for which the orbit of z_0 (that is, the sequence of values of z_n starting from $z_0 = 0$) does *not* run to infinity. By plotting these values on the complex c plane, I have discovered a certain shape, which proved incredible in its wealth of structure, and to which I was privileged to see by my name attached (black portion of Figure 10). It is the Logo of this workshop on chaos; you see it on the road signs indicating how to get to the conference hall. Some people even think that it is the picture of this speaker, but it is obviously not true at all. The Mandelbrot set looks like a very prickly cactus, while my own geometric contour has been converging very uncomfortably to that of a plump orange. (And, of course, my manners are not prickly at all, but very low-key, quiet and smooth!) The boundary line of the Mandelbrot set has many self-similarity properties: any part of it is similar to a much smaller segment of it.

Each of these smaller and smaller figures has a very special meaning. If you place c in the cardioid at the center of the main black "body" of M, you get a limit cycle of period 1. If you place c in the round "head" (looking to the left): cycle of period 2. In the "hands" (up and down): cycle of period 3. Farther to the left, in the little "island" on top of the head: cycle of period 4. And so on. If you are outside of the black region, there are no finite cycles at all.

Let us take the constant c to be real, and move it along the axis. If c is in the "body" of M, the orbits converge to a limit. By moving c from $c = 0$ to negative values (towards the left), you experience a cascade of bifurcations (Figure 11), finally the cascade runs into a "cycle of infinite length", and the system's evolution is called chaotic.

This phenomenon of cascading bifurcation has been discovered by the mathematician P.J. Myrberg in the 1950's, and later was rediscovered in physics by Edward Lorenz, one of the great figures in exploring chaotic phenomena in Nature, and applied by Robert May to ecology. Does its study belong to mathematics or to physics? If you focus on the regularities of the bifurcation points (see the papers of Grossmann & Thomae and of Feigenbaum), you may say it belongs to (applied) mathematics. If you observe these phenomena empirically, like Lorenz and May did, it seems to belong to physics (and to other sciences). In my mind, it belongs to both. I see no sharp boundary between the two, and when I move from one to the other, I experience no need to apologize.

Session 2

SEARCH FOR EXTRASOLAR PLANETS

Although theorists studying star formation predict that planets are the rule rather than the exception during the birth of a star, there exists no unambiguous evidence at this time for the any other planetary system beyond our own. There are, however, many tantalizing clues, including some of the papers presented in this session.

J. Scargle presented an overview of the four classical methods for attempting to detect planets. B. Campbell presented his results from radial velocity studies spanning 1981-1987. Minimum achievable accuracy is now 13 m/sec, and 7 stars show what may be true radial velocity variations at the 2.5 sigma level or above. These variations may be confirmed as being due to planetary companions. Victoria Lindsay touched similar topics. E. Levy outlined the capabilities of the proposed Astrometric Telescope Facility for Space Station. S. Tapia read a paper by R. McMillan's group about radial velocity program to study 15 stars, reporting intrinsic oscillations in K giants. J. Scargle read a paper by B. Borucki's group summarizing preliminary results from groundbased photometry studies. Fractional changes in luminosity of 5.10^{-4} can be measured. T. Ruzmaikina considered what initial values of angular momentum a collapsing cloud possess and still form a planetary system. Magnetic braking must be invoked to produce planets rather than a multiple star system. D. Backman used IRAS data to show that 18by analogy with Vega, dusty circumstellar disks. Three stellar disks studied in detail have physical voids at their centers that require a mechanism for maintenance (planets??). B. Smith reported that poor observing conditions have prevented the optical imaging of any of 75 additional potential disk candidates from the IRAS catalogue. Observations of Beta Pictoris show that its disk extends over at least 2500 AU. R. Brown delivered a rather pessimistic assessment of the HST's ability to directly image anything but giant planets around the very nearest tars. A real world error budget of the point spread function indicates that future optical systems must improve surface roughness by a factor of about 200 in order to attempt systematic explorations called for by the earlier speakers. R. Terrile discussed a proposal for a proposed future orbital instrument, the circumstellar imaging telescope. A coronograph may reduce diffracted light by a factor of 1000. An ideal error analysis suggests that a 1 meter class mirror could image Beta Pictoris disks to a distance of 100 parsecs, systematically survey for giant planets, and image very nearby terrestrial planets. A real world error budget may not be so optimistic, but this technology seems like a very promising next step.

Jill C. Tarter

PLANETARY DETECTION TECHNIQUES: AN OVERVIEW

J.D. Scargle

NASA Ames Research Center
Moffet Field, CA 94035, USA

ABSTRACT. Interest in the search for planets around other stars is due
to its connection with SETI, the desire to understand the formation of
stars and planetary systems, and curiosity about our place in the Cosmic
Neighborhood. Proposed techniques include direct imaging, detection of
transits, and detection of stellar radial velocity and positional
perturbations. Each method has advantages and is effective at detecting
different kinds of planets. The fundamental questions in this field
should be answered during the next decade.

1. THE MOTIVATION FOR PLANETARY SEARCHES

Most astronomers feel that planetary systems are probably common, and
yet there is currently no clear evidence for the existence of any planet
outside our Solar System. A number of groups are attacking this
difficult problem; objectives of searches range from circumstellar
matter in the solid state, through circumstellar ("protoplanetary")
disks and single planets, to systems of planets like the Solar System.
    There are many reasons to search for planets, all stemming from a
basic curiosity about possible cosmic neighbors. Both the theory of
star formation and the Copernican Principle suggest that planets exist
throughout the Universe. We wish to test this speculation and learn
whether the Solar System is indeed unique, rare, common or typical. The
search for planets is the first step in the search for extra-terrestrial
intelligence, as the surface of a planet is probably the only viable
location for the origin and evolution of life.
    Detection and study of other planets is needed to understand the
origin and evolution of stars and planetary systems. Initial progress
in this field is now frustrated by knowing only one example of a
planetary system. Either the discovery of other systems or the
opposite--proof that their abundance is much lower than theory
predicts--would change this frustrating situation. In addition planets,
composed of solid and liquid matter and capable of nurturing life, are
intrinsically very interesting objects. See also the reviews Black
(1980) and Tarter et al. (1986).

G. Marx (ed.), Bioastronomy – The Next Steps, 79–82.

2. A BRIEF HISTORY OF PLANETARY DETECTION

Perhaps the first serious interest in the search for extrasolar planets
arose in a series of six workshops on Interstellar Communication in
1975-1976, chaired by Philip Morrison. These workshops spawned two
others (in 1976, chaired by Jesse Greenstein and David Black) which
attempted to identify and assess possible planetary detection techniques
(see Morrison, Billingham, and Wolfe 1977). Project Orion (Black 1979)
followed up with a technical study of several planetary detection
methods, emphasizing a ground-based astrometric concept. Another
workshop series in 1978-79 (Black and Brunk 1980) identified, discussed
and compared ground-based techniques.
    Further developments at the NASA Ames Research center, under the
leadership of Black, focused on the astrometric technique as the most
feasible in connection with the project recommended by the science
workshops: a systematic survey, of approximately 100 stars in the solar
neighborhood capable of sensing planets as small as Uranus and Neptune.
A small program of NASA grants to universities has funded projects in
the areas of astrometry, radial velocity, and solar observations (to
evaluate non-planetary "signals" for solar-type stars).
    There is currently an awakening of interest in the subject,
encompassing several techniques and many institutions, as evidenced by
the variety of articles in this volume.

3. PLANETARY DETECTION METHODS

There are two kinds of search technique, depending on whether or not one
tries to detect light from the planet itself.

3.1. Direct Methods

Direct detection involves the attempt to extract a faint planetary image
from the bright image of the star. The main advantage of this method is
fast verification (one doesn't need data over a complete orbit). The
major difficulty is the need for extraordinarily efficient rejection of
diffracted and scattered light in the area surrounding the star image
(cf. paper by R. Brown and C. Burrows) because of the enormous
brightness contrast between the the star and planet. Proposals
discussed elsewhere in this volume include: an appodized coronagraphic
telescope (B. Smith and R. Terrile), an optical interferometer
(B. Burke), a special large space telescope (R. Angel), and imaging with
gravitational lenses (F. Drake). For completeness I mention the rather
exotic suggestion that radar be used to find and study nearby planets
(Williams 1985).

3.2. Indirect Methods

Indirect detection involves sensing an effect which the planet has on
the star. There are two kinds of indirect technique, depending on
whether one tries to sense the effect on the star's light or its motion.

3.2.1. The Photometric Planetary Detection Method. The light from a star will be diminished if a planet transits in front of the star as seen from the Earth (see paper by W. Borucki et al.). The main advantages are a relatively large signal and lack of reliance on the absence of very low-level systematic errors. The main disadvantages are the need to view the orbital plane edge-on, the need to observe at the time of the transit, and the fact that natural luminosity fluctuations will probably mask terrestrial-size and smaller planets.

3.2.2. Gravitational Perturbation Methods. The gravitational force of a planet causes the parent star's motion to be perturbed in a periodic way. There are two kinds of technique, depending on whether one monitors the radial or the transverse component of this motion.

3.2.2.1. The Radial Velocity Method. Measurements of stellar radial velocity, with accuracies in the neighborhood of 10-100 meters/sec, over intervals of time comparable to the orbital period, can indicate the presence of a planet. Several active programs are described elsewhere in this volume by B. Campbell, by V. Lindsay, and by R. McMillan P. Smith. Related projects are being carried out by Heacox (U. of Hawaii) and Cochran (U. of Texas). This is the only method that does not require a space platform to be fully effective. The main disadvantage is the potential confusion by natural motions in the stellar atmosphere.

3.2.2.2. The Astrometric Method. Measurements over time of stellar position, with accuracies in the neighborhood of 10-100 microarcseconds, can also indicate a planet. While moderately dependent on distance, this method is adequate for a survey of the volume within some 10's of light-years of the sun, which has been identified as key to scientific progress (see Section 2). A proposed NASA Astrometric Telescope Facility is described by E. Levy (this volume), Levy et al. (1986), and Sobeck (1987). Astrometric instruments using a Ronchi grating have been developed by Gatewood (Allegheny Observatory), Jones (Lick Observatory), and Buffington (U.C. San Diego). Interferometric techniques are being studied by Reasenberg (Harvard) and Shao (MIT). Monet (U.S. Naval Observatory) is carrying out astrometric measurements with a CCD. McAlister (Georgia State) is using speckle astrometric measurements in binaries to seek planets orbiting one of the components.

4. Comparison of Techniques

Each of these methods is most sensitive to a different kind of planet-star combination. How easy a given such combination is to detect can be specified by five parameters: (1) the size of the planet (2) the size of the star, (3) the distance between the planet and star, (4) the distance of the star from the earth, and (5) the orientation of the orbital plane. The 'size' parameter is different (mass, brightness, area), for each method, but we take these to be related (e.g., taking a common density for all planets relates area and mass). For standard values of parameters (2), (4) and (5), e.g. solar type star at 10 parsecs with orbit edge-on, the remaining two-dimensional space (planet mass and orbit size) can be easily visualized. Let SS be the area of this space corresponding to our Solar System. Then 10-microarcsecond astrometry is sensitive to a large section of SS, and radial velocities accurate to 10 m/sec to a somewhat smaller area. Direct imaging with a 1-m unapodized telescope is very far away from any part of SS, underscoring the importance of apodization and scattered light reduction. Photometry with an accuracy of .01% (which should be achievable in space) cuts across most of SS. However the real limit will be set by the considerations given in Section 3.2.1.

Some of the techniques are feasible only from space (astrometric and direct). The radial velocity approach is fully effective from the ground, while photometry is partially so. The rate at which detection capability drops with distance from the Earth ranges from very fast (direct), through fast (astrometric), to slow (radial velocity and photometric). Each method is troubled by natural, non-planetary signals, but astrometry and imaging are affected only in minor ways.

The detection of extra-solar planets is a difficult observational problem, but one of tremendous scientific and philosophical importance. The known techniques are complementary; all of them should be pursued.

REFERENCES

Black, D. (ed.), 1979, Project Orion, NASA SP-436.

---- 1980, 'In Search of Other Planetary Systems,' Space Science Reviews, Vol. 25, p. 35.

---- and Brunk, W., 1980, 'An Assessment of Ground-Based Techniques for Detecting Other Planetary Systems, Volume I: An Overview, and Volume II: Position Papers,' NASA Conference Publication 2124.

Levy, E., Gatewood, G., Stein, J., and McMillan, R., 1986, 'Astrometric Telescope of Ten Microarcsecond Accuracy on the Space Station,' S.P.I.E., Vol.628--Advanced Technology Optical Telescopes III.

Morrison, P., Billingham, H., and Wolfe, J. (eds.), 1977, The Search for Extraterrestrial Intelligence, NASA SP-419.

Sobeck, C., ed., 1987, Astrometric Telescope Facility: Preliminary Systems Definition Study, Volume I: Executive Summary, and Volume II: Technical Description, NASA Technical Memorandum 89429.

Tarter, J., Black, D., and Billingham, J. 1986, 'Review of Methodology and Technology Available for the Detection of Extrasolar Planetary Systems,' 1986, J. Brit. Interplanetary Soc. , Vol. 39, p. 418.

Williams., F., 1985, 'A Radar for the Exploration of Extrasolar Planets,' Proc. I.E.E.E., Vol. 73, p. 355.

A SEARCH FOR PLANETARY MASS COMPANIONS TO NEARBY STARS

B. Campbell

Dominion Astrophysical Observatory, Herzberg Institute
5071 W. Saanich Road, Victoria, B.C., Canada

G.A.H. Walker and S. Yang

Dept. of Geophysics& Astronomy, Univ. of British Columbia
2219 Main Mall, Vancouver, B.C., Canada

ABSTRACT. Seven of 16 solar-like stars observed with a precision radial velocity technique for the past six years show evidence of long term low level variations. The average external error of the velocities is 13 meters per second, and the observed variations are in the range 25-65 m s^{-1}. The trends in the velocities imply periods of more than ~10 years, but these cannot be due to brown dwarf-mass companions in very long period orbits, since these would have been previously detected by conventional astrometry. Companions of a few Jupiter masses are implied. Since brown dwarfs of 10-80 Jupiter masses are either rare or non-existent, these companions probably represent the tip of the planetary mass spectrum.

1. INTRODUCTION

Until very recently, it has not been feasible to search for extrasolar planetary systems by measuring line-of-sight velocities for stars, since radial velocity measurements have seldom been more precise than about 500 m s^{-1}. The orbital speed of the parent star due to a Jupiter-like planet, which would amount to a few times 10 m s^{-1} at most, would clearly be lost in the errors of measurement. An increase in precision of between one and two orders of magnitude would be necessary before radial velocity detection of planetary companions would be feasible.

We have developed a new technique for velocity measurement with just such an improvement in precision. We have used this technique to monitor the relative velocities of 16 stars for the past six years. The precision of the radial velocities, and the time span of our observations, are now sufficient that we can make some preliminary comments on the incidence of sub-stellar mass companions to solar-like stars.

G. Marx (ed.), Bioastronomy – The Next Steps, 83–90.

2. THE HYDROGEN FLUORIDE ABSORPTION CELL TECHNIQUE

This new technique was originally proposed by Campbell and Walker (1979). An absorption cell containing hydrogen fluoride gas is placed ahead of the slit of a conventional coudé spectrograph. Starlight passing through the cell before entering the spectrograph has absorption lines of HF superimposed. These lines provide wavelength fiducials which cannot be displaced relative to the stellar absorption lines by optical effects within the spectrograph. This is in contrast to conventional techniques, in which a separate beam of light is used to generate a comparison spectrum for wavelength calibration. In this conventional approach there is no guarantee that both the stellar and comparison beams traverse the spectrograph in exactly the same way. There are then unavoidable displacements of the comparison and stellar spectra in the focal plane, which translate into significant errors in velocity, even at the highest dispersions. All such systematic errors are eliminated by using an absorption cell.

Hydrogen fluoride was chosen for this application because (a) it produces strong absorption lines for a cell length of only 90 cm, (b) the lines are well spaced so that many unblended stellar lines fall between, and (c) there is no interference by weak lines of rare isotopes, or by telluric absorption lines. To be sure, HF is a noxious gas, capable of injury to personnel and damage to optical equipment, and so care in handling of the absorption cell is required.

Observations have been made with the HF cell at the coudé spectrograph of the 3.6 m Canada-France-Hawaii telescope on Mauna Kea. The detector is a Reticon photodiode array, which has sufficient dynamic range that spectra of very high signal-to-noise ratio, typically 1000:1, can be obtained. In such high quality spectra, the relative positions of individual HF and stellar absorption lines can be determined to about 0.2 microns, which for the ensemble of lines observed tranlates into a net velocity precision of about 10 m s^{-1}.

Details of the data reduction procedure are described by Campbell et al. (1986), and Campbell and Walker (1987). This involves alternately cancelling by spectral division the HF and stellar lines (observed independently), and measuring line positions using the difference technique of Fahlman and Glaspey (1973). The relative line positions are corrected for the slight effects of line blending and instrumental profile variations. Radial velocities are then derived in the usual way, and corrected to the barycenter of the solar system with the precision earth orbital routines of Stumpff (1980). Note that care must be taken to determine the appropriate <u>time</u> of each observation when correcting to the barycenter, since the acceleration of the earth's surface due to rotation is 114 m s^{-1} hr^{-1} in Hawaii. We therefore determine the <u>mean</u> time of each exposure, by monitoring the flux entering the spectrograph with an exposure meter.

The final step in the reductions takes account of the slight night-to-night zero-point variations in the velocities. (Such variations can be attributed to changes in the optical alignment of the spectrograph.) This is accomplished by using, for any given star, the mean of the velocities of all <u>other</u> stars observed in the same observing

run to form a correction. In this step we make use of the fact that
most of our program stars show no sign of real short term velocity
variations. Hence the deviations about any long term trends (see below)
can be used to establish zero-point corrections for each observing run.

3. ANALYSIS OF THE RADIAL VELOCITY CURVES

3.1. A Test for Velocity Variations

A typical velocity curve for one of our program stars, τ Ceti, is shown
in Figure 1. For each plotted point the associated error bar is the
standard error of the velocities for the typically 10-12 stellar lines
measured in each spectrum. These therefore represent the _internal_
errors of measurement.
        Clearly there are no significant velocity variations revealed by
this data on time scales of a few years. The mean internal error for
the τ Ceti velocities is 9.4 m s^{-1}, while the standard deviation of the
velocities themselves (henceforth called the _external_ error) is 13.9 m
s^{-1}. We do not consider the difference in internal and external errors
to be significant, and so we conclude that there are no real variations
present for periods up to about 6 years. Similarly we find no evidence
for rapid variations in any of the the remaining program stars.

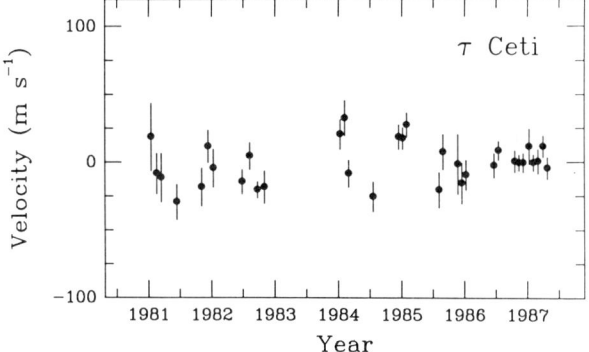

Figure 1. Relative velocities for τ Ceti.
Error bars are 1 sigma internal errors.

        To assess objectively whether there might be long term trends in
the velocities, we have adopted the following procedure. Each data set
has been fit with both a straight line, and a parabola, using multiple
regression techniques. We have thus determined the first derivative
(slope) and second derivative (curvature) of the data, and the formal 1
sigma uncertainties in each of these quantities, which depend on the
external errors. The ratios of the absolute values of the slope and
curvature to their respective errors ($|S|/\sigma_S$ and $|C|/\sigma_C$) is a measure of
the significance of long term velocity variations.

Figure 2 shows a histogram of the observed $|S|/\sigma_S$ and $|C|/\sigma_C$, 28 values in total for 14 stars. (Two stars, χ^1 Orionis and γ Cephei, are not considered here, since they have large velocity variations probably due to stellar companions.) The continuous line in Figure 2 is the distribution which results from Monte Carlo simulations of 2000 sets of velocity data. Each such data set consists of 27 points (the average number of observations per star), and the rms deviation of the velocities at each point is just the actual scatter in the velocities for all stars observed in each of 27 observing runs. Thus the simulated data accurately mimics that for the real stars, on average, except that no velocity trends are imposed. The distribution of $|S|/\sigma_S$ and $|C|/\sigma_C$ derived from the random data is, not surprisingly, a gaussian with sigma = 1.

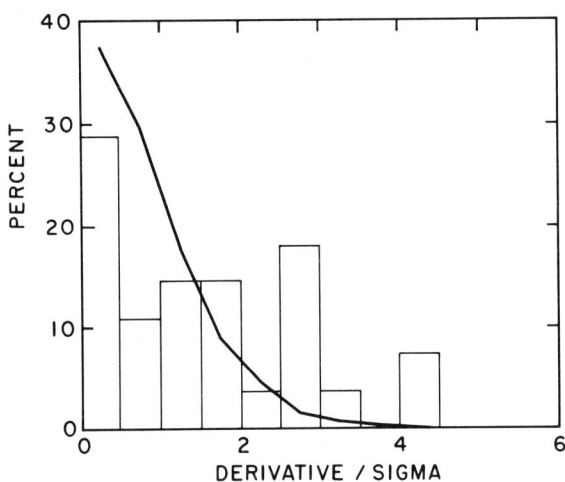

Figure 2. Histogram of observed $|S|/\sigma_S$ and $|C|/\sigma_C$. Continuous line shows the distribution from 2000 simulations.

The observed distribution in Figure 2, when compared to that from the simulations, appears to have an excess of values above 2.5 sigma. For 7 stars $|S|/\sigma_S$ or $|C|/\sigma_C$ is greater than 2.5, and these significant values are shown in the second column of Table 1. From the simulated distribution we would predict only 0.6 out of 28 cases more significant than 2.5 sigma. Clearly γ^2 Delphini has the lowest significance of this group, and so we do not consider its variation to be real. On the other hand, the variations for ε Eridani (see Figure 3) and ξ Bootis A are significant at the >4 sigma level, and so we classify these as "probable" variables. For the remaining stars, the variations appear to be significant, but not at a high degree, and so we classify these as "possible" variables.

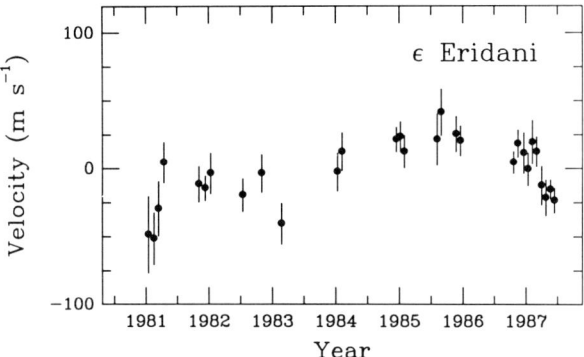

Figure 3.   Velocities for ε Eridani, a probable variable.

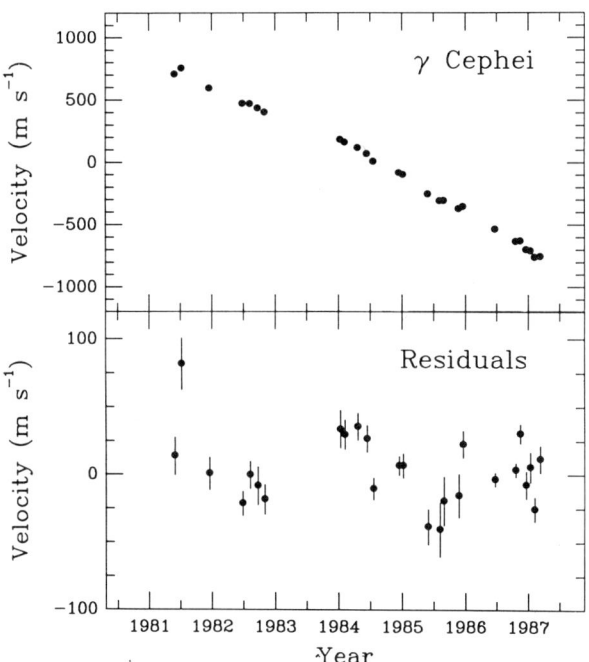

Figure 4.   Velocities for γ Cephei (upper panel) and residuals
            after subtracting a second order fit (lower panel).

3.2. γ Cephei - A Triple System?

The upper panel of Figure 4 shows the velocity curve for the K subgiant γ Cephei. The very obvious velocity variation revealed by the precision radial velocities apparently escaped detection with conventional techniques. This variation is probably due to a stellar secondary, and so this object is not of particular interest.

However, close examination of the velocity curve for γ Cephei reveals that it is not smooth, but shows evidence of low amplitude variations on top of the general trend. To show these variations, we have subtracted a second order fit to the velocities, and the resulting residuals are shown in the lower panel of Figure 4. This reveals the low level variation, which we estimate has a period of about 2.7 years, and an amplitude of about 25 m s^{-1}. Since our data cover two full periods of this variation, we believe this is a "probable" variable, and that there is very likely a third body present in the system. However, a more precise determination of the perturbation due to the third body must await a better definition of the orbit due to the secondary.

4. LIMITS TO COMPANION MASSES

To place upper limits on the mass, M_c, of possible companions to our program stars, we can make use of the slopes and curvatures derived above. The 1 sigma upper limits for these are simply $S_{max} = |S| + \sigma_S$, and $C_{max} = |C| + \sigma_C$. Assuming circular orbits, it can be shown that by combining these limits one obtains an upper limit to the companion mass in Jupiter masses as a function of period (P) as follows:

$$M_c \sin \underline{i} < 5.59 \times 10^{-3} \, P^{4/3} \, M_*^{2/3} \left[S_{max} + \frac{P}{2\pi} C_{max} \right] \tag{1}$$

where M_* is the mass of the primary star in solar masses. Note that this limit goes directly with period.

An additional constraint comes from old parallax surveys, which according to Gatewood (1986) should have revealed perturbations by unseen companions of $\alpha = 0.01$ arc seconds or more. In this case, again assuming circular orbits, we have

$$M_c < 1.05 \times 10^3 \frac{\alpha}{\pi_p} \frac{M_*^{2/3}}{P} \, , \tag{2}$$

where π_p is the parallax. This limit goes inversely with period, and so the combination of equations (1) and (2) yields an upper limit on the companion mass which is independent of period. These limits are given in the last column of Table 1 for 14 program stars. Note, however, that perturbations of period longer than about 50 years would not have been detected in parallax surveys, and so these limits apply to companions within about 14 AU of their parent stars. It should be emphasized that these limits are directly proportional to distance of the star from the Sun, which accounts for the wide range in these values. The median value of these upper limits is 8 Jupiter masses, which we consider the representative upper limit for companions to these stars.

Also shown in Table 1 are lower limits to the companion masses for
the 7 stars classified as "possible" or "probable" velocity variables.
These limits are derived from the observed velocity ranges, which cannot
be greater than twice the velocity amplitude. The lower limit to the
period is 12 years, twice the span of our observations, since none of
the variables appears to have been observed for more than half of a
period (with the possible exception of ϵ Eridani). For γ Cephei the
companion mass derived from the actual period and amplitude is 1.7
Jupiter masses.

TABLE 1. Mass limits for companions in orbits with
12 $<$ P $<$ 50 years.

Star	$\|S\|/\sigma_S$ or $\|C\|/\sigma_C$	$M_C \sin \underline{i}$ (Jupiter masses) Lower Limit	Upper Limit
τ Ceti	–	–	4.2
ι Persei	–	–	12.9
ϵ Eridani	4.12	2.4	4.8
o^2 Eridani	–	–	5.2
36 Ursae Majoris	2.94	1.2	15.4
β Virginis	2.98	1.1	11.1
β Comae	–	–	9.9
61 Virginis	2.93	1.8	7.6
ξ Bootis A	4.37	1.6	8.0
σ Draconis	–	–	6.3
β Aquilae	3.27	1.1	15.1
γ^2 Delphini	2.56	–	61
η Cephei	–	–	16.7
61 Cygni A	–	–	3.5
γ Cephei	–	1.7	

5. THE COMPANION OBJECTS – BROWN DWARFS OR PLANETS?

It is abundantly clear that none of our program stars has a sub-stellar
mass companion in the usual brown dwarf range of ~10–80 Jupiter masses.
This is consistent with many other recent surveys for brown dwarfs,
which have so far failed to turn up a single confirmed case. The report.
by McCarthy, Probst, and Low (1985) of a brown dwarf companion to Van
Biesbrock 8 was apparently erroneous (Skrutskie, Forrest, and Shure
1987; Perrier and Mariotti 1987). Infrared searches for companions to
nearby stars by Jameson, Sherrington, and Giles (1983), Skrutskie,
Forrest, and Shure (1986), and Kumar (1987), have turned up no
substellar companions to brightness limits corresponding to roughtly 40
Jupiter masses. A search of the IRAS catalog by Chester et al. (1986)
has failed to turn up any brown dwarf candidates in the field. A survey
for low-amplitude velocity variations in a sample of 65 M dwarfs by

Marcy et al. (1986) yielded only one possible brown dwarf companion, but Marcy (1987) now believes that this object is very close to the limit for stable hydrogen burning. In summary, the evidence to date suggests that brown dwarfs are rare, or perhaps non-existent. (Possibly this is fortunate, since if brown dwarfs were ubiquitous, there might be few stable planetary systems!)

If "massive" brown dwarfs (~10-80 Jupiter masses) are so rare, how are we to classify the companion objects that we have found? We have 7 cases out of 15 in which there is a "possible" or "probable" companion in the range of ~1-8 Jupiter masses. Such a preponderance of very low mass companions, combined with the apparent absence of intermediate mass objects, implies that there is a gap in the distribution of sub-stellar mass objects. If this gap is real, the implication is that the objects we have detected are at the tip of the planetary mass spectrum. And if our tentative detection of 7 companions to 16 stars is confirmed, this implies that the fraction of solar-like stars with planetary systems is probably $f_p > 0.45$.

We are most grateful to the staff of CFHT for their continuing assistance with the HF precision radial velocity program. We acknowledge the support of the National Research Council of Canada, and the Natural Sciences and Engineering Research Council of Canada.

REFERENCES

Campbell, B. and Walker, G. A. H. 1979, Pub. A.S.P., **91**, 540.
————. 1987, in preparation.
Campbell, B., Walker, G. A. H., Pritchet, C., and Long, B. 1986, in Astrophysics of Brown Dwarfs, M. C. Kafatos, R. S. Harrington, and S. P. Maran, eds. (Cambridge: Cambridge University Press), p. 37.
Chester, T. J., Fullmer, L. D., Beichman, C. A., Gillet, F. C., Low, F. J., and Neugebauer, G. 1986, Bull. A.A.S., **18**, 961.
Fahlman, G. G. and Glaspey, J. W. 1973, in Astronomical Observations with Television-Type Sensors, J. W. Glaspey and G. A. H. Walker, eds. (Vancouver: University of British Columbia), p. 347.
Gatewood, G. 1986, private communication.
Jameson, R. F., Sherrington, M. R., and Giles, A. B. 1983, M.N.R.A.S., **205**, 39P.
Kumar, K. C. 1987, preprint.
Marcy, G. W. 1987, private communication.
Marcy, G. W., Lindsay, V., Berengren, J., and Moore, D. 1986, in Astrophysics of Brown Dwarfs, M. C. Kafatos, R. S. Harrington, and S. P. Maran, eds. (Cambridge: Cambridge University Press), p. 50.
McCarthy, D. W., Probst, R. G., and Low, F. J. 1985, Ap. J. (Letters), **290**, L9.
Perrier, C. and Mariotti, J.-M. 1987, Ap. J. (Letters), **312**, L27.
Skrutskie, M. F., Forrest, W. J., and Shure, M. A. 1986, in Astrophysics of Brown Dwarfs, M. C. Kafatos, R. S. Harrington, and S. P. Maran, eds. (Cambridge: Cambridge University Press), p. 82.
————. 1987, Ap. J. (Letters), **312**, L55.
Stumpff, P. 1980, Astr. and Ap. Suppl., **41**, 1.

A RADIAL VELOCITY SEARCH FOR EXTRASOLAR PLANETS

R.S. McMillan and P.H. Smith
Lunar & Planetary Laboratory, University of Arizona
Tucson AZ 85721 USA

We are measuring small changes in the line-of-sight velocities of
stars to detect the oscillatory reflex acceleration induced by large
planets. Our instrument is an optical spectrometer for which
wavelengths are first calibrated by transmission through a tunable
Fabry-Perot etalon interferometer. The intrinsic stability of the
etalon and an image-scrambling fiber optic light feed provide great
sensitivity to line-of-sight accelerations and immunity to systematic
errors. The interferometer is being frequently calibrated to $\pm$ 6
meters/sec in Doppler shift. The standard deviation of a one-hour
exposure on a typical solar-type star of blue magnitude 4.0 is $\pm$ 12
m/s. This random error "averages down" through an observing season,
giving adequate accuracy for the search for Jovian-mass planets.
McMillan et al. (1985, 1986) describe the instrument, principle of
operation, test results, calibration, observing methods, and
sensitivity to planets as a function of mass and period.

Formal observations began in September 1985 and as of 1987 July 27,
5284 observations have been made of 17 stars on 173 nights. With our
first 1.5 years of observations we already have discovered a 1.84-day
oscillation of the radial velocity of the K1 III star Arcturus (Smith
et al. 1987), two modes of short-period oscillations in the K0 III
star Pollux (Smith and McMillan 1987a), and non-variability of the
radial velocity of the G0 V star Eta Cas A, previously suspected to be
a spectroscopic binary (McMillan and Smith 1987). Our data also rule
out previously-suspected 97-minute oscillations in the radial velocity
of Arcturus.

We have acquired considerable experience with this instrument and
its operation and calibration, and are making observations with
uniform procedures every month. The lack of disturbance to our optics
and the frequent sampling of the velocities of the program stars
distinguish our observing program from that of Campbell et al.

G. Marx (ed.), Bioastronomy – The Next Steps, 91–92.
© *1988 by Kluwer Academic Publishers.*

(1987). We intend to continue the observing program for at least three more years, a minimum duration for finding planets massive enough to cause detectable changes in the Doppler shifts of stars. Currently there are 17 stars in the observing program and we plan to expand the list to about 32 during the next year. Smith and McMillan (1987b) provide a list of the stars, tabulate the number of observations and number of nights for each star to date, make comments on the characteristics of each star, and indicate which stars are in common with other planet detection surveys.

REFERENCES

Campbell, B., Walker, G. A. H., and Yang, S. 1987, in Proc. IAU Colloq. 99, Bioastronomy - The Next Steps, G. Marx, ed., in press.

McMillan, R. S., Smith, P. H., Frecker, J. E., Merline, W. J., and Perry, M. L. 1985, in Proc. IAU Colloq. 88, Stellar Radial Velocities, A. G. Davis Philip and D. W. Latham, eds., (Schenectady: L. Davis Press), p. 63.

McMillan, R. S., Smith, P. H., Frecker, J. E., Merline, W. J., and Perry, M. L. 1986, Proc. SPIE, 627, 2.

McMillan, R. S., and Smith, P. H. 1987, Pub. A. S. P., 99, No. 618, in press.

Smith, P. H., and McMillan, R. S. 1987a, Proc. IAU Symp. 132, in press.

Smith, P. H., and McMillan, R. S. 1987b, in Proc. 27th Liege International Astrophysical Colloquium, Observational Astrophysics with High Precision Data, (Dordrecht: Reidel), in press.

Smith, P. H., McMillan, R. S., and Merline, W. J. 1987, Ap. J. (Lett.), 317, L79.

IRAS STATISTICS ON MAIN SEQUENCE STELLAR IR EXCESSES AND MODELS OF CIRCUMSTELLAR PARTICLE CLOUDS

D.E. Backman and F.C. Gillett
Kitt Peak National Observatory
National Optical Astronomy Observatories++
P.O. Box 26732, Tucson AZ 85726 USA

ABSTRACT. IRAS coadded-survey data of volume-limited samples of main sequence stars have been examined for evidence of excess far-IR flux. Of 134 systems, 25 (19%) show significant excesses at 25, 60, or 100 μm that may be due to emission from clouds of orbiting particles. The fraction with excess more luminous than $2 \times 10^{-5} L_*$ is about 15%, roughly independent of spectral type. Several of the stars with excess appear to be older than 2×10^9 yrs, implying that the particle cloud phenomenon is not solely a feature of young objects.

Models of three prominent clouds that have been spatially resolved (β Pic, α PsA, and α Lyr) indicate central depleted regions with radii of order 20 AU. One possible explanation for maintenance of the depleted regions is that a planet orbits at and defines each cloud's inner boundary, sweeping up particles entering this region.

1. Introduction

The strong far-IR flux from α Lyrae, α Piscis Austrini, β Pictoris, and ϵ Eridani (Aumann *et al.* 1984, Gillett *et al.* 1984, Gillett 1985) is prototypical of a phenomenon which may be associated with planetary system formation. Several discoveries of similar excesses have been made using IRAS Point Source Catalog (PSC) data (Aumann 1985, Sadakane and Nishida 1986, Coté 1987). This paper will (1) summarize analysis of coadded IRAS survey data on nearby stars, with a potential detection completeness limit approximately 3 times more sensitive than the PSC, and (2) report some results of modeling of three cases where the emitting region has been resolved, namely β Pic, α PsA, and α Lyr. The importance of spatial resolution in this context is that it has provided strong evidence that the excess emission is from particles substantially larger than interstellar grains in orbit about the stars (Aumann *et al.* 1984, Gillett 1985).

Section 2 presents a study of volume-limited stellar samples resulting in statistics of far-IR excesses found to be similar to the prototypes in terms of color temperature and fractional luminosity (L_e/L_*). Until the cases of excess presented here are spatially resolved and more precise IR photometry is performed, they should be

++Operated by the Association of Universities for Research in Astronomy, Inc. under contract with the National Science Foundation.

G. Marx (ed.), Bioastronomy – The Next Steps, 93–99.

considered candidates rather than definite examples of orbiting particle clouds.

The modeling results presented in §3 for the resolved systems are intended to indicate possible normal properties of the larger set of stars with excesses. A connection to planetary systems is suggested by particle removal time scales calculated from characteristic particle sizes and cloud inner boundary radii.

2. Investigation of Nearby Stars Using Coadded IRAS Survey Data

Coaddition of the all-sky survey data yields the most sensitive photometric information available from IRAS for most of the sky. The sample of 134 stellar systems selected here is a subset of the Gliese Catalog (Gliese 1969; Gliese and Jahreiss 1978). The sample was limited to stars of spectral types A, F, G, and K, luminosity classes V and IV-V, and trigonometric parallax $\pi_t \geq 0.045$ (d $\lesssim$ 22 pc). A crucial selection criterion was the flux in the shortest wavelength IRAS band (12 μm), which was limited to $f_{12} \geq 1.45$ Jy before color correction. The 12 μm flux was assumed to define the flux from the stellar photosphere for extrapolation to longer wavelengths.

Binary systems were counted as single stars, identified by the name of the primary. Stellar companions do not measurably affect net system colors because all stellar photospheres have nearly the same color across the IRAS bands. The summed luminosity of components listed by Gliese (1969) was used in calculating fractional cloud luminosities.

The volume searched decreases with later spectral type as a result of the 12 μm flux selection criterion. Table I lists as a function of spectral type the approximate distance limit of the sample (reciprocal of 25th percentile π_t), the total number of stars qualifying, and the fraction of stars with excess above limits in significance and fractional luminosity.

TABLE I - STATISTICAL PROPERTIES OF THE SAMPLE

type	approx. sample radius (pc)	total number	excess >3σ	excess $\tau \geq 2 \times 10^{-5}$
A	20	22	45%	23%
F	18	51	12%	12%
G	15	39	13%	10%
K	7	22	18%	18%

The amount of excess e_λ in a band was defined by the formula:

$$e_\lambda = [f_\lambda - f_{12}/x] \pm [\sigma_\lambda^2 + (\sigma_{12}/x)^2 + (yf_{12}/x^2)^2]^{1/2},$$

where $f_{12} \pm \sigma_{12}$ is the 12 μm flux, $f_\lambda \pm \sigma_\lambda$ is the flux in the band, and $x \pm y$ is the assumed photospheric flux ratio between 12 μm and the given band.

The photospheric flux ratios were determined from the observed 12/25 micron ratios. There are few examples of significant excess at 25 μm, so the median 12/25 μm ratio (after removal of cases of clear excess) was assumed to be a good measure of true stellar properties. The median ratios for each spectral type were then extrapolated at constant color temperature to determine the photospheric 12/60 μm and 12/100 μm flux ratios.

There are only 4 examples of excursions greater than 3σ "blueward" of the median ratios (negative e_λ), providing evidence that positive excesses larger than 3σ are significant and that "σ" values derived from measurement uncertainty are approximately correct. An extrapolation uncertainty of ±3% was assumed for each band.

The color temperatures of the significant excesses are shown in the histogram in Figure 1. The figure is divided into two sections: excesses more significant than 10σ, and excesses with significance between 3 and 10σ. The source color temperature was calculated from the 25/100 μm flux ratio if there was 25 μm excess or from the 60/100 μm ratio if there was not. Temperatures in cases of 60 or 100 μm flux upper limits were evaluated using 3σ limits as fluxes.

COLOR TEMPERATURES OF EXCESSES

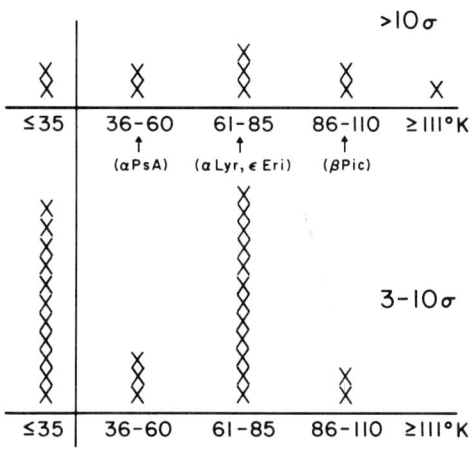

Figure 1

The stars in the set with moderate significance are mostly new candidates because of the increased sensitivity of this study. References are given in the last column of table II for stars previously recognized as having far-IR excess. The similarity of the two histograms for T > 35 K gives some confidence that the set with moderate significance probably represents excesses caused by the same physical mechanism as the set with high significance. The likeliest interpretation of the cold excesses (T ≤ 35 K) is source confusion rather than particle clouds around the stars because: (1) the objects with T ≤ 35 K are concentrated toward the galactic plane as one would expect for cirrus contamination, whereas the objects with T > 35 are not; and (2) the warmest well-studied cirrus source has a color temperature of 34 K (Low *et al.* 1984). Source counts at appropriate flux and color temperature ranges in the IRAS Serendipitous Survey Catalog (Kleinmann *et al.* 1986) imply that ~2-3 of the 25 warm excesses and ~6 of the 13 cold excesses may be due to source confusion.

Twenty-five stars with significant far-IR excesses at 25, 60, or 100 μm comparable in temperature to β Pic, α PsA, α Lyr, and ε Eri are thus left from a sample of 134. The stars, their spectral types, excess flux by band (not color corrected) and significance, color

temperature of the excess, and fractional cloud luminosity are listed
in table II.   In a few cases, stellar ages (see below) are also given.
TABLE II - STARS WITH 25, 60 or 100 μm EXCESSES >3σ AND T > 35 K

Gliese		spec	25 μm excess (Jy)	σ_{15}	60 μm excess (Jy)	σ_{60}	100 μm excess (Jy)	σ_{100}	T_c	τ	age	IRAS refs
68.0	DM+19 279	K1	0.10	1.7	0.11	3.4	0.38	2.9	85 K	7.1e-5		
71.0	TAU CET	G8	0.06	0.7	0.08	2.1	0.42	6.0	76	1.2e-5	3.5	B
111.0	TAU(1) ERI	F6	0.17	3.9	0.89	35.6	3.65	22.7	62	2.2e-4	2.2±1.6	A2
121.0	TAU(3) ERI	A4	0.02	0.8	0.04	1.6	0.15	3.4	75	7.5e-6		
144.0	EPS ERI	K2	0.29	3.9	1.33	41.7	2.27	41.6	74	9.7e-5		G
167.1	GAM DOR	F0	0.05	2.2	0.21	9.3	0.21	4.1	83	3.0e-5		A2
217.1	ZET LEP	A3	0.68	15.4	0.40	12.9	<0.11	-	139	7.1e-5		C
219.0	BET PIC	A5	9.05	194.1	20.44	363.4	13.15	115.3	106	2.4e-3		G
245.0	PSI(5) AUR	G0	0.11	2.6	0.43	8.9	0.53	2.4	81	1.7e-4	3.8	
248.0	ALF PIC	A5	0.13	3.5	0.00	0.1	<0.19	-	79	9.2e-6		
292.0A	DM-34 4036	F5	0.03	0.9	<0.03	-	3.54	4.2	49	1.5e-4	<2.1	
297.1	B CAR	F5	0.04	2.0	1.76	51.8	3.05	16.9	52	3.5e-4	4.5±1.1	
321.3A	DEL VEL	A0	0.03	0.4	0.29	6.9	<0.02	-	63	2.8e-6		C
364.0	DM-23 8646	G0	0.12	2.8	0.13	5.4	0.24	3.0	98	7.2e-5	2.6	
448.0	BET LEO	A3	0.41	5.5	0.77	14.9	0.63	7.0	109	1.9e-5		A2,C
557.0	SIG BOO	F2	0.06	2.2	0.09	3.5	<0.06	-	83	2.3e-5	2.0±1.5	
580.1	BET CIR	A3	0.07	0.8	2.72	4.5	<9.58	-	42	1.7e-4		
584.0A	ETA CRB	G2	0.02	0.8	<-0.03	-	0.23	4.9	69	2.0e-5		
673.1	DM-24 13337	A9	0.15	1.3	1.02	2.9	11.48	5.9	52	2.9e-4		
691.0	MU ARA	G5	0.22	7.3	0.08	1.1	<0.96	-	67	1.0e-4		
721.0	ALF LYR	A0	1.14	3.4	7.85	114.7	8.89	48.1	74	1.5e-5		A1,G
764.0	SIG DRA	K0	0.04	1.5	-0.00	-0.1	0.43	3.5	70	2.0e-5		
820.0A	61 CYG	K5	0.38	5.1	0.57	5.0	2.38	1.9	77	1.5e-4		B
822.0A	DEL EQL	F8	0.15	3.8	0.07	2.2	0.48	4.9	88	5.4e-5	6.4±1.0	
881.0	ALF PSA	A3	0.34	2.3	8.66	157.2	11.95	109.3	58	5.0e-5		G

References: A1 = Aumann *et al.* 1984; A2 = Aumann 1985; B = Backman, Gillett, and Low 1986;
C = Coté 1987; G = Gillett *et al.* 1984

The fourth column of table I gives the proportion of excesses
above 3σ significance as a function of spectral type.   The previously
noted preponderance of A stars with far-IR excesses is partly due to a
luminosity effect (Aumann 1985; Sadakane and Nishida 1986).   If the
geometry and surface area of the clouds are similar from star to star,
the clouds around A stars in a given volume will be easiest to detect.

It is useful, therefore, to compare the excesses on the basis of a
property that is independent of the stellar characteristics.   An
appropriate quantity is the bolometric ratio of fluxes from cloud and
star ($\tau \equiv L_g/L_*$), listed in column 11 of table II.   The excesses
discovered so far have fractional luminosity ranging from $\sim 10^{-6}$ to
$\sim 10^{-3}$.   In contrast, the luminosity of the known zodiacal dust emission in our (inner) solar system is $\sim 8 \times 10^{-8}$ $L_\odot$ (J. Good, priv. comm.).

The last column of table I gives the number of excesses qualified
both by the 3σ significance limit <u>and</u> a limit in fractional luminosity
greater than 2×10^{-5} (cf. a value of 1.5×10^{-5} for α Lyr).   It appears
possible that far-IR excesses occur equally often, within the uncer-
tainties of small number statistics, in main sequence spectral classes
A through K.   The chosen cutoff in fractional luminosity corresponds
closely to the limit of detectable excess around G and K stars in our
sample but is well above the detectability limit for A stars.
Increased sensitivity for the same search volumes would likely
increase the fraction of G and K stars found to have excess without
increasing the fraction of A stars.

One is led to the possibility that most main sequence stars have particle clouds emitting in the far-IR. The evidence for this proposition is: (1) in cases where the illumination source is brightest, namely A stars, the proportion of stars with excesses detectable by IRAS approaches 50%; (2) stellar properties such as spectral type, rotation (v sin i), or status as a spectroscopic binary are not strongly correlated with the presence of an excess; (3) some stars as old as the sun have detectable clouds (see below); and, (4) there is no reason to believe that the chosen cutoff in significance corresponds to a lower limit in cloud luminosities.

The far-IR flux received at earth from a cloud around the perimeter of our solar system would be smaller than the uncertainty in emission attributed to the inner solar system zodiacal dust (Good, Hauser, and Gautier 1986) if the cloud had (1) the same emissivity and radial extent as the α Lyr cloud (§3), (2) a fractional luminosity $5x10^{-6}$, ~1/4 the value for α Lyr, and (3) a range of inclinations $>\pm5°$ about the ecliptic plane. Such a cloud would be apparent to an "outside" observer using an IRAS-type instrument but difficult for an "inside" observer to discern.

The last column of table II gives age estimates in units of 10^9 yrs for some of the F stars (Varsik 1987) and G stars (from Duncan 1981; also, using Simon, Herbig, and Boesgaard 1985) with significant excess. Some stars with ages comparable to the age of the solar system apparently have particle clouds. Thus it seems unlikely that the clouds are evidence of planetary accretion in progress.

3. Models of the Resolved Clouds

Simple models of the β Pic, α PsA, and α Lyr clouds were calculated and compared to special IRAS observations. Details of the data, modeling process, and results for a range of assumptions can be found in Backman et al. (1988). Characteristic angular radii for the clouds can be used to deduce particle sizes for the resolved clouds because particle radiative efficiency depends on size. The model assumptions used were: (1) the particles have absorptive/emissive efficiency $\epsilon = 1/(1+\lambda/\lambda_o)$ with parameter λ_o; (2) the projected particle number in the ring between r and r+dr follows a power law: $n(r)dr \propto r^{\gamma+1}dr$, with parameter γ. The models were fit to the observed fluxes and 60 μm scan profiles after subtraction of the contribution from the central star.

Table III gives model results for the case in which γ equals 0 and λ_o has a single value. Included are: wavelength of characteristic emissivity break λ_o, inner and outer cloud radii, corresponding small grain temperatures, total particle area, (radiative) Poynting-Robertson orbit decay time from the cloud inner radius (see below), and stellar main sequence lifetime.

TABLE III - MODEL RESULTS FOR γ = 0

	λ_o (μm)	r_1 (AU)	T_1	r_2 (AU)	T_2	A (cm^2)	t_{P-R} (yrs)	$t_{m.s.}$ (yrs)
β Pic	0.8	17	210 K	790	45 K	$5x10^{30}$	$1x10^4$	$2x10^9$
α PsA	80	26	115 K	450	32 K	$3x10^{28}$	$1x10^6$	$1x10^9$
α Lyr	1100	22	165 K	245	50 K	$4x10^{27}$	$3x10^6$	$5x10^8$

The values of the inner radii are model-dependent. If $\gamma = -1$ is used, r_1 increases over the $\gamma = 0$ solution by about 30%. The existence of a central region significantly depleted in particles in each case is clear, however. For example, the average density of material around α Lyr at $r < r_1$ cannot be larger than ~ 0.2 of its value at r_1 (for the same "γ" in both regions) or there would be significantly more 12 and 25 μm flux emitted than is observed. The existence of an inner depleted region in the case of β Pic was proposed by Smith and Terrile (1984) on the basis of lack of extinction on the direct line of sight to the star through the disk; they hypothesized that the inner region was cleared by incorporation of material into planets.

Water ice sublimation (Isobe 1970) would occur at temperatures close to 100 K. The inner radii of the α Lyr and β Pic clouds are at substantially higher temperatures and thus the central depletion is not a result of ice sublimation.

First-order calculations indicate that destruction of particles by mutual collisions should occur at least as rapidly as removal by Poynting-Robertson (PR) radiation drag, which causes particles to spiral toward the star. However, modest stellar winds would produce substantial additional "corpuscular" PR drag. For the three modeled cases, the mean particle size is progressively larger for smaller total particle area. A progression in particle size versus total particle area could indicate that the identity of large particles is maintained over time and that small particles removed are no longer replaced with fragments of larger bodies, i.e., that a PR-like effect is important. The progression in particle size is in the same sense as the stellar luminosities, so a connection to minimum stable particle size (Gillett 1985) may also be involved.

The last two columns of table III compare the stars' main sequence lifetimes (Gillett 1985) with the (radiation) PR decay time for the particles at the inner edges of the clouds. A density of 3 g cm^{-3} and a particle radius "a" related to the emissivity parameter by a = $\lambda_o/2\pi$ were assumed. The orbit decay time for the inner particles is short compared to the stellar lifetimes; the particles detected by IRAS should move to fill the depleted regions.

One possible explanation for the continued existence of the depleted regions is the presence of an object able to clear a sufficiently wide volume of particles by gravitational accretion. Particles moving in toward the star would be unable to cross through this volume. In the case of the β Pic system, calculations for (radiation) PR drift indicate that an object with mass and radius approximately that of Saturn would act in this fashion.

Acknowledgments: The authors would like to thank G. Neugebauer, F. Low, and G. Aumann for assistance with the IRAS slow-scan observations and reductions, S. Aoki for assistance with the Addscan reductions, M. Cook and J. DuHamel for preparation of the figure, J. Goad for a critical reading of the manuscript, I. Gatley for helpful discussions, the Data Management Team of IPAC, especially H. Hanson, for rapid production of the Addscan data, and the IPAC library chopper crew for heroic bundling and mailing.

REFERENCES

Aumann, H. H., Gillett, F. C., Beichman, C. A., de Jong, T., Houck, J. R., Low, F. J., Neugebauer, G., Walker, R. G., and Wesselius, P. R. 1984, Ap. J. Lett. **278**, L23.
Aumann, H. H. 1985, P. A. S. Pacific **97**, 885.
Backman, D. E., Gillett, F. C., Aumann, H. H., Neugebauer, G., Low, F. J., and Waters, R. 1988, in prep.

Coté, J. 1987, A. A. **181**, 77.

Duncan, D. K. 1981, Ap. J. **248**, 651.

Gillett, F. C., Aumann, H. H., Low, F. J., Neugebauer, G.,
Waters, R. 1984, paper presented at Protostars and
Planets II Conference, Tucson, Arizona.

Gillett, F. C. 1985, in Light on Dark Matter ed. F. P. Israel,
Reidel Co., Dordrecht, p. 61.

Gliese, W. 1969, Catalogue of Nearby Stars , Veröffentl. Astron.
Rechen-Instituts, Heidelberg, Nr. 22.

Gliese, W. and Jahreiss, H. 1979, A. A. Suppl. **38**, 423.

Good, J. C., Hauser, M. G., and Gautier, T. N. 1986,
Adv. in Space Res. **6** (7), 83.

Kleinmann, S. G., Cutri, R. M., Young, E. T., Low, F. J., Gillett,
F.C. 1986, Explanatory Supplement to the IRAS Serendipitous
Survey Catalog, U.S. Government Printing Office, Washington DC.

Isobe, S. 1970, P. A. S. Japan **22**, 429.

Low, F. J., et al. 1984, Ap. J. Lett. **278**, L19.

Sadakane, K. and Nishida, M. 1986, P. A. S. Pacific **98**, 685.

Simon, T., Herbig, G., and Boesgaard, A. M. 1985 Ap. J. **293**, 551.

Smith, B. A. and Terrile, R. J. 1984, Science **226**, 1421.

Varsik, J. 1987, Ap. J. (preprint).

PROGRESS IN DETERMINING THE SPACE ORIENTATION OF STARS

L.R. Doyle

SETI Institute, Space Science, NASA Ames Research Center
Moffett Field CA 94035 USA

ABSTRACT. *The theory for determining the space orientation of the rotation axis of certain stars has previously been outlined, and we investigate here the precision and observational procedure with which this theory can be tested. Presently available data is presented, and observing stars are outlined. The importance of this determination to the methods of detecting extrasolar planetary systems is further developed, and a few examples are given.*

A theory for determining the space orientation of the rotation axis of stars has been proposed (Doyle *et al.*, 1983; 1984) based, in part, on asymmetries detectable on the star by observing it in sunspot emission. Baliunas *et al.* (1983), have determined the rotation period of certain stars by noting the short-term variations in their sunspot (or starspot) emission in the Ca II H & K line. Essentially, the rotation of the sunspot regions (which cause the emission feature in the spectra), around the stellar limb gives a determination of the stellar rotation period. This value divided by the stellar circumference gives the stellar rotational velocity. Since measurements of Doppler-shifted line broadening due to stellar rotation gives $V \sin i$, this latter value can be divided by the former to give the stellar inclination to the line-of-sight.

A second component of the space orientation of the stellar rotation axis is its projection on the plane-of-the-sky (in a clocklike sense). This determination could also be made by essentially measuring the star's diameter in the Ca II K line, which, (at least for the Sun), shows an equatorial "sash" due to the confinement of sunspots to equatorial latitudes (see for example, Doyle *et al.*, 1984). Ca II K-line speckle interferometry performed on late-type (sunspot bearing) resolvable stars could give this component of the space orientation to within a few degrees for angularly resolvable stars. The application of this determination to the various methods of detecting extrasolar planetary systems has been pointed out (Borucki and Summers, 1984; Doyle, 1985) with special relevance to the photometric detection method. The assumption is that the rotation axis of the star will be perpindicular to the planetary orbital plane, and our own solar system tends to support such theories. (The inclination of the solar equator to the ecliptic plane is about $7°15'$).

Indirect detection techniques (detection of an extrasolar planetary system by it's effect on the parent star) are much more sensitive than most direct detection techniques. The indirect methods fall into three essential types: astrometric, radial

101

G. Marx (ed.), Bioastronomy – The Next Steps, 101–105.
© *1988 by Kluwer Academic Publishers.*

velocity, and photometric. Astrometric techniques try to detect a "wobble" of the star due to the offset around the barycenter of the star as planets orbit it. This method, to first order, is not confined by a given space orientation of the system (a system could be detected varying in a left and right fashion - if the inclination was edge-on - just as it could be detected moving in a sinusoidal fashion if the system rotation axis was pole-on). The equations governing the detection of this motion are given (after Black, 1980) by:

$$x = R_*[cos\lambda_*(cos\psi cos\phi - sin\psi sin\phi cosi) - sin\lambda_*(sin\psi cos\phi + cos\psi sin\phi cosi)] \quad (1)$$

$$y = R_*[cos\psi sin\phi + sin\psi cos\phi cosi) + sin\lambda_*(sin\psi sin\phi - cos\psi cos\phi cosi)] \quad (2)$$

where R_* is the distance of the system barycenter from the star, ψ is the longitude of periastron (zero for circular orbits), ϕ is the angle between the true orbital plane and the plane of the sky, and λ_* is the position angle of the planet (or planets) in the orbital plane of the system. The values of x and y would correspond, for example, to right ascension and declination, respectively, and finally i is the inclination of the true orbit to the plane of the sky ($90°$ being edge-on, $0°$ being pole-on).

The astrometric method may benefit, then, from a determination of both the inclination i, as well as the clocklike component (which we will here call n) since a determination of these could remove ambiguities of determination due to multiple planetary bodies, proper motion components, and perhaps even intrinsic stellar pulsations of a significant magnitude. The expected shape of the variation could also be previously determined and the above equations would essentially reduce to only R_* and λ_* terms.

Detecting a planetary system by measuring regular radial velocity variations in the star would benefit from the rotation axis determination for the same reasons. In addition, this method's sensitivity goes to zero as the inclination to the line-of-sight drops. This method would benefit, then, by knowing the stellar orientation because sufficiently high inclination stars could be previously selected for observation.

The photometric method, which would detect a slight drop in the light from a star due to a planetary transit in the line-of-sight, significantly relies on planetary orbital inclinations being very close to $90°$. This method is not really statistically feasible unless such systems can be pre-determined. However, with such a determination it can become the most sensitive of extrasolar planetary detection techniques.

In Table 1 we give some examples of some presently known stellar inclinations to the line-of-sight determined from this technique. As can be seen, accuracy is a significant problem, but even with this technique in its initial development certain detection methods can already be suggested. P_{rot} values are those determined by

Baliunas *et al.* (1983), by measuring the variation in the CaII H & K emission, a measure of which is the value $< S >$ known as the Wilson index (see for example, Vaughan *et al.*, 1981). As has previously been pointed out (Doyle *et al.*, 1984) among major sources of error are uncertainties in $V \sin i$ which may not be measured at the same location in the stellar atmosphere as the period determined from Ca II H & K emission, which comes from the chromosphere.

Table 1

Stellar Rotation Axes Line-of-Sight Inclinations With Suggested Detection Method

HD No.	Name	Spec Type	$V \sin i$	$< S >$	P_{rot}(days)	$\sin i$	i°	Method
1835	9 Cet	G2 V	7.0 ± 0.8	0.336	7.7 ± 0.1	≈ 1	≈ 90	P, RV, A
2454		F2 V	8	0.160		1.0	90	P, RV, A
6903	ψ^3 Psc	G0 III	≈ 97	0.294	6.2 ± 0.1	≈ 1	≈ 90	P, RV, A
10700	τ Cet	G8 V	0.8 ± 0.8	0.168	31.9 ± 3.6	0.6 ± 0.6	37^{+53}_{-37}	A
16160		K4 V	< 3	0.231	45	≈ 1	≈ 90	P, RV, A
19373	ι Per	G0 V	3.5 ± 0.7		22.3 ± 2.5	1.0 ± 0.3	90^{+0}_{-46}	P, RV, A
20630	κ Cet	G5 V	> 15	0.348	9.4 ± 0.1	≈ 1	≈ 90	P, RV, A
22049	ϵ Eri	K2 V	2.1 ± 1.0	0.533	11.3 ± 0.2	0.5 ± 0.3	30^{+23}_{-18}	A
25998	50 Per	F7 V	22	0.282	2.6 ± 0.1	0.96	73	RV, A
26913		G3 V	< 6	0.380	7.17 ± 0.06	< 0.86	< 59	RV, A
39587	χ^1 Ori	G0 V	9.4 ± 0.4		5.1	0.8 ± 0.1	53^{+11}_{-9}	RV, A
61421	α CMi A	F5 IV	2.8 ± 0.3		2.4 ± 0.3	0.06 ± 0.01	$3^{+1}_{-0.2}$	A
102870	β Vir	F8 V	3.2 ± 0.4		12.1 ± 0.6	0.5 ± 0.1	30^{+7}_{-7}	A
114710	β Com	G0 V	3.9 ± 0.4		12.4 ± 0.2	1.0 ± 0.2	90^{+0}_{-33}	P, RV, A
115617	61 Vir	G6 V	2.0 ± 0.4		32.7 ± 3.7	1.2 ± 0.3	90^{+0}_{-26}	P, RV, A
131156	ξ Boo A	G8 V	2.7 ± 0.6		6.2	0.4 ± 0.1	24^{+6}_{-7}	A
154417		F8 V	5.5 ± 0.7	0.259	7.58 ± 0.02	0.78 ± 0.1	52^{+10}_{-9}	RV, A
165341	70 Oph A	K0 V	2.6 ± 0.8		19.7 ± 0.1	1.2 ± 0.4	90^{+0}_{-32}	P, RV, A
185144	σ Dra	K0 V	1.5 ± 1.0		30.0 ± 3.4	1.0 ± 0.7	90^{+0}_{-53}	P, RV, A
190406	15 Sge	G1 V	$3 - 5$	0.190	13.5 ± 0.2	$0.75 - 1.0$	$49 - 90$	RV, A
201091	61 Cyg A	K5 V	≈ 3	0.694	37.9 ± 1.0	≈ 1	≈ 90	P, RV, A
201092	61 Cyg B	K7 V	< 3	1.088	48	≈ 1	≈ 90	P, RV, A
206860		G0 V	10.2 ± 1.1	0.319	4.64 ± 0.02	0.93 ± 0.1	68^{+22}_{-12}	RV, A
222107	λ And	G8 IV-III	< 20	0.931	56.4 ± 0.8	≈ 1	≈ 90	P, RV, A
A = Astrometric,		RV =	Radial	Velocity,	P =	Photometric	Detection	Methods

The study is presently being extended to giant and supergiant stars and preliminarily determined rotation periods appear to be much longer than solar-type stars (Baliunas, 1986, personal communication). A few of these are listed in Table 2, where the stellar disc is sufficiently large to be theoretically resolved (the star image is not an Airy disc), by the telescope size given, so that a good determination of the clocklike component, n, can be made, as well. Applying our equation for the determination of the accuracy of this component, n, (after Doyle et al., 1984) we have:

$$\delta n = \sqrt{\frac{\left(10^{\frac{m+26}{2.5}}\right) N_s E_p N_I^2}{N_p N_e \left(\Delta\nu/\nu\right) Q S A \tau}} \tag{3}$$

We will take some representative values: $A = 8100 cm^2$ (telescope aperature), $N_s = 10^2$ (number of speckles in the specklegram), $E_p = 4 \times 10^{-12} ergs$ (photon energy), $N_I = 10^2$ (number of pixels/speck in image space), $N_p = 10^6$ (number of pixels per Fourier space speck), $N_e = 10^2$ (number of speckle exposures), $\Delta\nu/\nu = 2 \times 10^{-4}$ (bandpass of K filter), $Q = 0.1$ (quantum efficiency), $S = 10^6 ergs/sec - cm^2$ (solar constant), and $\tau = 10^{-3} sec$ (exposure time). For Arcturus (α Boo), then, which has apparent magnitude $m = 0.06$, we obtain $\delta n \approx 0.°1$.

Table 2

Clocklike Orientation Stars - Initial Proposed Observing List					
Name	apparent radius (")	telescope (in.)	Spec type/class	m_v	Comments
ω Cet	0.057	70	M6/III	2.0	To be added to H&K program
α Ori	0.066	61	M2/Iab	0.8	Probable circumstellar shell
R Leo	0.032	124	M8/III	5.0	Probable intrinsic variable
α Boo	0.022	181	K2/III	0.06	H&K period well determined
α Sco	0.042	95	M0/Iab	1.2	H&K determined, shell
α Her	0.031	128	M5/III	3.1	H&K being determined
β Peg	0.016	248	M2/II-III	2.56	H&K being determined

Of course, the determination of the inclination to the line-of-sight parameter would be the most essential, although it is the more difficult to determine accurately, (see Soderblom, 1985; Campbell and Garrison, 1985; as well), especially since the sine function becomes less sensitive to inclination angle just where the photometric method needs the determination made, around 90°. For example, a 1% error in the determination of $\sin i$ at around 90° translates to an almost 10° error in i, more than enough to prevent the accurate prediction of a transit. Clearly many more accurate inclinations need to be determined.

A relationship may be expected between the amplitude of the sunspot variations and the inclinations to the line-of-sight, as edge-on stars with a sufficiently asymmetric distribution of sunspots may be expected to vary in Ca II H & K emission to a greater extent; in some way related to the projected Ca II H & K emission area. However, a comparison would best be made between similar type stars at the onset of their sunspot cycles. The determination of the parameter $< S >$ involves a continuum scaling, among other things, that would also have to be scaled in a normalizing fashion. No relationship, as yet, emerges when the $< S >$ values are plotted against sin i values from Table 1, but if such a relationship could be found, as more accurate determinations of $< S >$ and sin i can be made, a great number of additional stellar inclinations could then be estimated and the number of systems with determined space orientations for extrasolar planetary detection would be significantly extended.

BIBLIOGRAPHY

(1) Baliunas, S.L., Vaughan, A.H., Hartmann, L., Middelkoop, F., Mihalas, D., Noyes, R.W., Preston, G.W., Frazer, J., and Lanning, H., 1983, 'Stellar Rotation in Lower Main-Sequence Stars Measured From Time Variations in H and K Emission-Line Fluxes II. Detailed Analysis of the 1980 Observing Season Data', **Ap.J. 275**, 752-772.

(2) Black, D.C., 1980, 'In Search of Other Planetary Systems ', **Space Sci. Rev. 25**, 35-81.

(3) Borucki, B. and Summers, A., 1984, 'The Photometric Method of Detecting Other Planetary Systems', **Icarus 58**, 121-134.

(4) Campbell, B. and Garrison, R.F., 1985, 'On the Inclination of Extra-solar Planetary Orbits', **P.A.S.P. 97**, 180-182.

(5) Doyle, L.R., Wilcox, T.J., and Lorre, J.J., 1983, 'The Space Orientation of Stars, (abstract), **P.A.S.P. 95**, 588.

(6) Doyle, L.R., Wilcox, T.J., and Lorre, J.J., 1984, 'The Space Orientation of Stars', **Ap.J. 287**, 307-314.

(7) Doyle, L.R., 1985, 'Assisting the Detection of Extrasolar Planets Through the Determination of Stellar Space Orientations', in **Proc. I.A.U. Symp. 112, Boston, MA., June 1984**, 97-102.

(8) Soderblom, D., 1985, 'Determining the Spacial Orientation of Stellar Rotation Axes', **P.A.S.P. 97**, 57-59.

(9) Vaughan, A.H., Baliunas, S.L., Middelkoop, F., Hartmann, L.W., Mihalas, D., Noyes, R.W., and Preston, G.W., 1981, 'Stellar Rotation in Lower Main-Sequence Stars Measured From Time Variations in H and K Emission-line Fluxes I. Initial Results', **Ap.J. 250**, 276-283.

A PHOTOMETRIC APPROACH
TO DETECTING EARTH-SIZED PLANETS

W.J. Borucki

NASA Ames Research Center
M.S. 2455-3, Moffett Field CA 94035, USA

L.E. Allen and W.S. Taylor

SETI Institute
101 First Street 410, Los Altos, CA 94022, USA

A.T. Young

Department of Astronomy, San Diego State University
San Diego, CA 92182, USA

A.R. Schaefer

National Bureau of Standards
Washington D.C. 20234, USA,

ABSTRACT. The photometric method of searching for planets around stars depends on observing the decrease in light flux produced by the transit of a planet across the stellar disk. The magnitude of this reduction is proportional to the ratio of the planet's area to that of the star. For the solar system, the decrease in light amounts to 0.01 per cent for terrestrial-sized planets. To overcome the effects of scintillation and variable extinction in the Earth's atmosphere, it will be necessary to operate the photometric system on a space platform. The photometric method works only for planets whose orbital plane is near our line of sight. Thus many stars must be monitored to insure that some stars with appropriately-oriented orbital planes are observed. If every solar-type star has a planetary system similar to our solar system, then a photometer that monitors 1000 stars with the requisite precision should detect at least 10 transits per year of observation. Thus a 3-year observation period should allow meaningful statements to be made about the frequency of solar-type planetary systems. A state-of-the-art photometer is being developed to test components and concepts. The goal is the development of a photometer that can routinely measure the relative brightness of stars to a precision of 1 part in 100,000. Such precision should be achievable using "quantum perfect" detectors. Results of field tests of a prototype photometer are promising.

1. INTRODUCTION

Detection of extrasolar planetary systems is important to the understanding of star formation, the formation of

107

G. Marx (ed.), Bioastronomy – The Next Steps, 107–116.

planetary sytems, and to the estimation of how common life
is throughout the universe. Black (1980), Tarter et al.
(1985), and Scargle (this conference) present summaries of
the various techniques that have been considered for
detecting other planetary systems. Efforts are underway to
implement the radial velocity and astrometric techniques. As
presently envisioned, these techniques will be able to
detect major planets at least as massive as Uranus and
Neptune, but will not be able to detect planets as small as
the Earth. Rosenblatt, 1971, and Borucki and Summers, 1984,
discuss photometric methods that have the potential of
detecting Earth-sized planets if a technique with the
required precision can be developed.

The simplest photometric method of searching for planets
around stars depends on observing the decrease in starlight
produced by the transit of a planet across the stellar disk.
The magnitude of this reduction is proportional to the ratio
of the planet's area to that of the star. For the solar
system, the decrease in light amounts to one percent for
giant planets such as Jupiter and Saturn, 0.1 percent for
planets like Uranus and Neptune, and 0.01 per cent for
terrestrial-sized planets. To assure that a transit would be
clearly distinguished from noise fluctuations, it is
necessary to operate with a high signal to noise ratio
(SNR). Consequently, the photometer should produce a SNR of
approximately 10^5 for the dimmest stars that will be
monitored.

2. PROBABILITY OF OBSERVING A TRANSIT

Note that the photometric method works only for planets
whose orbital plane is near our line of sight. As a planet
orbits its star, the planet casts a shadow on the celestial
sphere. The ratio of the area of this shadow band to the
area of the sphere is the probability that an observer will
observe a transit in one orbital period. Figure 1 sketches
the geometry used to derive the probability p, that a
transit will be observed. Let d* be the stellar diameter, r
the orbital radius, S the width of the shadow on the
celestial sphere, Θ the angle subtended by the star an the
planet, and y the distance from the planet to the celestial
sphere. Then p is given by:

$$p = (2\pi y S)/4\pi y^2 \tag{1}$$

$$S = y\Theta \quad \text{implies } p = \Theta/2 \tag{2}$$

and for small Θ; $\Theta = d_*/r$. $\tag{3}$

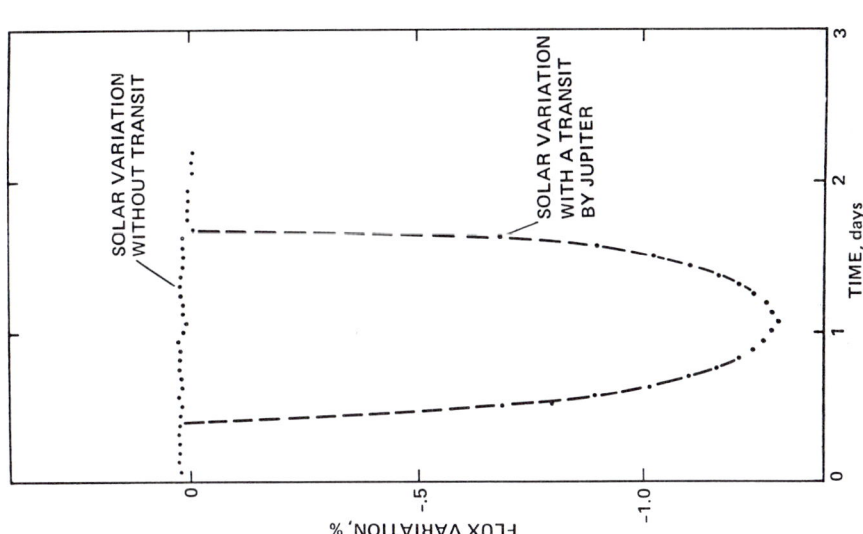

Fig. 2. Simulation of the flux variation of the Sun due to a transit by Jupiter as seen by a distant observer.

GEOMETRY

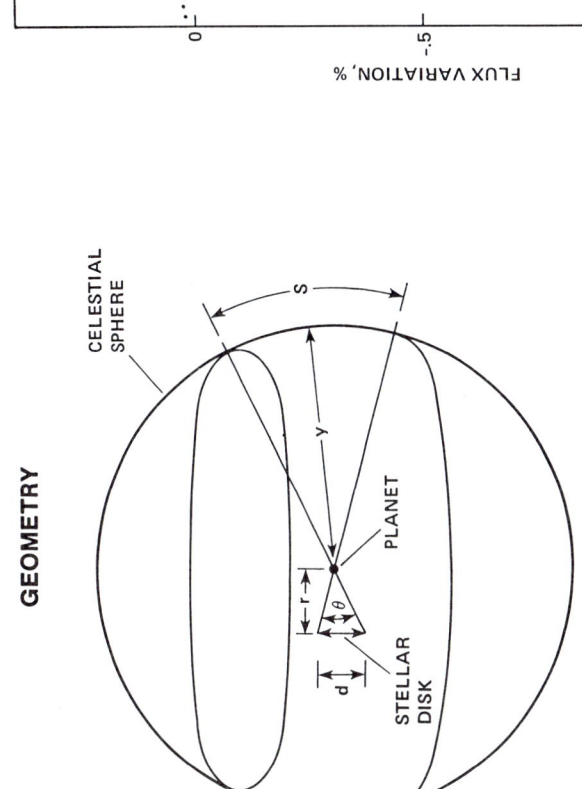

Fig. 1. Sketch of the geometry used to calculate the probability of an observation of a planetary transit. The planet is at the center of the sphere and is a distance r from a star of diameter d_*. The width of the shadow band cast on the celestial sphere is S.

Therefore; $\qquad p = d_*/2r$ $\qquad\qquad\qquad\qquad$ (4)

The number of transits expected to be observed in one terrestrial year T, is:

$$n = pT/T_p \qquad\qquad\qquad (5)$$

The planet's orbital period can be computed from;

$$T_p = 2\pi r^{3/2}/[GM_*]^{1/2} \qquad\qquad (6)$$

where G_* is the universal gravitational constant and M_* is the mass of the star. Therefore n can be written as;

$$n = d_*[GM_*]^{1/2} r^{-5/2} T/4\pi \qquad\qquad (7)$$

Equation (7) shows that n is largest for close-in planets of large and massive stars. For equation (7) to be valid, observations must be carried out 24 hours per day, ie., 3 or 4 telescopes equally spaced in longitude must be used or the observations must be carried out from space. If observations are carried out for an average of 6 hours per day, then p and n should be reduced slightly for inner planets that have a transit-time of less than 18 hours(See Borucki and Summers, 1984). Table 1 gives n based on the assumption that every solar-type star has a planetary system similar to ours.

Table 1: Expected Number of Transits n, per Star as a Function of Orbital Radius

Orbital Radius (AU)		Number of Transits/year
0.39	(Mercury)	5.0×10^{-2}
0.72	(Venus)	1.0×10^{-2}
1.0	(Earth)	4.6×10^{-3}
1.5	(Mars)	1.6×10^{-3}
5.2	(Jupiter)	7.5×10^{-5}
9.5	(Saturn)	1.6×10^{-5}
19.	(Uranus)	2.9×10^{-6}
30.	(Neptune)	9.4×10^{-7}

$$N = \text{Sum of } n = 6.7 \times 10^{-2}$$

It is clear from Table 1 that the probability of observing a transit for a planetary system that has randomly oriented orbital planes is approximately 7 % if it is assumed that all program stars have planetary systems with detectable planets at the positions of planets in our solar system. However, it will be shown later that detection of planets significantly smaller than the Earth is unlikely because the natural variability of stars will make it very difficult to obtain a sufficiently large signal to noise ratio to get an unambiguous detection. If we assume that no other planetary system has planets as large as the Earth in orbits like that of Mercury or Mars, then N becomes 0.015. Although this value of N is used in further calculations, it should be remembered that we are ignorant of the actual distribution of planets in other planetary systems.

If we can construct a multichannel photometer that monitors approximately 1000 stars continuously, then we should detect 1 transit per month. Because the highest probability of an observation is associated with the planets with short orbital periods, a confirmation of a planetary detection could be made within a year by detecting a signal of the same amplitude and duration one orbital period later. An approximate orbital period can be predicted from the duration of the transit and an estimate of the stellar diameter. The results of a 3 year observation program should decisively answer the question of the frequency of Earth-sized planets around solar-type stars. If no transits are detected, the results would strongly imply that other solar systems like ours are very uncommon. On the other hand, the detection of 30 or more planetary systems would imply that solar systems are quite common. In short, no matter what the number of detections, the results would provide a meaningful estimate to the frequency of planetary systems.

Before we can detect the 1 part in 10^4 variation of the stellar brightness due to a transit by an Earth-sized planet, it is necessary to insure that signal variations due to other sources do not overwhelm the transit signal. Three sources of error can be identified; variability of the star itself, instrument problems, and variations of the stellar light flux due to atmospheric variability. Each of these noise sources will be discussed in the following sections.

3. ESTIMATE OF STELLAR VARIABILTIY

Aside from technical limitations imposed by the photometer and the fundamental limit of photon counting statistics, the intrinsic photometric variability of stars imposes a practical limit on the detectability of small planets. No star can be regarded as completely nonvariable. Small

variations of stellar flux are caused by convection and
starspots, modulated by the rotation of the star (Radick et
al., 1983).

The Sun gives us an opportunity to examine the
variations of at least one main-sequence dwarf star.
Radiometric observations from the Solar Maximum Mission
(SMM) have suceeded in achieving 1 part per 100,000 orbit-
to-orbit precision (Willson et al., 1981).   To simulate
what a signal would look like from a planetary transit in
the presence of stellar variability, data from the SMM were
combined with calculated flux reductions due to a planetary
transit of the Sun by Jupiter and by the Earth.   Figure 2
shows what a distant observer in Jupiter's orbital plane
would observe during such a transit.   The Jupiter signal is
clearly detectable.   Signals by Saturn, Uranus, and Neptune
would also be clearly detectable. A simulation of the signal
expected from the transit of a terrestrial-sized planet
across the sun is presented in Figure 3. Note the greatly
expanded vertical scale. The points on the upper curve
represent 1 hr samples of the solar flux measured by the SMM
radiometer.   The lower curve connects points that represent
the flux expected if an Earth transit had occurred during
the measurements.   The data give a S/N of about 4, but it is
clear that the variability of the Sun makes the detection of
terrestrial-sized planets marginally possible.   The SMM
observations shown in Figures 2 and 3 were made at the peak
of the solar activity cycle.   The solar variability for
other portions of the solar activity cycle is expected to be
much less than that shown.   Therefore, detection of
terrestrial-sized planets should be quite feasible when the
stellar activity levels are below the peak activity level of
the Sun. The detection of larger planets and low-mass
companion stars should present no problem even in the
presence of substantial stellar activity levels.   However,
it is clear that stellar variability is a fundamental limit
to the ability of the photometric method to detect small
planets. (A more comprehensive discussion is given in
Borucki et al., 1984.)

4. ATMOSPHERIC LIMITATIONS

Two major sources of signal degradation are fluctuations in
terrestrial atmospheric extinction and scintillation.
(Seeing does not appreciably affect differential photometric
measurements if large focal-plane apertures and careful
techniques are used.)

For bright stars, scintillation noise will dominate
shot noise. Young(1974) showed that scintillation noise can
be expected to vary with telescope aperture to the -2/3
power, integration time to the -1/2 power, wind speed, and
with the square of the air mass.   In particular, for a

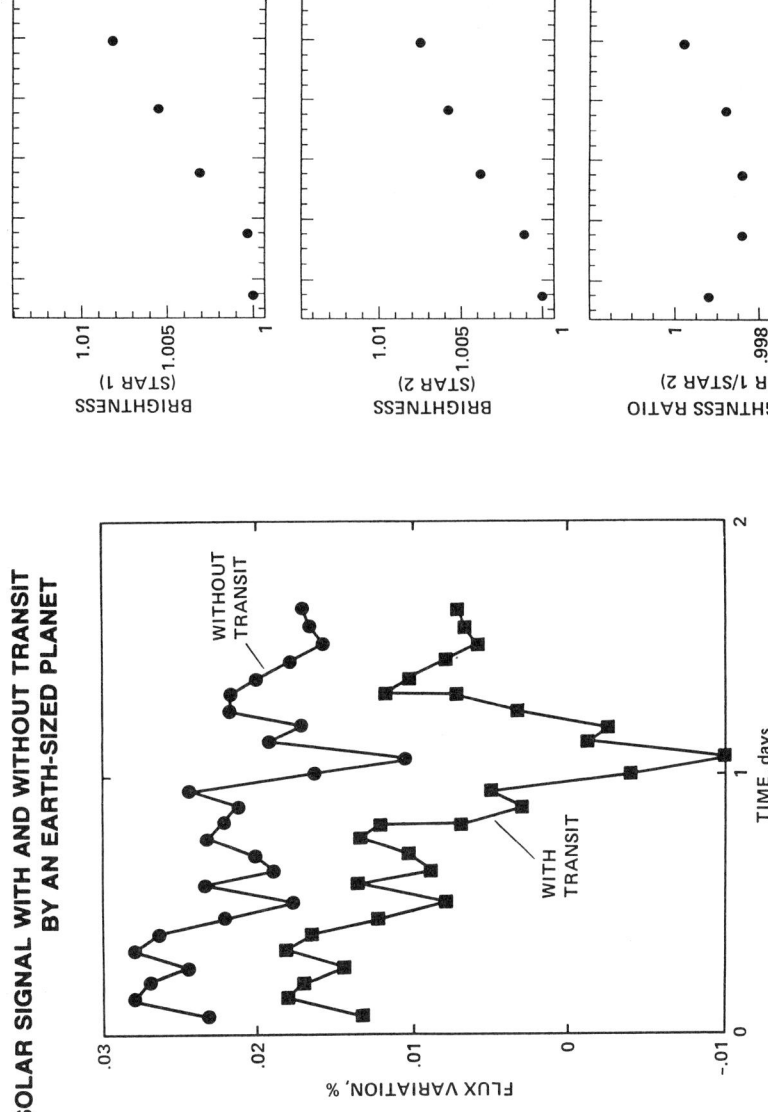

SOLAR SIGNAL WITH AND WITHOUT TRANSIT
BY AN EARTH-SIZED PLANET

Fig. 3. Simulation of the flux variation of the Sun due to
a transit by the Earth as seen by a distant observer. The
upper curve shows measurements from the active cavity
radiometer aboard the SMM satellite (Willson et al., 1981).
The lower curve shows simulated transit. Note that the scale
on the ordinate is greatly expanded compared to that used in
Figure 1.

Figure 4. The variation of the brightness of 2 stars and the
ratio of their brightness as a function of time. Star 1 is α
Aries and Star 2 is β Arietis.

sample time of 8 hrs and an air mass of 2, a telescope with a 9 m aperture is required to keep the scintillation noise at or below 1 part in 100,000. Because scintillation and extinction problems are not a problem for telescopes in space, a 1 to 2 m aperture telescope and a 30 minute observation period would be sufficient to obtain the necessary precision for stars brighter than $m_v = 11$. There are approximately 10^5 solar type stars brighter than $m_v = 11$ (Allen, 1963). Consequently, there should be no shortage of program stars.

5. INSTRUMENTATION FOR HIGH PRECISION PHOTOMETRY

Although space-based radiometric measurements of the sun have attained a relative precision for hour-long observations of 1 part in 10^5 (Willson et al, 1981), current stellar photometers that must observe much dimmer objects through the terrestrial atmosphere obtain a night-to-night precision of only 3 parts in 10^3. Current limitations are caused by a combination of instrument drift, variations in atmospheric extinction during the measurement sequence, and approximations used in data reduction. Instrument drift can be reduced with "quantum-perfect" detectors. The concept of "quantum perfect" detectors refers to detectors that closely approximate a quantum mechanical system that converts every photon in a detector to exactly one electron-hole pair. The physics of operating silicon diodes as self-calibrated "quantum perfect" detectors is desribed Zalewski and Geist, 1980, Geist et al., 1982, and Bruenig, 1987. Variations of atmospheric extinction are eliminated by operating from a space platform. A prototype photometer has been constructed for ground-based tests of different types of detectors, preamps, and optical components. Three channels simultaneously monitor two stars and the sky. The silicon diode detectors are followed by transimpedance preamps that produce 10^{11} volts/amp. Analog to digital conversion and signal averaging is done by HP 3457 digital multimeters. The measurements are transmitted to and stored by an IBM PC. The photometer is attached to the focal plane of the Carneige twin astrograph at Lick Observatory. Figure 4 shows some data obtained on α and β Arietis with the prototype photometer. The top 2 strips show the output from each of the two channels monitoring the stars after corrections for the sky background, the detector dark currents, and for the extinction due to the changes in the air mass with time. It is clear that during the approximately 3 hour observational period, the extinction per air mass has changed by about 1 %. The ratio of the fluxes from the two stars is shown in the bottom strip. The significant improvement in the measurement precision is clear. The top 2 records have a standard deviation of 4.5×10^{-3} whereas the ratio has a SD of 5×10^{-4}. For solar-type stars, the precision obtained during this test is sufficient to detect Jupiter-sized or Uranus-sized planets with SNRs of 20 and 2, respectively. Although the precision demonstrated by the breadboard photometer is too low to search for Earth-sized planets, the significant increase in precision over conventional photometers is encouraging.

6. DISCUSSION

Although it appears more practical to search for Earth-sized planets with a single 1 meter telescope in space than to use ground-based 9 meter telescopes, ground-based tests of prototype planetary detection photometers should be quite useful. A ground-based search for Uranus and Neptune-sized planets in short-period orbits should be quite practical once the capability to monitor 1000 stars at a precision of 1 part in 10,000 is reached. Determination of the frequency of such planets will add significantly to our knowledge of other planetary systems even if no planets are detected. Furthermore, improvements to photometric precision should significantly impact other areas of astronomical research; for example: measurements of low-level stellar variability, determination of asteroid properties, and the recognition of low contrast features in planetary atmospheres and on planetary surfaces.

PULSE DETECTION IN THE VAX

Searches for 3 consecutive pulses at 1 resolution

Statistical parameters:

* Probability of single hit = P_{cell} = 10^{-5}

* Threshold in units of average noise = T = 11.5

* N = 1000 seconds

* M = 10^7 channels

* Number of distinct pulse trains = Q = $MN^3/12$

* False alarms per observation = F = QP_{cell}^3 = 5/6

* F conserved at different resolutions

Average computational parameters:

* 1000 hits/second

* Pulse field = 10^5 hits

* Extrapolations/hit = $QP_{cell}^2/10^5$ = 5/6

* 250 Extrapolations in last second

TABLE 1

3 PULSE DETECTION AT LOW THRESHOLDS

Pairs in search cone

$$q = KN^3 P_{cell}^2$$

where $0.01 < K < 0.1$

Probability of third pulse

$$P_3 = 1 - (1 - P_{cell})^9 \approx 9P_{cell}$$

For high sensitivity

$$P \approx 0.1, \text{ given noise alone}$$

Implies long search with low detection probability.

INSTEAD make searches short and test total power.

Sensitivities within 0.5dB of search using no threshold.

TABLE 2

REFERENCES

Allen, C.W.(1963). <u>Astrophysical Quantities. 2nd edition</u>. University of London, The Athlone Press.

Black, D.C.(1980). 'In search of other planetary systems' <u>Space Science Reviews</u> **25**, 35-81.

Borucki, W.J., and A.L. Summers(1984). 'The photometric method of detecting other planetary systems' <u>Icarus</u> **58**, 121-134.

Borucki, W.J., J.D. Scargle, and H.S. Hudson(1984). 'Detectability of extrasolar planetary transits' <u>Ap. J.</u> **291**, 852-854.

Bruening, R.J. (1987). 'Spectral irradiance scales based on filtered absolute silicon photodetectors' <u>Applied Optics</u> **26**, 1051-1057.

Geist, J., W.K. Gladden, & E.F. Zalewski(1982) 'Physics of photon-flux measurements with silicon photodiodes' <u>J. Opt. Soc. Am.</u> **72**, 1068-1075.

Radick, R.R., M.S. Wilkerson, S.P. Worden, J.L. Africano, A. Klimke, S. Ruden, W. Rogers, T.E. Armandroff, and M.S. Giampapa(1983). 'The photometric variability of solar-type stars. II. Stars selected from Wilson's chromosphric activity survey' <u>Pub. Astr. Soc. Pacif.</u> **95**, 300.

Rosenblatt, F. (1971). 'A two-color photometric method for detection of extra-solar planetary systems' <u>Icarus</u> **14**, 71-93.

Tarter. J.C., D.C. Black, and J. Billingham (1985). 'Review of methodology and technology available for the detection of extrasolar planetary systems' Paper IAA-85-493; <u>Proceedings of the 36th International Astronautical Congress, held in Stockholm, October 1985</u>.

Willson, R.C., S. Gulkis, M. Janssen, G.A. Chapman, and H. Hudson(1981). 'Observations of solar irradiance variability' <u>Science</u> **211**, 700-702.

Young, A.T.(1974). 'Observational technique and data reduction' In <u>Methods of Experimental Physics</u> **12A**. Edited by N. Carleton.

Zalewski, E.F., and J. Geist (1980) 'Silicon photodiode absolute spectral response self-calibration' <u>Applied Optics</u> **19**, 1214-1216.

SYSTEMATIC ASPECTS OF DIRECT EXTRASOLAR PLANET DETECTION

Robert A. Brown

Space Telescope Science Institute

3700 San Martin Drive, Baltimore MD-21218 USA

ABSTRACT. Using the first optical observatory in space, the Hubble Space Telescope, images of possible extrasolar planets will have poor contrast against the background of diffracted and scattered starlight. The very long exposure time required to achieve an adequate signal-to-noise ratio will make their detection infeasible. For a future telescope, a 16-fold increase in either the smoothness or the collecting area of the optics would reduce the exposure time to a tolerable value, but the contrast would remain low and the required photometric precision high. In this situation, the feasibility of detection would be contingent on the careful identification and control of systematic errors.

1. INTRODUCTION

Diffraction-limited optical telescopes in space will perform unique and powerful studies of the environments of stars. To a far greater extent than is possible at ground-based observatories, the point-spread functions (PSFs) of the Hubble Space Telescope (HST) and its successors will discriminate between stellar and circumstellar light. This will enable the direct observation of phenomena related to our planetary system and to the processes that formed it — for example, accretion disks and gas jets near young stellar objects. Possibly, extrasolar planets will be discovered, although that is a challenge of extreme intrinsic difficulty: a planet would be faint ($m_v \gtrsim 26$), and the planet-to-star flux ratio would be small ($\lesssim 10^{-9}$). Nevertheless, taken as a whole, future space–based observations of stellar environments will advance our understanding of the Solar System and extend a frontier of planetology toward the science of stars.

There is a current debate about the qualitative performance required for extrasolar planet detection, and it bears directly on scientific intent. The first and absolutely indispensable criterion is that a proposed technique should be capable of detecting planets. The second is that a null result should be meaningful. This would be the basis for a survey to determine the statistical occurrence of planets, and it has more stringent technical implications because it requires that any (or at least a known fraction of) planets present *must* be detected — within some size, reflectivity, and geometrical constraints, of course. The view dissenting from the stronger criterion holds that, since no extra-solar planets have yet been imaged, *any* detection is a worthwhile goal in itself, and a statistically valid survey could come later. The current paper does not pursue this debate; it only discusses system- and experiment-level issues related to the minimum capability — a reasonable (but not known) probability that a relatively conspicuous planet would be discovered.

A critical instrumental factor for extrasolar planet detection is the contrast ratio of the peak surface brightness of the planetary image to the surrounding scattered starlight.

117

G. Marx (ed.), Bioastronomy – The Next Steps, 117–123.
© *1988 by Kluwer Academic Publishers.*

That ratio governs — in concert with the planet flux — the fundamental rate at which information about the planet can accumulate. Space-based telescopes, free of atmospheric "seeing," will improve the contrast ratio by offering narrower, diffraction-limited PSF central cores. The outer wing, however, determines the denominator of the contrast ratio, and relatively less practical experience and understanding has been achieved on this facet of optical PSFs — either on the ground or in space.

Although extrasolar planet detection was not a design objective of HST, there is interest in its potential contribution. Brown and Burrows[1] have analyzed its static optical performance using metrological data from manufacturing. They computed the expected HST PSF due to aperture diffraction, optical aberrations, and scattering by the residual surface roughness of the mirrors. The predicted contrast ratio between a hypothetical planet and the background starlight was found to be very small, even with extreme assumptions about the planet's size, reflectivity, and placement. Citing limitations in the HST detectors that would prevent achieving the signal-to-noise level required by the low contrast ratio, Brown and Burrows concluded that extrasolar planet detection would be very difficult with HST. The exposure time analysis below complements and fortifies that conclusion.

Future optical telescopes will be developed with higher performance requirements than HST, perhaps with extrasolar planet detection as a specific goal.[2] Because of the great effort that would entail, technical feasibility must be demonstrated by clarifying and validating proposed instrumental approaches. That process has begun by addressing the basis for improving the contrast ratio by enhancing both the core and wings of the telescopic image. However, the low information rate of the planetary signal — even with greatly improved static optical performance — indicates the need for a system-level conception of the experiment to identify and analyze systematic liens against its success. This paper attempts incremental progress toward that goal.

2. COUNTING STATISTICS AND THE REQUIRED EXPOSURE TIMES

2.1. The HST Case

We consider a Cassegrainian telescope that forms the image of a star and a companion planet onto a focal-plane detector. The upper limit to the planetary information in an exposure is set by the counting statistics of the recorded photons. Extending the notation of Brown and Burrows, we define:

$R \equiv$ radius of telescope aperture (HST: 120cm)
$\varepsilon \equiv$ fractional size of telescope central obscuration (HST: 0.33)
$\lambda =$ wavelength of light
$\Delta\lambda \equiv$ spectral bandpass
πF and $\pi F'' \equiv$ total stellar and planetary fluxes within $\Delta\lambda$ (photons cm^{-2} sec^{-1}).
θ and $\theta'' \equiv$ angles from stellar and planetary image centers
PSF $\equiv$ telescope point-spread function (sr^{-1})
$\Omega_1 \equiv$ solid angle inside first dark ring of the PSF
$\eta \equiv$ factor (~ 1 and implicitly defined in Eq. (1)) accounting for the finite size of Ω_1
$q_1 \equiv$ quantum efficiency of the detector
$q_2 \equiv$ system transmittance to the detector
$T \equiv$ duration of an exposure in seconds.

The stellar and planetary photons recorded within Ω_1 are given by

$$N_{star}(\theta) = \pi F \cdot PSF(\theta) \cdot \Omega_1 \cdot \eta(\theta) \cdot (1-\varepsilon^2)\pi R^2 \cdot q_1 q_2 \cdot T \qquad (1)$$

and

$$N_{planet}(\theta'') = N_{star}(\theta) \cdot \frac{\eta(\theta'')}{\eta(\theta)} \cdot \left[\frac{\pi F'' \cdot PSF(\theta'')}{\pi F \cdot PSF(\theta)} \right]. \tag{2}$$

The bracketed quantity in Eq. (2) is the previously discussed contrast ratio. Somewhat arbitrarily, we define the minimum information for detecting the planet to be a signal-to-noise ratio of five inside Ω_1 centered on the planet position:

$$\frac{N_{planet}(\theta''=0)}{\sqrt{N_{star}(\theta) + N_{planet}(\theta''=0)}} \geq 5. \tag{3}$$

(The telescope image must be at least critically sampled to avoid loss of spatial information. The angular subtense of a critically sized element is $\lambda/4R$, and no fewer than about twenty elements would receive the light within Ω_1. Critically-sized detector elements located near the center of the planet image would achieve nearly the same signal-to-noise ratio as specified for Ω_1. In practice, the minimum desirable size for a detector element would be determined by the detector noise characteristics.)

The basic condition in Eq. (3) requires long integration times. The Jupiter-Sun system provides an optimistic example because Jupiter is nearly a maximal planet[3] and the Sun is more luminous than most stars. Consider that HST attempts to detect Jupiter

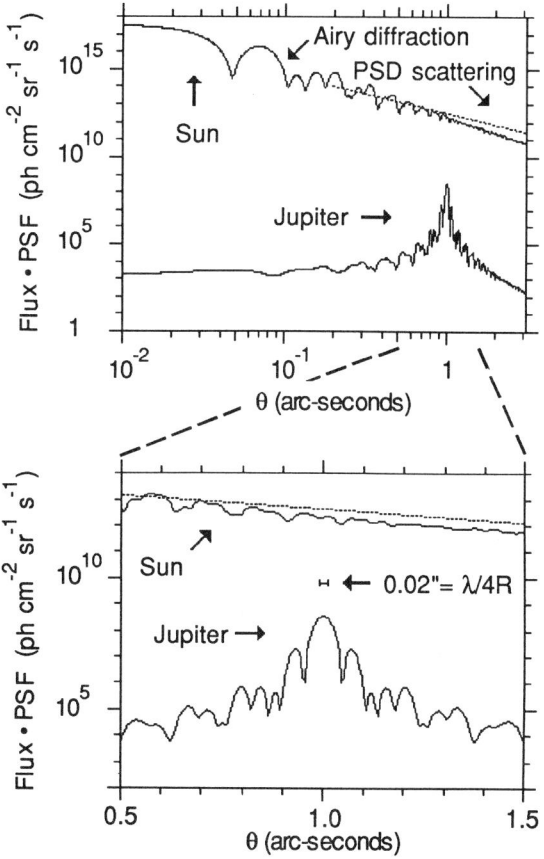

Figure 1. Surface brightness distributions illustrating the direct extra-solar planet detection problem. The HST flux-PSF products — the numerator and denominator of the bracketed quantity in Eq. (2) — are shown for a hypothetical Sun-Jupiter system viewed from 5pc in a 500Å bandwidth centered at a wavelength of 5000Å. The solid curve is the Airy diffraction pattern for a 120cm radius aperture with a 33% central obscuration (HST). The dashed line gives the PSF due to the power spectral density (PSD) of residual surface errors on the HST mirrors as estimated by Brown and Burrows. The contrast ratio between the central surface brightness of Jupiter and the local scattered and diffracted sunlight is about 6×10^{-5}. The size of a detector element for critical sampling is indicated.

from a distance of 5 parsecs (pc). (Only about 50 stars are closer than 5pc, and none is within 1pc.) At its greatest elongation of about 1 arc-second ("), Jupiter would be about 26th visual magnitude, and the ratio of planetary to stellar flux would be about 10^{-9}. We assume a 10% spectral bandpass centered on $\lambda =5000$Å. Fig. (1) shows separately the numerator and denominator of the contrast ratio, the peak value of which is about 1.8×10^{-4}. We find by numerically integrating the Airy PSF (see Brown and Burrows) that $\eta(0) \approx 0.25$ and $\eta(1") \approx 1$, so Eq. (3) requires the number of recorded planetary photons to be greater than 5.6×10^5. Assuming $q_1 q_2 = 0.2$, and using the values $\pi F" \cdot PSF(0) = 3.6 \times 10^8 ph$ cm^{-2} sr^{-1} s^{-1}, $(1 - \varepsilon^2) \pi R^2 = 4 \times 10^4$cm^2, and $\Omega_1 = 1.6 \times 10^{-13}$sr, then one planetary photon would be recorded about every nine seconds within the first PSF minimum. The exposure time from Eqs. (1-2) would be greater than $T(5\sigma) = 5 \times 10^6$s or 58 days.

Of all the instrumental factors governing the exposure time, only one is not a simple function of R and λ: the PSF wing. $T(5\sigma)$ is directly proportional to the PSF evaluated at the separation angle between the star and the planet. In addition to aperture diffraction, other scattering mechanisms can contribute to this wing brightness, and Brown and Burrows have investigated the effects of the power spectral density (PSD) of the residual figure errors on the telescope mirrors. The PSD PSF is proportional to $\lambda^{-4} \delta^2$, where δ is the root-mean-square mirror roughness in a specific range of spatial frequencies at the telescope exit pupil. That range is defined by the extrema of $2\pi\theta/\lambda$ for the θ- and λ-domains of interest. The functional form of the PSD PSF versus θ is the same as for the PSD spectrum itself versus spatial frequency, and Brown and Burrows found a -2.2 power law in the case of HST. They predicted that PSD scattering would produce scattered light 1.8 times the Airy diffraction pattern at $\lambda=5000$Å and $\theta=1"$, as illustrated in Fig. (1). This increases the expected exposure time by factor 2.8 to $T(5\sigma)=1.4 \times 10^7$s.

For *systematic* reasons, this exposure time is unacceptably long. In the first instance, the signal-to-noise calculation would be invalidated by the planet moving — Jupiter at 5pc would typically appear to travel 0.02" (the critical sample size at 5000Å) in about 10 days. Secondly, long exposures are vulnerable to operational problems. For example, rare but large noise fluctuations in pointing sensors could cause loss of pointing control, as could computer processing upsets due to cosmic ray hits. Based on experience with both space-based and ground-based observing, 10 hours is a reasonable goal for the duration of one astronomical observation, so reducing the required exposure time by a factor of about 400 would be a significant step toward feasibility for extrasolar planet detection, and this would require an optical system enhanced with respect to HST.

The telescope mirror smoothness is significant for the extrasolar planet detection problem because the PSD PSF limits the extent to which the exposure time can be reduced by apodizing the exit pupil to modulate the Airy diffraction pattern.

2.2. The Potential Benefit of Apodization

The theory of apodization for a Cassegrainian telescope has been presented by Mauron,[4] and the apodizer he analyzed in detail has been implemented in HST's Faint Object Camera (FOC).[5] Apodization can shape — and suppress — the wing of the aperture diffraction pattern, but the composite PSF wing cannot be reduced below the contributions of other scattering mechanisms, PSD scattering in particular. On account of the relatively high level of mirror roughness, little benefit is expected from reducing the Airy wing on HST. Nevertheless, the FOC apodizer will effectively remove the diffraction signatures of the secondary mirror support and the primary mirror mounting pads.

A future optical telescope may be built with smoother optics or greater collecting area or both. To estimate the potential exposure time reduction relative to the HST-Jupiter-Sun case, the parametric dependencies of $T(5\sigma)$ on δ, R, and λ must be introduced.

Assume the Airy wing is removed by apodization to a level below the PSD PSF — but without reducing either the system efficiency, $q_1 q_2$, or the PSF evaluated at $\theta=0$ (idealizations). The product $PSF(\theta=0) \cdot \Omega_1 \cdot \eta(\theta=0)$ is independent of R, λ, and δ (ignoring the Strehl factor of order unity that was fully treated by Brown and Burrows). The scattering characteristics of the planet are assumed to be independent of λ. Then Eqs. (1-3) indicate $T(5\sigma)$, which would have the value 9×10^6s for the apodized HST PSF, is proportional

    (a) to R^{-2} through the collecting area,
    (b) to $\lambda^{-4}\delta^2$ through the denominator of the contrast ratio,
    (c) to $R^{-2}\lambda^2$ through the numerator, and
    (d) to $\Phi(\lambda)^{-1}$, the inverse of the relative stellar flux.

The relative stellar flux is normalized to the value used in the HST-Jupiter-Sun case, namely 2.25×10^4ph cm^{-2} s^{-1}, which is valid for a 10% bandwidth centered on 5000Å and a blackbody at 5850K with radius $r_s=6.96\times10^{10}$cm (i.e. the Sun) at distance of 5pc. Tab. (1) provides $\Phi(\lambda)$ for the Sun and for M5 and A1 dwarfs at 5pc in a 10% spectral bandpass. (More than half the stars closer than 5pc are M4-6 dwarfs and the brightest, Sirius, is an A1 dwarf.)

Table 1. $\Phi(\lambda)$, the relative photon flux in a 10% spectral bandpass from stars of various spectral types at a distance of 5pc. The values are normalized to the solar value at 5000Å, 2.25×10^4ph cm^{-2} s^{-1}. The calculations assumed a blackbody spectrum from an object of the indicated temperature and radius.

$\lambda(Å)$	$\Phi(\lambda)$		
	Sun (5850K, 1r_s)	M5 (2800K, 0.32r_s)	A1 (9700K, 2.2r_s)
2000	9.8×10^{-3}	1.7×10^{-9}	6.4
3000	0.17	2.4×10^{-6}	23.
4000	0.57	7.1×10^{-5}	34.
5000	1.0	4.7×10^{-4}	37.
6500	1.4	2.3×10^{-3}	35.
8000	1.6	5.4×10^{-3}	31.
10,000	1.6	1.0×10^{-2}	25.

Let us define:
    $\Lambda \equiv \lambda/5000Å$
    $\mathfrak{R} \equiv R/120cm$
    $\Delta \equiv \delta/\delta_{HST}$ (see Brown and Burrows for a discussion of δ_{HST})
    $T \equiv T(5\sigma)/9\times10^6$s
and then

$$T = \Lambda^{-2} \mathfrak{R}^{-4} \Delta^2 \Phi(\lambda) \qquad (4)$$

This formula can be used to explore design and observational possibilities, such as:

(i) If the only change is smoother optics, a 10 hour exposure time would be achieved with Δ=0.06 at 5000Å for the Sun. This would be further reduced to 2.4 hours for λ=8000Å.

(ii) A telescope with half the aperture radius of HST ($\mathfrak{R}$=0.5) would have to be four times smoother to give the same performance. With 4 times the aperture radius and no smoother optics than HST, the 10 hour exposure time could be achieved.

(iii) Choosing a more typical, less luminous star for study would make the exposure time much longer. Selecting the brightest star would shorten the exposure time by less than about a factor forty.

So, if a telescope like HST were developed that had either 16 times smoother optics or 16 times the collecting area, then the Airy and PSD PSFs would not prevent accumulating information about the existence of a maximal planet near an unusually bright, nearby star in a 10-hour exposure. However, the signal would still be weak — contrast ratio below 3% — and vulnerable to other systematic effects. The planetary information must be regarded as latent until we understand how to protect it fully and to transform it competently into a scientific result.

3. SYSTEMATIC ASPECTS

In their optical analysis of the feasibility of planet direct detection, Brown and Burrows introduced a "figure of merit" for the optical system, which is the essentially the contrast ratio discussed above — the central surface brightness of the planetary image divided by that of the local stellar wing intensity. Without elaborating, they stated that if the contrast ratio were much less than unity, then planet detection would be "difficult", and "systematic sources of error would have to be carefully analyzed and controlled". While Brown and Burrows did not specifically address photon counting errors, their maxim is illustrated by the information rate/exposure time analysis of Sect. (2): poor contrast necessitates long exposure times, which would be limited by problems that – while they may be either deterministic (Jupiter moving) or stochastic (operational disruptions) – are essentially systematic, beyond experimental control, and not statistically reducible.

Determining *all* the systematic influences and susceptibilities in this experiment calls for a systems analysis and a synoptic viewpoint. Fig. (2) provides a conceptual template for the planet detection experiment. The seven squared boxes identify major tasks.

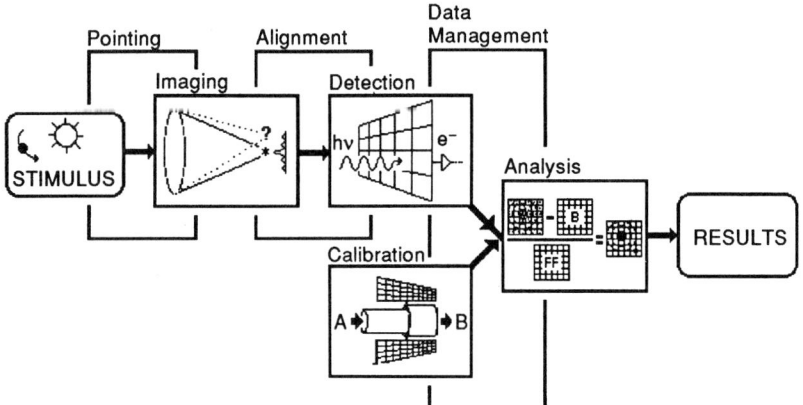

Figure 2. A systems analysis template for the direct extrasolar planet detection experiment. Seven tasks are identified that must be analyzed to assure feasibility.

Briefly, pointing and alignment must hold the planetary image stationary on the apodizer and the detector; imaging must form a stable, sharp PSF with no extra light in the wing; detection must provide a signal containing information about the numbers and locations of incident photons; analysis converts the detector signal into a measurement of the surface brightness distribution near the star using calibration data documenting the instrumental response; and insofar as the data base will be complex — multi-dimensional and time-dependent — a data management task has been separately identified.

It is the burden of this paper that technical liens against direct planet detection must be *fully* identified and their remedial requirements derived and assessed. A complete and appropriate set of questions about the experiment must be posed that will establish its feasibility. Those questions have the following generic form: how good does the pointing (imaging, alignment, detection, calibration, data management, and analysis) need to be? The feasibility of this project is equivalent to the concatenated feasibilities of all the discovered requirements. Beyond that, an *assurance* of successfully implementing the experiment demands extra margin in the specifications of all critical tasks — the technical equivalent of fiscal contingency reserves.

Thanks to C. J. Burrows, M. W. Werner, and G. D. Illingworth for critical readings of the original manuscript. Support for this work was provided by NASA under Contract NAS5-26555 through the Space Telescope Science Institute, which is operated by AURA, Inc.

REFERENCES

1. Brown, R. A. , and Burrows, C. J. 'On the feasibility of direct planet detection using Hubble Space Telescope.' *Icarus* (submitted).

2. Terrile, R. J. 'Direct imaging of extra-solar planetary systems with a low-scattered light telescope.' *Bioastronomy — The Next Steps: Proceedings of IAU Colloquium No. 99* (G. Marx, ed.) 1988.

3. Hubbard, W. B. *Planetary Interiors*. Van Nostrand Reinhold: New York, 1984, pp. 236-240.

4. Mauron, N. 'Haute Resolution Angulaire et Coronographie en Astronomie Spatiale: Etude Theorique et Experimentale du Mode Coronographique de la "Faint Object Camera." ' Unpublished thesis at Université de Droit, d'Economie et des Sciences d'Aix-Marseille, 1980.

5. Macchetto, F. 'The Faint Object Camera.' *The Space Telescope Observatory: Proceedings of the Special Session of Commission 44, IAU 18th General Assembly* (D. N. B. Hall, ed.) 1982, pp. 40-54.

DIRECT IMAGING OF EXTRA-SOLAR PLANETARY SYSTEMS WITH A LOW-SCATTERED LIGHT TELESCOPE

R.J. Terrile

Jet Propulsion Laboratory, CalTech
Pasadena CA 91109, USA

ABSTRACT: In a joint study conducted by the Jet Propulsion Laboratory and the Perkin-Elmer Corporation it was found that an Earth orbital, one meter class low scattered light coronagraphic telescope can achieve a broad range of scientific objectives including the direct detection of Jupiter-sized planets around the nearby stars. Recent major advances in our understanding of coronagraphic performance and in the field of super smooth mirror fabrication allow such an instrument to be designed and built within current technology.

INTRODUCTION

In order to achieve the most ambitious goals of remotely detecting, and characterizing (temperature measurements, detection and composition of atmospheres and oceans, detection of biogenic elements, etc.) extra-solar planets, the direct light from the planet must be examined without contamination from the light of the parent star. Only direct methods of planet detection are suited for obtaining these observations. One such method, which could be implemented with current state of the art technology, is the Circumstellar Imaging Telescope (CIT). This is an Earth orbital, low-scattered light, coronagraphic telescope currently under study at the Jet Propulsion Laboratory (Ref. 1-3).

The largest obstacle in making observations of extra-solar planets is the separation of the planet light from the scattered and diffracted light from the parent star. In the case of our own solar system seen from the nearby stars, Jupiter would be the brightest planet and would, in visible light, be about 1 billion times fainter than the sun (Ref. 4,5). Without the use of special optical techniques to reduce the contribution of starlight, detection of the planet would be impossible.

Currently, one of the most promising methods of direct detection of extra-solar planets uses a combination of a coronagraph to reduce the diffraction of starlight by the telescope and a super-smooth mirror to reduce scattered light. This combination of technologies is currently under study in the Circumstellar Imaging Telescope. This telescope is different than any other astronomical telescope in that it can reduce diffracted light in the wings of an image of a point source (Airy Pattern) by a factor of 1000 over the entire field of view of the

125

G. Marx (ed.), Bioastronomy – The Next Steps, 125–130.
© 1988 by Kluwer Academic Publishers.

instrument. Such an instrument would have to orbit the Earth in order to get above the detrimental effects of the atmospheric seeing. A 2 meter diameter instrument in Earth orbit, would allow the direct detection of a Jupiter-size planet around a solar type star out to a distance of about 10 parsecs. These same techniques could be employed on a larger instrument of 10 meters in diameter to detect the Earth out to these distances.

SCIENTIFIC OBJECTIVES

Recent discoveries in the field of extra-solar planetary systems (Ref. 6,7) have led to the formulation of a series of science objectives which can be addressed by the proper instrumentation. Ideally, it is important to conduct a survey of the nearby stars in order to determine the frequency of circumstellar disks as a function of the spectral class of the parent star. Important parameters to be measured include the disk morphology, mass distribution, composition (through spectroscopy and narrow bandwidth imaging), particle size distribution (through polarization studies) and possibly dynamics if appropriate spectral regions can be found. Furthermore, imaging of other disks at various viewing geometries and ages will allow morphological studies and perhaps lead to an understanding of the origin and evolution of planetary systems. Finally, direct detection of Jupiter-sized planets around the nearby stars would be of fundamental importance to our view of planet formation. Additionally, depending on the distance to the parent star, filter photometry and phase measurements could be done over the orbital period and Neptune-sized objects could be detected around the nearest stars.

In addition to the extra-solar planetary science objectives there are many astrophysical science objectives as well. These include the study of faint material near quasars and the nuclei of active galaxies and imaging of galactic jets. Also the study of novae and supernovae remnants as well as interacting binary stars and early type stars (T-Tauri). Generally, regions near bright astrophysical objects can be studied for faint material.

RESULTS

The diffraction control technology employs a principle which was first used in the 1930's by Lyot to look at the corona of the sun (Ref. 8). The instrument is a Lyot coronagraph and works by blocking out the center light from a bright object in the field of view with an occulting spot in the focal plane. Down steam from this point the pupil of the telescope is imaged and masked off to remove the contribution of light which is diffracted from the edges and secondary support structures of the primary and secondary mirrors. This allows the reimaging of the focal plane onto a detector with the diffracted component of light reduced by about a factor of 100 from that of a conventional system. The main disadvantage is the loss of the field of view near the bright central object. In a JPL - Perkin-Elmer study it was found that an additional factor of 10 in rejection of diffracted light can be achieved by optimizing the coronagraph design. This optimization also allows imaging arbitrarily close to the central object without loss of diffraction rejection. Figure 1 shows the intensity of light in the focal plane as a function of distance from the center of the Airy disk for several methods of apodization. The hybrid coronagraph reduces diffracted light by 3 orders of magnitude. However, this factor of 1000 improvement in diffracted light can only be utilized if the narrow angle scattering due to figure errors in the primary mirror is at least 1000 times less than the diffraction of a conventional mirror. This requirement imposes very strong constraints on the mid-spatial scale frequency errors of the telescope.

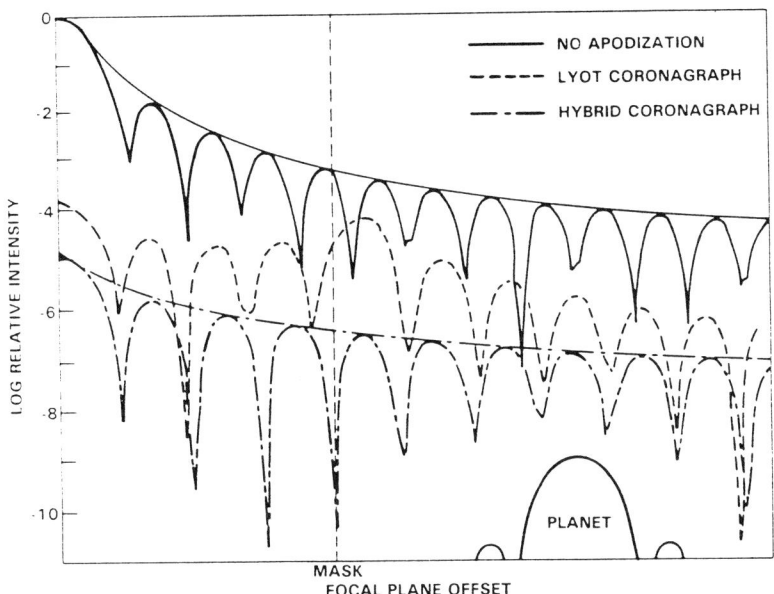

Figure 1: The intensity of starlight in the focal plane as a function of distance from the center of the Airy disk. Solid lines show the diffracted star light (for a solar type star) normalized at the center of the Airy disk. Also in solid lines is the pattern from a Jupiter-sized planet shown at 5 au around a star about 10 parsecs away. The dashed lines show the reduction of diffracted light from a classical Lyot coronagraph and for our hybrid design. The reduction in diffracted light for the hybrid design is about 3 orders of magnitude.

The other key technology which allows the use of a low-scattered light telescope is the fabrication of a super-smooth mirror. One of the most important findings of the JPL - Perkin-Elmer study was that mirror fabrication technology has not been driven by astronomical mirrors but by other optical uses, such as in the fabrication of micro-electronics.

For example, micro-lithography mirrors (0.5 meter diameter spherical mirrors) have been characterized as being about a factor of 5 times smoother than the Hubble Space Telescope mirror at mid-spatial frequencies. Furthermore, these mirrors are mass produced at the current rate of about 20 per month and are made to specifications. During fabrication figure errors on these mirrors decrease monotonically until the desired specification is reached and then polishing stops and the mirror is delivered. The metrology used to measure the figure of the mirrors can go beyond this specification and there is no inherent reason why much smoother mirrors could not be fabricated. A primary mirror between a factor of 2 to 3 times better than the currently produced mirrors is needed for the direct detection and characterization of extra-solar planets. This level of mirror smoothness is currently within the capabilities of existing fabrication techniques. Other non-conventional manufacturing techniques currently under study, could produce even smoother mirror surfaces and would further improve the performance of an extra-solar planet detecting and characterizing telescope.

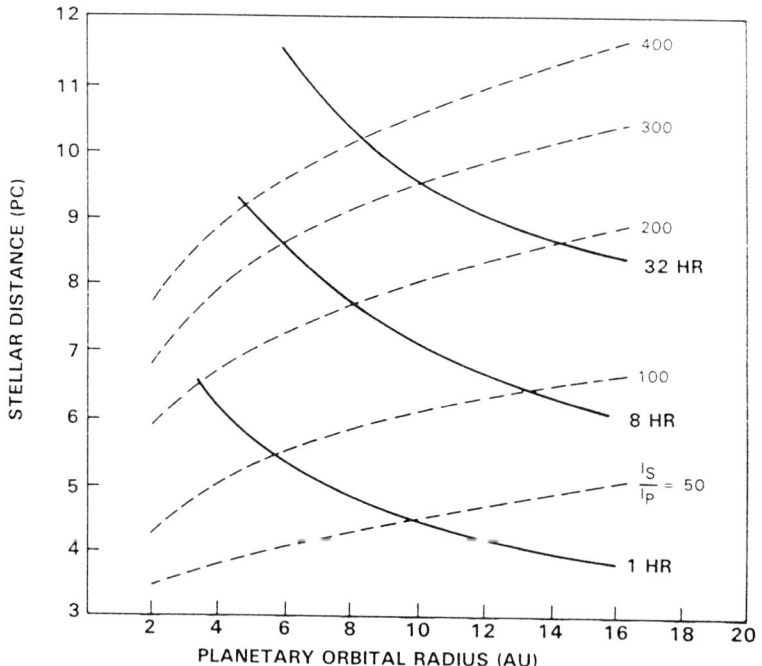

Figure 2: Detection of Jupiter-sized planets with a 1.5 meter diameter Circumstellar Imaging Telescope. Solid lines indicate integration times required to achieve a signal-to-noise ratio of 5 for a Jupiter-sized planet around a solar type star. A 400 nm bandpass at 600 nm was assumed and a mirror 15 times better than HST was used with a 2.5 slope of the power spectral density curve of the mirror. Also shown are

trajectories of constant flux ratio where flux ratio is defined as the number of scattered star photons in the Airy disk of the planet divided by the photons from the planet.

DISCUSSION

The practicality of detection of a Jupiter-sized planet depends on integration time and flux ratio. Integration time is governed by the diameter of the primary mirror and the scattered light in the optics (mirror smoothness at mid-spatial frequencies and coronagraphic efficiency). In our calculations we assume a required signal-to-noise ratio of 5 for a Jupiter-sized planet around a solar type star using a 400 nm bandpass at 600 nm. The practical maximum length of the system integration time is governed by the timescales for changes in the point-spread function of the system which can not be calibrated. However, point-spread function changes which restrict the detection of extra-solar planets will generally arise from length scales on the order of 10's of cm on the primary mirror. It is these length scales which will contribute to a fixed speckle pattern at small angles in the focal plane. An analysis of the mirror figure stability at these length scales is currently under way, but for the present, we assume that a practical integration timescale is on the order of hours to tens of hours. Flux ratio is defined as the number of star photons collected for every planet photon in the Airy disk of the planet. It is dependent on the coronagraphic efficiency, the mirror smoothness and the slope of the power spectral density curve of the mirror. We assume that conventional ground-based techniques using charge-coupled-device (CCD) detectors for integration and frame adding can achieve detection with flux ratios of several hundred. Figure 2 shows the relationship of integration time and flux ratio for the detection of a Jupiter-sized planet with a 1.5 meter Circumstellar Imaging Telescope.

CONCLUSION

Future larger systems, employing primary mirrors with diameters of 10 meters or more would allow full characterization of the temperature, pressure, atmospheric composition etc. of extra-solar planets. These instruments will need to employ the same diffracted light rejection and super-smooth mirror technology that is used in the smaller instruments. However, because they will most likely be composed of multiple elements, additional technologies involved in the stabilization of the point spread function of a multiple element telescope will be needed. This additional technology will also allow phasing of individual elements so that interferometric observations could be made.

The logical first step before building a large extra-solar planet characterizing system is a modest sized (1 to 2 meter class) low-scattered light orbital telescope. The CIT is such a system and would not only address the technologies necessary for this complete remote characterization, but would also provide important data on circumstellar disks, zodiacal components, and proto-planetary material around nearby the stars. Such an examination and inventory of circumstellar material is necessary to evaluate the likelihood of conditions suitable for the formation of planets and evolution of life around the other stars.

ACKNOWLEDGMENTS

This work would not have been possible without the valuable contributions from C. Ftaclas and E. T. Siebert of the Perkin-Elmer Corporation. We also thank R. A. Brown for many useful discussions. The work described in this paper was carried out by the Jet Propulsion Laboratory, California Institute of Technology, under a contract with the National Aeronautics and Space Administration and was supported by the Director's Discretionary Fund at the Jet Propulsion Laboratory.

REFERENCES

1.   Terrile, R. J. (1986) "Direct Detection of Extra-Solar Planetary Systems From the Ground and Space." Aerospace America, in press.

2.   Ftaclas, C., Siebert, E. T. and Terrile, R. J. (1987) "The Circum-stellar Imaging Telescope – Direct Detection of Extra-Solar Planets., Bull. Amer. Astron. Soc., in press.

3.   Terrile, R. J., Ftaclas, C. and Siebert, E. T. (1987) "Direct Imaging of Extra-Solar Planetary Systems With A Low-Scattered Light Telescope." Bull. Amer. Astron. Soc., in press.

4.   Lockheed Palo Alto Research Laboratory (1979) "Systems Level Feasibility Study for the Detection of Extrasolar Planets, Part 2, Apodized Telescope (APOTS)." LMSC-D676425, Lockheed.

5.   Brown, R. A. (1987) "Systematic Aspects of Direct Extra-Solar Planet Detection." IAU Colloquium 99, Bioastronomy, Balaton, Hungary, this volume.

6.   Aumann, H. H., Gillett, F. C., Biechman, C. A., de Jong, T., Houck, J. R., Low, F. J., Neugebauer, G., Walker, R. G. and Wesselius, P. R. (1984) "Discovery of a Shell Around Alpha Lyrae." Astrophys. J., 278, L23.

7.   Smith, B. A. and Terrile, R. J. (1984) "A Circumstellar Disk Around Beta Pictoris." Science, 226, 1421.

8.   Lyot, B. (1939) "A Study of the Solar Corona and Prominences Without Eclipses." Mon. Not. Royal Astron. Soc., 99, 580.

DISCOVERY AND STUDY OF PLANETARY SYSTEMS USING ASTROMETRY FROM SPACE

E.H. Levy and R.S. McMillan
Lunar Planetary Laboratory, University of Arizona
Tucson AZ 85721 USA

G.D. Gatewood, J.W. Stein and M.W. Castelaz
Allegheny Observatory, Univesity of Pittsburg
Pittsburg PA 15215 USA

A. Buffington
Astrophysics and Space Science, University of California
San Diego, La Jolla CA 92098 USA

N, Nishioka and J.D. Scargle
NASA Ames Research Centerl Space Sciences Division
Moffett Field CA 94035 USA

Introduction

Much attention has been given over recent years to the question of discovering planetary systems around other stars. The interest in this question is provoked by both scientific and human imperatives. In this paper, I want to discuss the real intellectual issues that must be confronted in any attempt to discover planetary systems and the implications that these have for the performance of instruments built for that task. And, finally, I want to discuss the progress that potentially can be made with precise astrometric measurements. I will focus here on the general question of planet system discovery and study, and I will not take up other fascinating questions posed by more specialized—and perhaps more futuristic—approaches to the detection of possible markers of extraterrestrial life (Angel 1987; Burke 1987).

From a scientific perspective, we now have a theoretical picture of the simultaneous formation of stars and planetary systems which seems natural in the sense that it follows in a relatively straightforward fashion from the basic physical behavior of a collapsing, rotating interstellar cloud. This picture is also natural in that the dominating characteristics of the one planetary system that is known, the Solar System, are seen to emerge easily from the basic physical ideas, without strenuous and fanciful excursions of theory. I do not mean to imply that all of the details are understood—indeed, the totality of what we truly understand is probably dwarfed by that which remains mysterious—but, rather, that our current understanding provides a general and stable framework within which future advances in the subject of star and planet-system formation are likely to remain.

Looked at as a whole, our own planetary system exhibits a number of essential regularities, which are crucial to our understanding of the system's origin and which have strong implications for our thinking about planetary systems associated with other stars. Crudely speaking, all of the planets orbit the Sun in concentric circles, and in the same direction. All of the orbits are confined roughly to the same plane, and that plane coincides with the plane extending the solar equator. The planetary spins are aligned with one another and with the other angular momenta in the Solar System to a degree too great to be accidental.

In addition to these geometrical regularities, the planetary masses and compositions are distributed in a correspondingly regular fashion. The four planets close to the Sun are the smaller of the major planets and are composed predominantly of rock and metal—compounds which condense from a cosmic mix of material or remain solid at high temperatures, ranging toward 2000K. The planets in the outer Solar System include the four giants, which contain much larger amounts of

131

G. Marx (ed.), Bioastronomy – The Next Steps, 131–136.

cosmic gas, hydrogen and helium, as well as large amounts of ice—compounds which condense or remains solid in cosmic matter only at relatively low temperatures, below a few hundred Kelvins. In this gross architectural view of the Solar System, I will leave aside consideration of Pluto.

It is worth recalling that *the central dogma of planetary cosmogony* accounts for all of these regularities rather gracefully. The Sun and planets formed at about the same time from a single precursor object: the protoplanetary or protosolar nebula. This nebula had evolved substantially under the influence of viscous and other dissipative forces before planet formation, to the extent that memory of the original collapse was erased from the system. Before planet formation, the nebula had settled into a flattened disk with a single sense of Keplerian revolution. The temperature and pressure in the disk declined with distance from the center, reaching temperatures in the range of 1500-2000K within a fraction of an astronomical unit from the center, and falling below a few hundred Kelvins beyond several AU. In the higher temperature region, within two or three AU of the center, only rock and metallic compounds existed in the solid phase, while all of the more volatile components of cosmic matter remained vapor. The terrestrial planets apparently formed from a process of accretion of this solid matter. In the low temperature zone, beyond a few AU from the nebula's center, most condensible species—including the cosmically abundant water—existed in the solid form. Solid matter accretion in this region had a far richer source of material with which to work, producing large protoplanets with sufficiently strong gravity to capture the surrounding nebular gas. Thus, the present cosmogonical dogma holds that all of the planets formed through processes controlled by solid matter accretion, and that the contrasts in composition and mass between the four terrestrial planets of the inner solar system and the four giant planets of the outer solar system reflect the differences in the composition and the abundances of the compounds that could condense or remain solid in the inner and outer parts of the Solar System.

Altogether, the existing cosmogonical dogma is generally successful in explaining that which we already know about this planetary system, and it is probably correct in its crude outline. However, one important attribute of the existing dogma is its aspect of physical inevitability. This inevitability then has implications for the prevalence and character of other planetary systems that provide an important empirical challenge. The strength of those implications is enhanced by the fact that many observed protostars are seen to be associated with protostellar disks similar to that which is inferred to have been the birthplace of our own planetary system (Strom 1987; Sargent and Beckwith 1986). Therefore, the Copernican view that the Solar System is unremarkable in the universe is supported by comparing the conditions inferred to have existed at the Solar System's birth with the observed conditions of star formation more generally.

In order to sharpen the issues connected with discovering and studying other planetary systems, it is instructive to compare this archetypical planetary system, and its canonical mode of formation, with a binary star system; planetary systems have often—and most likely erroneously—been thought of as low-mass, end members of the distribution of multiple star systems. Stars in multiple component systems are distinguished from planets in several respects. Inasmuch as most stars show bulk elemental abundances closely matched to the cosmic values—more closely so than even the most cosmically representative of the known planets, Jupiter—it is apparent that these stars formed not from processes dominated by the accretion of condensed solid matter, but rather through hydrodynamic collapse processes which gathered the gas and dust together as a fluid. Moreover, the orbits of binary stars are generally highly elliptical in comparison with the circularized orbits on which the known planets move. There remains great uncertainty in our understanding of the formation of multiple star systems. It is conceivable, for example, that many stars form in bound multiplets, with, say, three to six members, and evolve by sequentially ejecting members, or throwing them into eccentric orbits, until only a bound binary is left (Van Albada 1968). However, inasmuch as the number of binary pairs seems to be greater than the number of single stars (Abt 1983, 1987), it seems clear that most extant binary pairs must have formed as binaries, and are not the reduced remnants of higher-order multiplicities. Their elliptical orbits then suggest that binary stars form as hydrodynamic subcondensations in a larger collapsing system. The individual stars apparently formed early and were isolated from the general infall, before dissipative processes had circularized the motion. The elliptical orbits represent memory of the infall velocities of the subcondensations (Bodenheimer 1978).

Within the constraints of our present, and admittedly sparse, knowledge, one can postulate the existence of a bifurcation in the possible evolutions of a protostar system. One the one hand, there may be collapse to a dissipative disk, followed by the formation of a central star (most likely?) surrounded by a system of planets. On the otherhand, there may occur hydrodynamic separation of subcondensations during the early stages of collapse, leading to a multiple star system. Clearly these are not entirely mutually exclusive processes. One or more members of a multiple star system may itself form with a sufficiently well-developed accretion disk to make its own planetary system. It is possible, too, that in some "planetary" systems, sufficiently massive components could form out of accretion disks—either as a result of purely hydrodynamic processes or precipitated by solid matter accumulation—to make a star heated by nuclear fusion, although the long-term stability of a planetary system with such a massive component is arguable. There could exist a wide variety of stellar and planetary systems, with only a fuzzy boundary separating them. This is, for now, an open empirical question that will need very detailed, precise, and widely applied observations in the future.

In this respect, at least one very important data point is already growing in its implications. There is an apparent astounding absence of the so-called brown dwarf stars, as revealed both in the infrared survey made with the Infrared Astronomical Satellite and in astrometric and radial velocity studies of stars (Gatewood et al. 1986; Campbell et al. 1986; Marcy et al. 1987). Dark stellar companions with masses comparable to several tens of Jupiters should easily be detectable from the ground with these latter two techniques. Although more complete studies are clearly needed, the presently apparent paucity of these objects, with masses intermediate between the known planets and the lowest mass stars, seems to support the idea that there is a real bifurcation separating the pathways of planet system formation from the pathways of star system formation. Either the discovery of many brown dwarfs and the determination of the orbits of those in multiple systems, or the confirmation of their absence, would mark a major step toward an understanding of the varieties of star and planet system formation and the relationships between these two phenomena.

The human perspective. Planets, more than most cosmical objects, evoke a deep and broadscale human interest. Of course we cannot claim to know what varieties of life are possible in the universe. The mind can—and has—run wild with fancies of living interstellar clouds and micron-scale life on neutron stars. However, it is a defensible, even if conservative, view that real life requires a relatively "benign"—though not necessarily Earth-like—planet in order to arise, persist, and flourish in an evolutionary way. There is at least no obvious evidence of life elsewhere in our Solar System. Even those of us who believe life to be a commonly occurring phenomenon are thus constrained to the view that life does not automatically occur and thrive in arbitrary environments. Interest in the prevalence of planets is inextricably tied to our interest in the prevalence of life. Ascertaining the prevalence of planets and the characters of typical planetary environments throughout the universe is a step toward understanding the possible distribution of life. It is also a step toward understanding our relation to the broader cosmos, to the things that make us, as living beings, unique or ordinary. Indeed, barring an astounding discovery in the impending searches for extraterrestrial intelligence, ascertaining the prevalence and character of planetary systems is likely to be the only unambiguous step in that direction we can take in the near future.

Heretofore, inquiry into the formation and character of planetary systems has suffered from the problem of singularity. Having only one planetary system to study creates an essential difficulty in separating those aspects of the Solar System that are accidental and specific from those aspects that are characteristic of planetary systems more generally. In this respect, planetary cosmogony has suffered from the same difficulty as has cosmology, which has—at least for the time being—only one observable universe on which to focus attention. However, in the study of planetary systems, this singularity is not one of principle, but rather one of past practice and practicality. By taking advantage of modern technology to discover and study other planetary systems, we can overcome the philosophical singularity and turn planetary cosmogony into a more complete science. By learning the general rules that govern the formation of planets, we will come to know—by inference, at least—the prevalence of planetary systems, the probable nature of planets, and the frequency with which habitable systems occur throughout the universe.

Discovering and Studying Planetary Systems

The remainder of this short paper will be occupied with the question of discovering and studying other planetary systems through the use of high-accuracy stellar astrometry.

The detection of an isolated planet orbiting some star would be a very important step forward, inasmuch as there still is no firm evidence as to the existence of any planet outside of the Solar System. However, in order to advance our scientific understanding of planetary systems and their formation, it is essential to set an intellectually more ambitious program, with more ambitious measurement objectives. The goals of a *planetary-system* search program should seek to:

- Advance our understanding of planet-system and star formation.
- Test the ideas that form the basis of the existing central dogma of planetary cosmogony.
- Explore the "systematics" and regularities of planetary systems.
- Achieve an understanding of planetary system regularities that is sharp enough to advance our perceptions about the occurrence of life-sustaining environments.
- Mount a search for planetary systems that is sensitive enough so that a negative result can be interpretable in an important way, having a profound impact on our understanding.
- Mount a search for planetary systems that is capable of either confirming our present ideas or revealing departures from our expectations.
- Identify targets for more detailed studies using a variety of measurement techniques.

As noted above, our present understanding suggests that planetary systems form through processes involving the strong influence of dissipation acting before the formation of large condensed objects, and that the accumulation of planets is dominated by the accretion of solid matter. A planetary system might be expected to show the signature of these processes in the form of circular, corevolving, and coplanar orbits, and with a distribution of planetary masses having the higher mass planets in the outer parts of the system, with lower mass planets closer to the central star.

Setting measurement objectives for such an observational program requires the application of present knowledge and ideas in order to define the information necessary to provide a substantial advance in understanding, regardless of the outcome. One important question concerns the sensitivity to planetary mass that would be needed in order to insure that a negative result would have a profound impact on our ideas. Jupiter is the largest planet in our planetary system. Finding that Jupiter-mass objects were absent around a large number of stars would set only a weak constraint on our ideas about the prevalence of planets. On the otherhand, all four of the giant planets in the Solar System seem to have accreted cores of condensible matter with similar masses in the range of 10 to 20 equivalent Earth masses. The absence of planets at least as massive as these giant-planet cores, around a large number of other stars, would dictate a substantial change in our ideas about the formation of this planetary system or about the wider prevalence of the processes that produced this system (see the more detailed discussion in Black and Rathjen 1984). *Thus, a necessary measurement objective, for a planetary-system search that would inevitably have a major impact on our understanding, is sensitivity to planets with masses approaching the masses of the giant planet cores, approaching the range of 10-20 earth masses.*

The most massive planets are expected to form in the outer parts of a system, where the temperatures in the protoplanetary nebula were below that at which the abundant water ice would persist in solid form. Generally then, *it is necessary to be able to detect planets at distances beyond about 3 astronomical units from the central star and, thus, having orbit periods in the range of 10 to 20 years.*

Finally, in order to profoundly impact our ideas about the formation and prevalence of planetary systems, it is necessary to include a sufficiently large population of stars in a search program. The specific number of stars needed to have such an impact is not logically deducible. However, *a search population of 100 stars* would surely be sufficient. The failure to find planets with masses in the range of 10-20 earth masses, and orbits periods of one or two decades, around some 100 stars could not fail to have a profound impact on our ideas. On the otherhand, with the more expected outcome, in which many planetary systems were discovered in the search population, there would be a sufficient number to enable the discovery of their varieties and regularities. This requires that *the accessible search space extends to about 10 parsecs.*

Astrometric Telescope Facility

A planet disturbs the inertial motion of the star about which it orbits, as both the planet and the star orbit the system's center of mass. Precise measurement of the star's orbit reveals both the orbit and the mass of the accompanying planet. In a system with several planets, accurate measurements over a long period of time permit separate determinations of the several planetary masses and orbits. As we have pointed out earlier, the masses and orbits of planets convey much of the information that is essential for testing our concepts about planet-system formation and for expanding our general understanding of planetary systems. Moveover, the discovery of systems with large outer planets arrayed with the expected regularities similar to those in our own system would create strong presumptive evidence for the presence of terrestrial-type planets nearer to the central star.

Astrometric Accuracy. Consider a 15 earth mass planet orbiting at 5 astronomical units from a solar-type star at 10 parsecs from the Solar System. The amplitude of the star's deviation from inertial motion would be about 2.5×10^{-5} arcseconds. Thus an astrometric telescope with an accuracy of about 10^{-5} arcseconds would accomplish the desired discovery and initial study program.

The discovery and study of other planetary systems will require measurements made from space, above the observational disturbances produced by the atmosphere. The diffraction limit of a one meter telescope is

$$\frac{\lambda}{D} \approx \frac{5 \times 10^{-5}}{100} \approx 5 \times 10^{-7} \, \text{radians} \approx 0.1 \, \text{arcseconds}.$$

Ideally then, with such a telescope, the photon limit to the measurement accuracy of a star's position is

$$\Delta x \approx \frac{0.1}{\sqrt{N}} \, \text{arcseconds},$$

where N is the number of collected photons. In principle, collecting about 10^8 photons will allow measurement of a star's position to the required 10 μarcseconds. In order to achieve this accuracy in a real measurement also requires the control or calibration of systematic errors to a final accuracy better than the desired 10^{-5} arcseconds.

Now consider that a solar type star at 10 parsecs distance delivers about 10^8 photons/second to a one meter telescope. Assuming a 1% photon efficiency in the instrumentation, 10^8 photons can be collected in a few minutes from a solar luminosity star at 10 parsecs distance. From a more typical 11th magnitude star, the collection of 10^8 photons will require several hours. Studies carried out in connection with a development project for an Astrometric Telescope Facility to be operated from the U.S. Space Station, and taking into account the many additional liens against effective photon efficiency, indicate that an observational program could readily be carried out that would achieve the desired 10 μarcsecond accuracy per year for some 100 stars.

Astrometric Telescope Facility. The Astrometric Telescope Facility is a specially designed astrometric telescope that is currently being developed in a joint project by the NASA-Ames Research Center and the University of Arizona's Lunar and Planetary Laboratory, in collaboration with the University of Pittsburgh's Allegheny Observatory (Levy *et al.* 1986; Sobeck 1987). The principle of measurement that is being developed for the Astrometric Telescope Facility is based on the *Multi-*

channel Astrometric Photometer (MAP) developed at the Allegheny Observatory (Gatewood *et al.* 1980).

The MAP's measurement principle involves scanning a high-precision Ronchi ruling in the telescope's focal plane and photometrically measuring the light from each star that passes through the ruling. The physical separation between the target star and the several reference frame stars is then encoded in the phase differences of the modulation curves. With this approach, the primary metric for the measurement is the precise Ronchi ruling. Among the advantages of this approach is the relatively high accuracy with which the relative phases of the modulation curves can be determined. Moreover, each phase measurement is averaged over some 10^4 lines on the Ronchi ruling; thus random errors in the ruling lines are averaged over a large number of lines, greatly relieving the demands of fabrication and stability that need to be levied on the ruling.

At present, a MAP-based astrometric system is in routine operation on the Allegheny Observatory's Thaw Telescope, and making astrometric measurements at an accuracy of several milliarcseconds per hour of observation—and limited by Earth's atmosphere (Gatewood 1987).

Other Astrometric Science Objectives. It is worth noting that high-precision astrometric measurements, of the kind needed for the planetary system search described here, will also have important applications to a wide variety of other astrophysical problems. Among the astrophysical problems that would yield important new information to astrometric investigations at 10^{-4} arcseconds or better are the determination of absolute stellar luminosities, direct calibration of cosmical distance-scale standards, calibration of stellar masses, the dynamics and mass distribution of external galaxies, the dynamics of star clusters, and large-scale Galactic structure and dynamics.

ACKNOWLEDGEMENTS. This work was supported by the U.S. National Aeronautics and Space Administration under grants Ames/NCC 2-348(ATF), Ames/NCC 2-349, and in part by NSG-7419. The conceptual design and feasibility studies for the Astrometric Telescope Facility also rely on contributions from many people at the NASA-Ames Research Center, too numerous to list here.

REFERENCES

Abt, H. A. 1983, *Ann. Rev. Astr. Astrophys.*, **21**, 343.

Abt, H. A. 1987, *Ap. J.*, **317**, 353.

Angel, J. R. P. 1987, *Proc. IAU Colloq., BioAstronomy—The Next Steps*, George Marx, ed., **99**, (Reidel Publ., The Netherlands).

Black, D. C. and Rathjen, S., Editors 1984, *Report from NASA-Ames Science Workshop*, **Toward Detection and Study of Other Planetary Systems**, (NASA-Ames Research Center, Moffett Field, California).

Bodenheimer, P. 1978, *Ap. J.*, **224**, 488.

Burke, B. F., 1987 *Proc. IAU Colloq., BioAstronomy—The Next Steps*, George Marx, ed., **99**, (Reidel Publ., The Netherlands).

Campbell, B., Walker, G. A. H., Pritchet, C., and Long, B. 1986, **Astrophysics of Brown Dwarfs**, (Kafatos, M. C., Harrington, R. S., and Maran, S. P., eds.; Cambridge Univ. Pr., London), 37.

Gatewood, G. D., 1987 *Astron. J.*, **94**, 213.

Gatewood, G. D., Breakiron, L. A., Goebel, R., Kipp, S., Russell, J. L., and Stein, J. W. 1980, *Icarus*, **39**, 205.

Gatewood, G. D., de Jonge, J. K., Stein, J., Han, I., and Breakiron, L. 1986, **Astrophysics of Brown Dwarfs**, (Kafatos, M. C., Harrington, R. S., and Maran, S. P., eds.; Cambridge Univ. Pr., London), 104.

Levy, E. H., Gatewood, G. D., Stein, J. W., and McMillan, R. S. 1986, *Proc. S.P.I.E.*, **628**, 181.

Marcy, G. W., Lindsay, V., and Wilson, K. 1987, *Pub. A. S. P.*, **99**, 490.

Sargent, A. I. and Beckwith, S. 1986, *B.A.A.S.*, **18**, 959.

Sobeck, C. (Editor) 1987, *NASA TM-89429*.

Strom, S. E. 1987, *B.A.A.S.*, **19**, 710.

Van Albada, T. S. 1968, *Bull. Astron. Inst. Netherlands*, **20**, 57.

Session 3

SEARCH FOR PRIMITIVE LIFE

Two next-generation-technology concepts for orbital telescopes capable of spectroscopic analyses of the atmospheres of planets circling nearby stars were presented. These systems could be used to search for presumptive evidence of primitive life in the form of some non-equilibrium trace gas chemistry that can only be explained by a biological source. Since detecting and directly imaging the planets is a formidable task in itself, it should not be surprising that the more difficult task of assay chemistry requires instrumentation that strains credibility with respect to required positional alignments and absolute surface and figure accuracies over hundreds of hours of on-orbit observation. B. Burke proposed a multielement interferometer working at optical wavelengths to study extrasolar planetary ozone and oxygen. The system would have an effective collecting area of 10 square meters with apodization of each of the individual elements, whose surface accuracies would require new optical mirror techniques. Three hundred hours of integration time would be required to detect a terrestrial planet at a distance of 10 parsecs and another factor of 10 more hours to detect the oxygen and ozone bands (assuming terrestrial models). Since the most difficult technical spec is the precise measurement and maintenance of all the baselines, this instrument may be more suitable for the Lunar farside, rather than low Earth orbit. R. Angel presented a concept for an orbiting monolithic mirror system that works at infrared wavelengths where the planet/star contrast ratio may be more favorable. This system must have a diameter sufficiently large so that the diffraction limit can resolve the planet from the star. It too requires apodization of the primary, but in addition it must be actively cooled. It would be constructed and polished on-orbit, but could suffer rapid degradation if the orbit is too low and filled with debris. Detection of terrestrial planets with this system should be possible, to a distance of 4 parsecs, in less than 1 hour of integration, but any chemical analysis will take much longer.

Studies need to commence immediately if either or both of these advanced technologies is to be ready in the coming decades. At that time current generation instruments will hopefully have detected larger planets around nearby stars and have pinpointed planetary systems within which terrestrial planets are likely.

Jill C. Tarter

OPTICAL INTERFEROMETRY
AND THE DETECTION OF EVIDENCE OF LIFE

B.F. Burke

Massachusetts Institute of Technology
Cambridge MA 02139, USA

ABSTRACT. An examination of the possibility of detecting planetary systems of other stars indicates that a suitably designed optical interferometer, following principles that have been developed in radio astronomy, can detect major planets of solar-type stars easily at a distance of ten parsecs from the sun. Out to at least six parsecs distance, images of earth-like planets of the dozen or so solar-type stars should be detectable. Spectroscopic study, with the same interferometric systems, offers a good chance of detecting life on such planets if the atmospheric signature is as pronounced as it is in the terrestrial case.

1. A SEARCH STRATEGY

1.1 The Atmospheric Evidence

The earth's atmosphere bears witness to the effects of life, since it is far from chemical equilibrium. Cloud (1974) first summarized the striking evidence that biological agents—most likely the blue-green algae or their primitive ancestors—generated most of the oxygen now found in the earth's atmosphere. One might expect that similarly dramatic nonequilibrium chemical mixtures, if found in the atmosphere of a planet orbiting another star, would constitute strong evidence for life there. The search possibilities have been examined by a number of conferences and groups, but in most discussions the problems associated with direct communication with higher life-forms generally received most of the attention. The 1971 Byurakhan Conference (Sagan, 1973), the series of workshops held in 1975–76 by the Ames Research Center (Morrison, Billingham, and Wolfe, 1977) and the 1979 Montreal Joint Discussion (Papagiannis, 1980) are examples of the early gatherings at which these questions were discussed. An official recommendation was made to the US Congress (Library of Congress Congressional Research Office, 1977) recommending that atmospheric studies of other planetary systems be undertaken, and a resolution to that same effect was passed by Commission 57 of the International Astronomical Union (Papagiannis, 1982) but in neither case was there specification of the means to accomplish this end.

The earth's atmosphere could be a guide to the more general possibility that all life patterns tend toward the production of an oxygen-rich atmosphere. Owen (1980) examined the spectroscopic possibilities, arguing that the expectation for water-based organisms and

139

eventual oxygen-producing life was firmly based in chemistry, and he concluded that the A-band of oxygen at $\lambda7600\overset{\circ}{A}$ was the best tracer, but he did not say how this task should be accomplished. He also discussed the prospects for examining the region $\lambda2000 - 3000\overset{\circ}{A}$, again without discussing the means, but concluded that the ozone absorption $\lambda3000 - 3100\overset{\circ}{A}$ (the Hartley bands) could not be distinguished from the Metropoulos-Beutler bands of sulphur dioxide. This could be taken as a slightly disappointing conclusion; some time earlier, Berkner and Marshall (1965) had argued that the existence of the ozone layer was a precondition for the evolution of higher terrestrial life forms, and Kasting and Donohue (1981) had discussed the information that one gains from the ozone/oxygen balance. Burke (1985c, 1986a), in a re-examination of the planetary detection problem, showed there were reasons to expect that application of modern interferometric techniques could result in a space-borne system for the visible and ultraviolet part of the spectrum that not only could detect the oxygen A-band, but could discriminate between ozone and sulphur dioxide in the ultraviolet. Shortly afterwards, Angel, Cheng, and Woolf (1986) showed that ozone and oxygen might well be detectable in the infrared part of the spectrum by a suitably designed monolithic telescope in space.

These more recent studies suggest that if a planet can be detected, its principal atmospheric properties can be analyzed. A particularly forceful statement of the inevitability of significant chemical anomalies in the atmosphere of life-bearing planets was made by Margulis and Lovelock (1974) in their Gaia hypothesis: that the atmosphere of our own planet is part of a feedback loop maintained by the earth's living system to support its continual existence. This principle may be sufficiently general to apply to all planets where life has arisen, and the implications of dramatic departures from chemical equilibrium in other planetary atmospheres (not necessarily oxygen overabundance) are sufficiently interesting to warrant study.

1.2 A Parallel Approach

It is generally accepted that planetary systems are not a rare phenomenon among the later-type stars; the abrupt decrease in the angular momentum of stars later than the A-type and the obvious case of our own solar system provide a reasonable expectation for finding planetary systems among the F, G, and K dwarfs and perhaps among the later types also. The most advanced technique for search at present is the astrometric approach, looking either for transverse or radial motion of a star about its barycenter, induced by the periodic motion of large planets (terrestrial planets have too little mass to cause observable effects with present techniques). The astrometric projects described by Shao et al (1984), Reasenberg (1984), Black (1984) and York et al (1984) are all examples of this approach. Radial velocity studies are also giving encouraging results, as reported by Campbell, Walker, and Yang (1987). Both the radial velocity and astronometric approaches, however, take a long time to give definitive results. If several major planets, analogous to Jupiter and Saturn, are the principal agents, several decades of work could well be needed to sort out the actual planetary components.

When one is faced with a difficult scientific problem of great interest, the use of mixed strategies can be a prudent course of action (the demonstration of parity non-conservation in both beta-emission by a complex nucleus and the muon decay system is a good example). The recent proposals by Burke and by Angel et al to detect planets by direct imaging has reopened an alternate approach, and the present study is a review of the prospects for direct optical imaging. It is, in a sense, a reopening of a discussion that occurred between Moroz, Burke, and Pariisky at the Byurakhan Symposium (Sagan, op.cit., pp 31–36).

The principal instrumental development that has occurred in the intervening years, and which changes the nature of the discussion, is the full development of the aperture synthesis concept in radio astronomy. The following section gives a brief summary of the concepts, conceived by Ryle (1960, 1972) and brought to full realization in the Very Large Array (VLA) of the National Radio Astronomy Observatory (Napier, Thompson, and Ekers, 1983). The concepts should carry over to optical wavelengths, and various concepts have been discussed at the 1984 Tucson Workshop (Reasenberg, 1984), the 1985 Cargése Symposium (ESA, 1985), and the 1986 Cambridge Workshop (Battelle, 1987). It appears that one can assume that space-borne instruments of this class will be built in the future for various astronomical purposes. The work can, in some instances, be done from the ground, but in all cases one should be aware of the characteristics that should be in the design specifications if the instruments can be made to serve for the study of other solar systems.

The proposal, therefore, would be to concentrate first on interferometric instruments that could detect major planets, such as Jupiter or Saturn, orbiting nearby stars. It will be seen that the instrument might be relatively modest. The next step, to detect the existence of earth-like planets in such solar systems, is more demanding. The search might be made easier because the large planets would determine the inclination of the principal plane, but the weak signature of reflected light from a small planet dictates the construction of a large instrument, almost certainly space-borne. The final step, the measurement of the atmospheric characteristics of such a planet, is demanding but possible and can be carried out with the postulated instrument for detection of earth-like planets.

2. PLANETARY IMAGING

2.1 Signal-to-Noise Considerations

An early proposal for detecting planets belonging to other stars was made by Bracewell and MacPhie (1979), who proposed a cryogenically cooled infrared interferometer in space, operating at $\lambda 40$ microns to pick up the thermal emission form a Jupiter-like planet. The more recent proposal, by Angel et al, to use a large monolithic space telescope, is also being discussed at this symposium. The suggestion to look at optical and ultraviolet wavelengths, where one would detect the planet by reflected light, was revived by Burke and is the subject of this discussion.

Consider a planet of diameter d and mean albedo η, assuming isotropic scattering, situated in a circular orbit of radius a and observed at maximum elongation. The ratio of planetary to stellar flux, S_p/S_s, will then be

$$S_p/S_s = (\eta/16)(d/a)^2 \tag{1}$$

Initially, the star τ Ceti, well known to SETI investigators as a prime nearby candidate star, will be used as an example for two cases: a planet similar to the earth, with $\eta = 0.4$ and $a = 1$ a.u., and a planet like Jupiter, with $\eta = 0.6$ and $a = 5$ a.u. The distance of τ Ceti from the earth is 3 parsecs, and its spectral type is G5 V, making it similar to our own sun in its characteristics. This gives the flux ratios

$$S_p/S_s = \begin{cases} 1.8 \times 10^{-10} & (\text{earth} - \text{like}) \\ 1.4 \times 10^{-9} & (\text{Jupiter} - \text{like}) \end{cases}$$

Since the total photon flux of τ Ceti is 8.5×10^9 photons/m^2/sec this means that the photon flux from an earth-like planet would be 1.5 photons/m^2/sec (11 photons/sec/m^2 in the Jupiter-like case). Since the contrasts are 97 db and 88 db, respectively (24 and 22 magnitudes) the combination of low flux and high dynamic range has inhibited discussion of optical imaging of planets. Nevertheless, Oliver (unpublished memorandum, referred to by Greenstein and Black, 1977) pointed out that the detection of Jupiter at a distance of 5 parsecs could be accomplished by the Hubble Space Telescope if it were properly apodized. It is the new promise offered by the aperture synthesis methods of radio astronomy that allows reopening of the subject.

The contrast between the optical and infrared detection problem can be summarized easily: at infrared wavelengths, the contrast between planet and star is more favorable, but with instruments of realistic size, both planet and star are within the primary resolution of the instrument. At optical wavelengths, the signal-to-noise problem is more severe, but the planet can be placed outside the primary instrumental resolution, and attention can be paid to sidelobe reduction.

An optical instrument will be postulated, having effective collecting area A, and diffraction sidelobes of intensity G with respect to the primary response (i.e. G is the relative power gain, normalized to units at maximum response). For an integration time t, the uncertainty in the flux level of the planet will be determined by the photon statistics. If the stellar flux is S_o, for a star at distance D_o, then for a similar star at distance D a signal-to-noise ratio of r standard deviations will require the following instrumental characteristics

$$A/G = r^2 (S_s/S_p)^2 (D/D_o)^2/(tS_o) \tag{2}$$

for the present case of τ Ceti, with $D = D_o$, $t = 10$ hours, and a $10 - \sigma$ detection

$$A/G = \begin{cases} 10^7 m^2 & (earth - like) \\ 2 \times 10^5 m^2 & (Jupiter - like) \end{cases}$$

These parameters give a good indication of the instrumental requirements; it still remains to describe the instrument that will satisfy them.

2.2 Aperture-Synthesis Interferometry

The requirements imposed on Equation (2) can be easily summarized. An optical instrument having several square meters of effective collecting area— $10m^2$ can be taken as a representative number—must yield at least 60 db rejection of the light from the nearby star if an earth-like planet is to be detected in orbit about τ Ceti; a more modest instrument, perhaps having $2m^2$ effective area, must yield at least 50 db rejection in order to detect a Jupiter-like planet. The assumed integration time is ten hours, so the system must be capable of retaining stability over many hours of observing time.

Similar requirements were met in radio astronomy by using Michelson interferometry. At radio wavelengths, the building of high-angular-resolution telescopes by building simple apertures quickly encounters physical and budgetary constraints. Interferometry overcomes these constraints, and gives other advantages as well. The full development of Michelson interferometry into aperture synthesis has been described in the comprehensive monograph of Thompson, Moran, and Swenson (1986), and a discussion of the applicability of the radio concepts has been given (Burke, 1987). Except for signal-to-noise considerations the concepts carry over from the radio to the optical case completely. The method, in brief outline, can be described as follows: a two-dimensional brightness distribution on the sky, $B(x, y)$, can be described by its Fourier transform $V(u, v)$. A

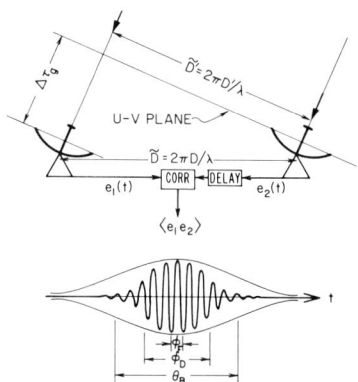

Figure 1: Two-element Michelson Interferometer

two-element Michelson interferometer, schematized in Figure 1, gives an output of the form shown when a source on the sky is passed by the receiving patterns of the apertures. Alternatively, this can be regarded as the gain pattern of the interferometer. The sinusoidal output is characterized by an amplitude and a phase, i.e. by a complex number (the complex amplitude) and this is proportional to the Fourier transform $V(u,v)$ of the source brightness distribution. The dual coordinates (u,v) are given by the projection on the plane of the sky, $\tilde{D}'$, of the interferometer baseline (measured in wavelengths). (See Thompson, Moran, and Swenson for a detailed treatment.)

The transition from simple interferometry to aperture synthesis consists of realizing that a set of N apertures, properly arranged, gives $N(N-1)/2$ independent baselines and thus the number of independent samples of the Fourier transform grows approximately as the square of the number of apertures when all possible cross-combinations are taken. The practical question of maintaining phase stability for the system has been thoroughly treated for the radio case, and the optical case appears to be soluble, although the technical questions are non-trivial (cf Burke, 1987 for a more complete discussion). One should note that, for the source-limited case, the effective area of the system is equal to the total area of the N elements.

One can postulate, therefore, that an N-element aperture synthesis interferometer can be built for optical wavelengths, and that it has the required phase stability (further comments on the phase stability question can be found in Section 3). The elements will be assumed to be identical; therefore each has a gain pattern $G(x,y)$ with respect to the instrumental axis, where (x,y) are the angular distances off axis (θ in the circularly symmetric case). The principal response determines the maximum field of view of the interferometer, shown as the angle θ_b in Figure 1. If the instrument is pointed at a planet, the parent star will be outside the primary field of view, with its flux suppressed by the gain G. These sidelobe photons, however, add noise to the output and it is these noise photons that must be stably averaged, resulting in the requirement of Equation (2).

The parametric dependence of the system can be displayed by re-writing Equation

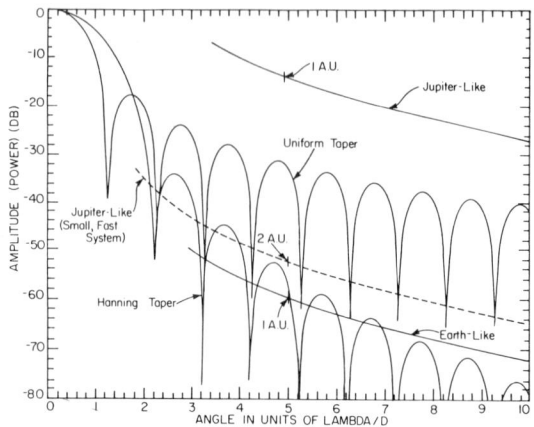

Figure 2: Relative Gain for a Circular Aperture with Uniform (upper curve) and Hanning (lower curve) Apodization. The detection limits on G for a Jupiter-like and earth-like planet are shown for $A = 10m^2$ and $D = 2m$. The dashed curve is for a Jupiter-like planet for $A = 2m^2$ and $D = 1m$. Note that the horizontal scale in this case is compressed by a factor of two.

(2) in the form

$$(Atf)/(r^2G) = (\eta/16)^2(d/a)^4(D/D_o)^2/S_o \qquad (3)$$

with the instrumental parameters on one side and the source parameters on the other. The parameter f represents the fraction of the stellar photons that are detectable by the instrument, acknowledging the fact that the photon efficiency and instrumental passband are finite. When f becomes smaller, t must become larger to keep the same signal-to-noise ratio.

The sidelobe gain, G, must be controlled stringently to bring the extraneous light from the star down to an acceptable level in the vicinity of the companion planet. The requirements are illustrated in Figure 2, which shows the relative gain of a circular aperture for two illustrative cases. The upper curve is the ideal Airy pattern for a uniformly illuminated circular aperture. The two slanting lines show the required gain for detecting a Jupiter-like (upper-curve) or earth-like (lower curve) planet orbiting τ Ceti for the following observing parameters: effective area of the system $A = 10m^2$, 10 hours for the product ft, and a 10σ signal-to-noise ratio, with primary mirrors $2m$ in diameter; using the planetary parameters given in Section 2.1.

One can see that the sidelobe level is not too high to prevent the detection of a Jupiter-like planet, if it is relatively close to the star (i.e., within 2 a.u.). At 5 a.u., however, the curve extrapolates in an unfavorable way, and detection of Jupiter or Saturn would take a considerably larger integration time than the assumed value. Detection of an earth-like planet would be impossible.

The solution is well-known to both radio and optical engineers. The illumination of the aperture must be tapered - apodized, in optical terminology (Burke, 1986a). The Hanning taper, a raised cosine function, is easy to calculate and demonstrates the principle adequately; other tapering illuminations can be designed for special purposes. The sidelobe pattern for the Hanning taper is shown in the lower curve of Figure 2. The consequences of apodization can be seen: the primary field of view is broadened, with a consequent diminishing of effective area by a factor of two in this case. What is more important, the sidelobe level is not only lower but diminishes more rapidly with angle. If a Jupiter-like planet were in the field of view, it would be detectable in less than a minute if $a \approx 1$ a.u.. At 5 a.u. it would require a comparable integration time in a formal sense; the signal strength would diminish at about the same rate as the sidelobe level. In Section 3, however, it will be seen that there are practical considerations that limit the expectable sidelobe level to -70 db or so. For a real system, therefore, a Jupiter-like planet would require $G < -45$ db, so a 25 db excess, implied if $G = -70$ db, translates to 10σ detection in a few minutes.

One notes, further, that the earth-like detection case becomes a reality. If the earth-like planet is orbiting τ Ceti at 1 a.u., the $10 - \sigma$ detection is marginally possible ($\theta = 5\lambda/D$), and longer integration is indicated. If a is slightly greater than 1 a.u., ten hours suffices for detection with the adopted parameters.

2.3 Spectroscopy: Dispersed Fringes

The discussion of aperture synthesis in the previous section did not address the problems that are introduced by wide bandwidth. Since one needs to have the largest possible photon count in the optical case, a bandwidth of 10 per cent or greater may be desirable, and this introduces a complication that is illustrated in Figure 1. Wide receiving bandwidth implies a short coherence time for the photons, and with the time delay adjusted for optimum response (i.e. when it compensates exactly for the geometrical time delay $\Delta\tau_g$, plus all differential instrumental effects) the compensation will be correct only for a single angle. The loss of coherence with angle is called the "delay beam" effect; the effective field size imposed by the delay beam is often smaller than the primary resolutoin of the individual elements. The angle ϕ_D in Figure 1 illustrates the effect of the delay beam. The order-of-magnitude is easily expressed in terms of the fringe spacing ϕ_F: a fractional bandwidth of $1/N$ will give N fringes, so ϕ_D is of order $N\phi_F$.

The effect is fundamental, but the solution is well known. The dispersed fringe method, first proposed by Koechlin (1977) and then discussed more fully by Stachnik and Labeyrie (1984), allows one to open up the field of view to the full aperture size, θ_B. In its most commonly discussed configuration, one images the converging beams to a common focal plane, merging the Airy discs. At this point, the image is the diffraction pattern of the primary aperture, crossed by N interference fringes which are imaged in turn through a dispersive system such that, at any given wavelength, the circular Airy pattern is imaged into a narrow ellipse, with the dispersion arranged so that the ellipses are dispersed in wavelength in the direction of the minor axis. Thus, the fringes are displayed as functions of angle in one coordinate, and frequency in the perpendicular coordinate. The effective bandwidth at any given wavelength can be made sufficiently narrow to broaden the delay beam to match the field of view of the individual receiving elements. (Note: fringe dispersal is discussed here for focal-plane interferometry but it works equally well for an interferometer working in the aperture plane.)

Thus, one can see that an optical interferometric aperture-synthesis system will be a spectrometer as well. If a planet can be detected, its major atmospheric absorption features can also be detected. A larger number of photons will be required in any spectral range, of course, and so the integration time must be lengthened. The factor ft in Equation (3) shows that the relation is reciprocal.

3. INSTRUMENTAL REALIZATION

3.1 Interferometry in Space

The state of optical interferometry on the earth's surface has been advancing, and a comprehensive review was held in 1984 in Tucson under the auspices of the American Astronomical Society (Boyce and Reasenberg, 1984). The most advanced project at present is probably the MIT-SAO-NRL interferometer described by Shao et al (1986) which is designed for astrometric work. At the same time, there has been increased interest in extending the techniques to space. In addition to the review edited by Boyce and Reasenberg, there are extensive treatments published in the proceedings of the 1984 Cargèse Symposium (European Space Agency, 1985). The problems associated with ground-based optical interferometry are now well appreciated, and stem from the disturbing effects of the earth's atmosphere. In order to meet the demanding specifications developed in the previous section, it appears probable that planetary detection will require a space-based aperture-synthesis array, specifically to avoid atmospheric effects. Such an instrument might well be developed for more general astrophysical purposes, and it should be valuable to understand the extra requirements that arise if it is to be useful for research on planetary systems.

The recent study cited earlier (Burke, 1987) can be taken as an example. An array of nine reflectors, each in the $1.5 - 2m$ size range, could be arranged in an optimum configuration on a circular locus, following the prescription of Cornwell (1987). The structure would be some tens of meters in diameter, but must not be so large that planets would be resolved. The base structure could be deployable, and is only present to provide a secure kinematic framework, and not to bear significant loads. Thermal effects must be compensated for, and accurate metrology would have to receive careful technological attention. The conclusion of the study, that radio astronomy methods could be applied directly to optical aperture-synthesis interferometers, was qualified only by the observation that heterodyne techniques would not be useful. This is not a significant limitation, since there are optical analogs for all transmission-line components.

The signal-to-noise analysis of the preceding section assumed that phase-coherent integration was performed. Even at radio wavelengths, this is seldom the case in the strictest sense. Even the VLA suffers some phase shift in its elements with time, and phase-calibrating reference sources must be observed, at intervals ranging from a few minutes to several hours, depending upon the specific needs of the project. Standard procedures have been developed that permit extraordinarily long coherent integration of the interference fringes. A recent observation at the VLA by Kellermann et al (1986) achieved the theoretical signal-to-noise ratio in a 74-hour integration, and there is no reason to expect that the limits have been reached, or even approached.

Coherent integration in the optical case is probably less of a problem than it first appears to be. The short-term stability of the instrument must be adequate to give a good measurement of fringe amplitude and phase for a strong source. Two cases can be distinguished: phase referencing to a source within the primary resolution, and phase referencing to a source that is nearby but outside the primary field of view. The second case is the relevant one. The parent star, although outside the field of view, is within an arc-second, or a few arc-seconds at most, of the planet. This means that it is well within the useable field of almost any optical device, so the starlight can be intercepted and diverted to a correlating system that derives the amplitude and phase of the star fringes, and uses this information to adjust the interferometric delay lines to the proper value. One thus has two correlator systems, a fringe-control correlator that monitors the star and a signal-processing correlator that is used for the planetary search. With the instrumental parameters controlled, the search for planetary fringes can proceed, provided that the fringe-control correlator system is kinematically rigid with respect to the signal-processing correlator. Laser-controlled servo systems, similar to those being developed for ground-controlled interferometer systems, would be required.

The fringe-control correlator could be similar to the type described by Shao et al (1986). Not all baselines would be processed; indeed, if the total size of the interferometer were $50m$ there would be problems using the longest baselines because a star like τ Ceti would be partly resolved. Baselines $10m$ and less would be acceptable for control purposes, and this requirement would have to be part of the specifications for an aperture-systhesis system capable of carrying out a planetary search.

3.2 Control of Extraneous Light

The signal analysis expressed by Equations (2) and (3) emphasizes the dominant role played by the light from the parent star. Even a small quantity of extraneous light can have a devastating effect, a conclusion that is evident from inspection of Figure 2. The starlight that enters throught the diffraction pattern is inescapable but controllable through proper illumination tapering of the input apertures, discussed in Section 2.3. Two further contributions, the effects of internally scattered light and of optical defects, must also be addressed.

The internal light scattering is reminiscent of the same problem encountered in a solar coronagraph, and the solution is similar. There are two choices that can be made: one can intercept and divert the starlight (as in a coronagraph) or one can intercept and divert the field of view in which the planetary search is to be made. Either choice appears to be feasible; and the final decision would be made on engineering grounds. The dual optical trains would then be conducted to the fringe-control and signal-processing correlators. The optical-defect contamination needs close attention. Geometrical optics assumes that the wavefronts in an optical system are regular, and derivation of the diffraction pattern at the focus assumes spherical, converging wavefronts. Defects in the optical system will always be present, and these will induce phase deviations by some *rms* value δ, and these have the effect of raising the sidelobe level. The problem has been treated by Ruze (1966) for the case of a paraboloidal reflector, and this treatment suffices for the present discussion.

First, let us review briefly the causes of the noise sidelobes. The relative gain contribution from the noise sidelobes at an angle θ off the optical axis will be denoted by $G_n(\theta)$. Intensity in the image plane will be the Fourier transform of the auto-correlation of the amplitude error in the aperture-plane phase front. The precise form of the sidelobes will depend upon the error distribution, but as a representative case let the errors have a random distribution, with an auto-correlation of the form

$$\langle y^2(\tau) \rangle = \langle \delta^2 \rangle \left(1 - e^{-(2\tau/\ell)^2}\right) \tag{4}$$

where τ is the separation, ℓ is a correlation length, and δ is the phase error. If the *rms* surface error is ϵ, then $\langle \delta^2 \rangle = (4\pi\epsilon/\lambda)^2$. It then follows from the work of Ruze that, for a uniformly illuminated aperture of diameter D with $\delta \ll \lambda$, the relative gain of the noise sidelobes will have the distribution

$$\langle G_n(\theta) \rangle = 8\pi^2(\ell/D)^2(\epsilon/\lambda)^2 e^{-(1/2\pi\ell\theta/\lambda)^2} \tag{5}$$

Physically, the gain distribution will be a speckle pattern extending over an angle $\approx (\lambda/\ell)$ with individual speckles of size $\approx (\lambda/D)$.

Interferometry overcomes the speckle problem with great effectiveness. The speckle pattern is derived from spatial frequencies less than (λ/D), whilst the interferometer fringes came from spatial frequencies much larger than (λ/D). Hence the interferometer fringe amplitude comes from the planet and not from the speckles, which only contribute the noise that was accounted for in the signal-to-noise calculation. Equation (4) above shows that the distribution of errors in the real case must be considered. The total power scattered from the main beam depends only upon ϵ, and if, for a given ϵ, ℓ is small, the scattered power is distributed over a large angle, and the close-in sidelobe level will be low. For the present estimate, $\varepsilon = \lambda/100$ and $\ell = 1$ cm is chosen as the correlation length, giving a sidelobe suppression of -70 db. Note that, since close-in sidelobes are of central importance, the spatial frequency spectrum must be controlled. If the only errors in the phase front were from large spatial frequencies, the instrumental sidelobe level would be troublesome. A beautifully polished surface, whose only errors in figure were on large scales, would be unsuitable for finding planets since all the noise power would be in sidelobes clustered close to the main beam. Mirror diameters of the order of 1.5 to $2m$ probably should be chosen for the receiving elements, because of the extensive industrial experience in making high-quality mirrors in this size range. Industrial suppliers, in informal inquiries, have indicated that achieving this precision of the element surfaces is an economic, not a technical, question.

3.3 Practical Considerations

Given an N-element aperture-synthesis space-borne interferometer, with its apertures properly apodized, and dual correlators (fringe-control and signal-processing), there are several additional features that might well be incorporated. The system envisaged above would have its mass distribution arranged so that there would be no net quadrupole moment, in order to minimize gravitational torques. The mirrors would have to have clear apertures in order to minimize sidelobes; this can be done by the circular array of mirrors described above, if they are off-axis paraboloids, reflecting their signals to a central station on the axis of the circle. The circular array of apertures has an additional advantage, since pointing the array at the target source equalizes the time delays to a first approximation. Pointing errors and structural deformations will still lead to a need for time delay correction, but the corrections should be small—of the order of a few millimeters at most.

One should also realize that the planet would be moving with respect to the parent star. For a 100-hour integration, this could mean a motion of several fringes (the earth orbiting τ Ceti would reach a maximum angular velocity of nearly five milliarc-seconds per day). This is not a problem, however, if the data are recorded and processed properly, and there are standard methods developed by radio astronomers for VLBI processing that are directly applicable. Instead of solving for fringe phase and amplitude only, one also searches in fringe-rate space for the maximum correlation. The principal burden is carried by the data reduction computation, but the VLBI experience is favorable (cf Thompson, Moran, and Swenson, Ch. 10 and 12). Possible concerns from zodiacal particles, photon efficiency, and read-out noise have been shown to be tractable (Burke 1986a). Note that the apodization need not be applied in the aperture plane and can equally well be accomplished using small masks close to the correlator.

4. PROSPECTS FOR PLANETARY IMAGING

4.1 The First Step

A two-level instrumental program appears to be indicated. The parametric expression, Equation (3) and the graphs in Figure 2 show that Jupiter-like planets are far easier to search for. The instrumental needs being less demanding, and space interferometry being even less than a fledgling field at present, a relatively modest first step would appear to be prudent. A system with several apertures, to take advantage of phase-closure techniques (Readhead and Wilkinson, 1978), would be indicated. If the total effective collecting area were to be $2m^2$, and if Hanning apodization (or other related tapering function) is used, the physical area of each aperture would be twice the effective area. A five-element system, with individual elements one meter in diameter, would be a reasonable possibility; the elements would be arranged on a structure ten to twenty meters in diameter. The collecting area would be one-fifth the area assumed in plotting the detection loci in Figure 2; furthermore, one might consider a fairly rapid synthesis schedule, with one hour per field. A realistic estimate must acknowledge that only a fraction of the photons will be detected; a value $f = 0.1$ is probably achievable. Inspection of the left side of Equation (3) shows that the product Aft/Gr^2 is reduced by a factor of 0.002, or 27 db. The locus for detection of the Jupiter-like planet in Figure 2 must be moved to the left by a factor of two (because the element diameter is reduced by two) and must be dropped down vertically by 27 db; the resulting locus is shown by the dashed line in Figure 2.

The "small" system, which might well become a reality over the next decade or so, is capable, therefore, of detecting major planets of nearby stars with relative ease. Note that the observing parameters ($r = 10\sigma, t = 1$ hour, $f = 0.1$) are realistic, and from the dashed curve in Figure 2, one sees that it is comfortably above the limiting gain for Hanning-tapered primary elements over a range that corresponds, for the τ Ceti example, to separations of 2 a.u. and larger. A Jupiter-like planet at 1 a.u. is still easily detectable, with an increase in observing time from 1 hour to 4 hours. The equivalent locus in Figure 2 for a Uranus-like planet ($\eta = 0.6, d = 0.34d_J$) would be dropped 18 db below the dashed line; it would require, therefore a longer integration time. The dominant sidelobes beyond 3 a.u. would be the noise sidelobes described in Section 3.2, which are estimated to be of the order of -70 db. The conclusion, therefore, is that at least 10 hours per field would be needed for 10σ detection of Uranus-like planets. One should note, however, that 5σ detection of planets requires only one-fourth as much observing time. A prudent procedure, therefore, would consist of briefer examinations, looking for 5σ evidence, and then spending longer integration times to verify the detection.

4.2 Systems for Detecting Earth-Like Planets

The initial parameters used in Section 2, all of which are technically feasible, describe an ambitious system that will probably be built some day. In addition to the proposals discussed at the Tucson Workshop and Cargèse Symposium, there are concepts that are still more ambitious. If a national or international project to establish a permanently manned lunar base becomes a reality, a large optical interferometer might well be a major scientific instrument that could be supported there (Burke, 1985b). The timescale is a long one, whether the instrument is planned to be in orbit or on the lunar surface, but there seem to be no technical barriers to realizing the necessary parameters: clear aperture of $2m^2$, surface tolerance of $\lambda/60$ over scales of $1cm$ to the full diameter, and a total area of at least $10m^2$. Such a system would be capable of detecting the Earth, Mars, Jupiter, Saturn, Uranus and Neptune orbiting τ Ceti, if it had a planetary system like our own.

5. SUMMARY

A relatively simple space-borne optical aperture-synthesis interferometric system, composed of (for instance) five one-meter apertures with a maximum baseline of ten to twenty meters would be capable of detecting major planets in orbit about nearby stars. In the case of τ Ceti, Jupiter-like or Uranus-like bodies would be detectable at distances from one to ten astronomical units from the parent star. Jupiter-like planets orbiting any of the dozen or so F, G, and K stars within six parsecs of the sun would be easily detectable. There are reasonable possibilities of detecting such major planets out to ten parsecs distance, and so thee would be approximately fifty candidate late-type stars within the range of study of such an instrument.

A larger interferometric instrument, with ten or more times the area and composed of apertures of at least two meters diameter, could detect an earth-like planet orbiting τ Ceti, and could look for absorption features in the atmosphere of such a planet. Both the Hartley bands of ozone and the A-band of oxygen could be detected since the interferometer system must also have spectroscopic capability.

It is clear that such a signature, indicating an oxygen-rich atmosphere such as life has generated on earth, would have immediate philosophical and scientific interest. A systematic program could start with the construction of a preliminary instrument capable of imaging major planets belonging to nearby stars. These observations would determine the orientation of the orbital plane, and would determine the size distribution of planets in such a planetary system. There are enough nearby stars to give a meaningful estimate of the incidence and forms of planetary systems, and the results could well encourage the construction of more ambitious instruments, capable of detecting atmospheric anomalies in planets as small as the earth, that could signal the presence of life. The results could influence, in turn, the SETI program that will be seeking signals from planets on which life has become sufficiently advanced.

References

[1] Angel, R., Cheng, A., & Woolf, N. *Nature* **322,** 341-343 (1986).

[2] Berkner, L.V. & Marshall, L.C. *J. Atmos. Sci* **22,** 225 (1965).

[3] Black, D.C. *Bull. Am. astr. Soc.* **16,** 767-769 (1984).

[4] Bracewell, R.N. & MacPhie, R.H. *Icarus* **38,** 136-147 (1979).

[5] Burke, B.F. *ESA* **SP-226,** 177-183 (1985a).

[6] Burke, B.F. *Lunar Bases and Space Activities of the 21st Century* (Mendell, W.W., ed.) 281-292 (Lunar and Planetary Inst., Houston 1985b).

[7] Burke, B.F. preprint 'Detection of Planetary Systems and the Search for Evidence of Life' (Cambridge) (1985c).

[8] Burke, B.F. *Nature* **322,** 340-341 (1986a).

[9] Burke, B.F. *Nature* **324,** 518 (1986b).

[10] Burke, B.F. *Reflective Optics* **SPIE 751,** 50-61 (1987)

[11] Campbell, Walker, & Yang *Bull. Am. astr. Soc.* **19,** 762 (1987).

[12] Cloud, P. *Am. Scient.* **62,** 54-66 (1974).

[13] Cornwell, T.J. 'Crystalline Antenna Arrays,' *NRAO Millimeter Array Memo* **38** (1987).

[14] Greenstein, J.L. & Black, D.C. *The Search for Extraterrestrial Intelligence* Morrison, P., Billingham, J., & Wolfe, J. **SP-419** 53-60 (NASA 1977).

[15] Kasting, J.F. & Donohue, T.M. *Life in the Universe* (1979 Ames Conf.) (Billingham, J. ed.) 149 (Cambridge, US 1981).

[16] Kellermann, K.I. *et al. Highlights of Astronomy* 367 (1986).

[17] Koechlin, L. *Optical Telescopes of the Future* (ESO Conf.) 475 (Geneva 1977).

[18] Lib. of Congress, Congressional Research Office, Science Policy Research Division; Committee on Science and Technology, US House of Representatives *Possibility of Intelligent Life in The Universe* **98-185 0** (US Govt. Printing Office 1977).

[19] Margulis, L. & Lovelock, J.E. *Icarus* **21,** 471-480 (1974).

[20] Morrison, Billingham, & Wolfe (Ames Conf.) (1977).

[21] Napier. P.J., Thompson, A.R. & Ekers, R.D. *Proc. IEEE* **71,** 1295-1320 (1983).

[22] Owen, T. in *Strategies for Search for Life in the Universe* (Papagiannis, M.D., ed.) 177-185 (Reidel, Dordrecht, 1980).

[23] Papagiannis, M.D. *IAU Trans.* 323-329 (1982).

[24] Readhead, A.C.S. & Wilkinson, P.N. *Astrophys. J.* **223** 25-36 (1978).

[25] Reasenberg, R.D. *Bull Am. astr. Soc.* **16,** 758-766 (1984).

[26] Ruze, J. *Proc. IEEE* **54,** 633-640 (1966).

[27] Ryle, M. & Hewish, A. *Mon. Not. R. astr. Soc.* **120,** 220-230 (1960).

[28] Ryle, M. *Nature* **239,** 435-438 (1972).

[29] Sagan, C. *Communication with Extraterrestrial Intelligence* (Byurakhan Conference) (MIT, Cambridge) (1973).

[30] Shao, M., Colavita, M.M., Staelin, D.H., Simm, R., & Johnston, K.J. *Bull. Am. astr. Soc.* **16,** 750-757 (1984).

[31] Shao, M., Colavita, M.M., & Staelin, D.H. *Proc.* **SPIE-628** 250 (1986).

[32] Stachnik, R. & Labeyrie, A. *Sky Telesc.* **67,** 205-209 (1984).

[33] Thompson, A.R., Moran, J.M. & Swenson, Jr., G.W. *Interferometry and Aperture Synthesis in Radio Astronomy* (Wiley, NY 1986).

[34] York, D.G. *et al. Bull Am. astr. Soc.* **16,** 775 (1984).

Conference Procedings:

[35] American Astronomical Society and National Aeronautics and Space Administration *Bull. Am. astr. Soc.* **16,** Workshop on High Angular Resolution Optical Interferometry from Space, (Tucson Workshop) (Reasenberg, R.D. & Boyce, P.B., eds.) 747-837 (1984).

[36] Battelle Cambridge Workshop on Imaging Interferometry (Columbus Division 1987).

[37] European Space Agency *ESA* **SP-226,** Colloquium on Kilometric Optical Arrays in Space, (Cargèse Symposium) (1985).

SEARCHING FOR EXTRASOLAR LIFE
WITH SPACE TELESCOPES

J.R.P. Angel

Steward Observatory, University of Arizona
Tucson AZ 85721, USA

ABSTRACT. Telescopes placed above the earth's atmosphere could obtain resolved images of planets around nearby stars. Given sufficiently large and accurate mirrors, even planets as small as the earth and as close to the star as the earth is to the sun could be distinguished. With a resolved image, the planets atmosphere could be studied by spectroscopic analysis. If, as on the earth, there are photo-synthesizing organisms, then oxygen features would be expected in the spectrum along with those of water and carbon dioxide.

Spectra of either the reflected light or the thermal emission would show features of oxygen if it is present. O_2 absorption is found at 0.76 microns wavelength in the optical spectru, O_3 at 10 microns in the infrared. The minimum size telescope needed to resolve planets close enough to a number of good candidate stars is 1.5m diameter at optical wavelengths, 16-20m at 10 microns wavelength. In practice the optical signal is so weak that, even with a number of 1.5m telescopes or a single larger telescope, several days of integration would be needed to obtain a useful absorption spectrum. The accuracy of the mirror surface needed in the optical far exceeds that of the Hubble space telescope. The problems of working in the infrared are the large size of the mirror, and the need to operate it at a temperature of 80K. If these can be accommodated, the accuracy requirements are not nearly so severe, and the telescope in short integrations would yield high quality spectra.

Do nearby stars that are like the sun have planetary systems? If so, do they have planets like the earth, with abundant liquid water? Have such planets developed life similar to that on earth? We do not now know the answers to these questions, but we could find them by placing sufficiently powerful telescopes in earth orbit.

The other planets in our solar system show no obvious signs of life, and their physical conditions are not favorable. If there is or was any life, direct exploration seems to be required to uncover it.

153

G. Marx (ed.), Bioastronomy – The Next Steps, 153–157.
© *1988 by Kluwer Academic Publishers.*

But a distant planet that is, like the earth, totally in the grip of life, could readily be identified, if we could obtain spectra of its atmosphere.

Spectroscopic analysis can show whether temperature and atmospheric composition favor the formation of life. More significantly, such analysis can show whether the atmosphere has been modified by the presence of life, as has the earth's. No oxygen was present in the earth's atmosphere until it was generated from carbon dioxide by living organisms in the process of photosynthesis. As far as we know, free oxygen will not exist except by biological action. Thus the presence of abundant oxygen in the atmosphere of an earthlike planet would constitute strong evidence for life.

Molecular constituents are detectable in the noisy spectra of very faint objects because of their strong and broad absorption features. An outstanding example is methane, which imprints a strong pattern of absorption in the light reflected from Titan and the outer planets. A low resolution spectrum with a signal/noise ratio of 10:1 would be adequate to give a clear identification. Oxygen is not as prominent, but nevertheless has some detectable features. In the earth's atmosphere it exists in two molecular forms, O_2 and O_3 (ozone), each with characteristic absorption features. At shorter wavelengths these are due to electronic transitions. Owen (1980) pointed out their value as indicators of the presence of life. The most prominent absorption of O_2 in reflected sunlight is in the near infrared at around 0.76 microns wavelength. This feature is over a band of 8 nm width. Ozone, although much less abundant than diatomic oxygen, causes complete absorption below 0.3 microns. However, the sun's radiation is quite weak this far into the ultraviolet spectrum, and most candidate stars are even weaker. Thus ozone may be more difficult to detect than O_2 in the optical spectrum.

Less than half of the sunlight falling on the earth is reflected. The remainder is absorbed as heat, and reemitted to space as thermal radiation, peaking at 10 microns wavelength. Telescopes operating in space at very low temperatures can be made very sensitive to this radiation. Detection of large, extrasolar Jupiter-like planets by their thermal emission has been proposed by Bracewell and Macphie (1979). The same emission can also be used to analyse atmospheric composition. For example, the thermal emission of Venus shows a strong absorption feature at 11.2 microns wavelength due to HSO_4-, from dissociated sulphuric acid. As we have pointed out (Angel, Cheng and Woolf, 1986, Paper I) the search for oxygen can be made in the thermal infrared spectrum. O_2 is not a strong absorber, but ozone as well as water and carbon dioxide all give strong features. In these bands the emitted radiation comes from higher, colder layers of the atmosphere and is less intense. The tenuous layer of ozone at high altitude is a strong absorber in the wavelength range 9-10 microns.

No existing or planned telescopes are powerful enough to detect either the reflected light or thermal emission of an earthlike planet of even the nearest stars. The Hubble space telescope and the new generation of 8m ground-based telescopes would just be able to make an optical detection if they were not blinded by starlight. The planned SIRTF infrared space observatory could detect the small amount of radiated heat, but not against the heat of the star and the background of heat emitted by zodiacal particles. The brightness ratio of the sun to the earth is nearly ten billion in visible light, ten million in thermal emission. Similar ratios would be expected for any planet supporting earthlike life assuming comparable size to the earth and distance to the host star. These similarities are needed to obtain the temperature of liquid water, and adequate gravity to preserve the atmosphere.

The huge differences in brightness compounded with the tiny angular separation are what make the detection problem so difficult. Suppose we wish to look not only at the nearest star, but at the nearest 100, which lie at distances up to 6.5 par secs (Allen, 1973). About half are in double systems, where conditions may have been quite different to those that led to the formation of the solar system. Most of the remainder are M stars which may be too faint and cool to support earthlike planets. But the sample includes nine brighter, single stars that are excellent candidates. Just as in our own solar system, planets warm enough to have liquid water will be relatively close to their star, and will not be nearly as prominent as more distant giant planets. At a distance of 5 pc, the radius of the earth's orbit would subtend an angle of 1/5 arc second. To detect a planet so close to a bright star, the stellar image must be very sharp. If any light spills out from the image core, it must not fall in the ring at 1/5 arc second radius, where the search must be made.

The image formed by a simple circular aperture shows a bright central maximum surrounded by a series of rings of decreasing intensity. In paper I we described a new method to mask or apodize a telescope so as to broaden and deepen the first dark ring around the central maximum. Provided the mirror surface is accurate enough, this method can suppress the starlight enough to allow detection of a planet. The mirror size needed for a dark ring at 1/5 arc second radius depends on the wavelength studied. At the 7600Å feature of O_2 the diameter must be 1.5m and at the 10 micron feature of O_3, 20m. The flux of photons received from a planet just like the earth at 4 pc distance would be anly one per minute by the 1.5m mirror in the O_2 band but 300 per second by the 20m mirror in the O_3 band.

A number of 1.5m telescopes would be needed for a detection of such a weak optical signal, and very long integration times. Radio interferometers are able to realize the theoretical advantage of long integrations, and Burke (1986) has suggested it may be necessary to operate the optical telescopes as an interferometer. In the infrared, there are higher photon fluxes, a stronger spectral feature and the telescope is necessarily big to make a sharp enough image. Good quality spectra then require only moderate integration periods. Paper I gives a detailed calculation of signal strength and noise background for a 16m telescope in the infrared, including zodiacal emission as well as stellar background. Less than an hour is needed to give a clear indication of ozone present at the same level as in the earth's atmosphere.

A very dark ring in the diffraction pattern of the star will not be obtained unless the imaging comes very close to the theoretical limit of perfection. Thus the distortion introduced by atmospheric turbulence will never be small enough to allow this type of observation from the ground. Even in space, telescopes made for normal diffraction limited imaging are not good enough. The Hubble telescope has the most perfect mirror yet made, but it is about 50 times rougher than needed for optical detection. Ripples in its surface spread out starlight into the dark ring. Normally the amount would be negligible, but it would totally swamp the faint light of an earthlike planet.

The spreading effect of ripples on infrared waves is much less than on light, and for the infrared, mirror segments of even slightly inferior quality than the Hubble telescope would work. NASA has under study a 20m space telescope, the Large Deployable Reflector. However, the surface accuracy for its proposed diffraction limited operation at 30 microns would be about 100 times worse than needed to detect earthlike planets at 10 microns. Furthermore, it is not presently envisaged that the telescope should be cooled. Cooling to 80°K is needed to detect faint 10µ emission, to eliminate the thermal background of the telescope itself.

How can telescopes of the same size but of much higher quality than today's billion dollar space instruments be built? If this advance is to be made, we must learn how to get higher accuracy without much increase in cost. The costs of launching and operating satellites will not necessarily be larger because of high accuracy. What we must do, though, is learn how to make very accurate components at acceptable cost, with good ways to ensure that the accuracy can be maintained in the space environment.

For optical detection, mirrors much smoother than the Hubble Telescope mirror could be made with better polishing and testing methods. Already there have been significant advances. The ripples characteristic of the short strokes of rigid tools should be eliminated by the method of stressed lap polishing (Angel, 1985). In this method the mirror is polished by full strokes with a relatively large tool, whose shape is changed as it moves so as always to fit the desired mirror surface. New ways to correct large scale errors are under development. These include glass removal by ion beams (McNeil, Wilson, and Barron, 1986). There is also the possibility of adjusting the thickness of the metallic reflecting surface by deposition in a controlled raster scan. This technique could be used for a final touch up of the surface figure in orbit, when the mirror is under the exact thermal and mechanical conditions of operation (Angel, 1986).

The large mirror needed for the infrared must be made of segments, as will be the Keck telescope. The challenge of achieving accuracy an order of magnitude higher than in the Keck telescope is not as difficult as might first be expected. In space no wind blows to distort and vibrate the assembly. Also, no weight is pulling in different directions, and the system can be optimised for one fixed operating temperature. A simple three point kinematic support is all that is needed for each segment. Precise alignment of the segment array will be facilitated by the light of the bright host star.

Many technical issues need to be worked through beyond simple surface accuarcy, but this paper is not the place for a full discussion. We favor the concept of a large infrared telescope, because it would yield planetary spectra of high signal to noise ratio, and be a universally powerful tool for astronomy at both optical and infrared wavelengths. However, it will require assembly in orbit, and must rely critically on good thermal design and shielding to allow for operation at 80K. For optical observations the technical challenges are to make the incredibly perfect optics needed to contain the ten billion times brighter star image, and to build up extremely weak signals by very long integrations.

In conclusion, it is now clear that there are practical possibilities for the detection and spectroscopy of extrasolar earthlike planets. We can look forward to studying the composition of the atmospheres of many planets around nearby stars, assuming the sun is not unique. It would now be valuable to develop models of atmospheric spectra corresponding to different scenarios for the origin and evolution of life. These are needed to help find the best telescope strategy for discriminating between planets with and without life.

References

Allen, W. C. 1973, Astrophysical Quantities, (Clowes, London).

Angel, J. R. P. 1984, Proc IAU Colloq. No. 79, 'Very Large Telescopes, Their Instrumentation and Programs', Garching, West Germany, M. -H. Ulrich and K. Kjar, eds., p. 519.

Angel, J. R. P., Cheng, A. Y. S., and Woolf, N. J. (Paper I) 1986, Nature, 322, 341.

Angel, J. R. P. 1986, Proc. S.P.I.E., 571, p. 40.

Bracewell, R. N., and MacPhie, R. H. 1979, Icarus, 38, 136.

Burke, B. F. 1986, Nature, 322, 340.

McNeil, J. R., Wilson, S. R., and Barron, A E. 1986, Workshop on Optical Fabrication and Testing, Technical Digest, Optical Society of America.

Owen, T. 1980, Strategies for the Search for Life in the Universe, ed. M. D. Papagiannis, (Reidel, Dordrecht), p. 177.

Session 4

ALTERNATE BIOLOGIES?
ALTERNATE PATHWAYS?

"We shall regars as alive a population of entities which has the properties of reproduction, heredity and variability" - *as* J. Maynard Smith *has been quoted to say by* Ronald D. Brown. *The heredity (a precondition for biological evolution, driven by natural selection) is attributed mostly to nucleic acids. According to this view (reviewed by* Joan Oró*) life started with self-replicating RNA, supported later by protein enzymes. The abiogenic synthesis of nucleic acids seems, however, to be difficult even on Earth, taking into account, that when oceans first appeared here, there was no reducing atmosphere present.*

A more general approach (advocated by R. Shapiro*) would request appropriate solvent (like water), some pattern forming materials (e.g. carbon chains) and free energy (as sunlight). Liquid water is plentiful on Earth, occasionally it might be present on Mars or Europa, or on some comets. Simple patterns may be offered by clay (*Cairns-Smith*) or rust (*G.Arrhenius*). The free energy may come from starshine (*Matthews*), from meteoric impacts, or from volcanism (fed by radioactivity or by tidal work,* Corliss*). Therefore (bio)chemical evolution might have started even outside of earth-like planets (on cometary debris, even in the atmosphere of red giants). Interplanetary transfer of organic materials in desiccated form cannot be excluded (*G.Horneck*). Even Earth might have been contaminated by cometary materials, to start here the grand scale (biological) evolution. Therefere it seems worthwhile to search for life-bearing compounds in dust clouds and on smaller bodies of the Solar System.*

George Marx

CONSTRAINTS IMPOSED BY COSMIC EVOLUTION TOWARDS THE DEVELOPMENT OF LIFE

J. Oró

Biochemical-Biophysical Dept., University of Houston
Houston TX 77004, USA

ABSTRACT.   The probability of terrestrial-týpe life emerging in any other place of the universe will depend on the constraints imposed by cosmic evolution on that particular place. A systematic examination of cosmic constraints, which in the case of the Earth must have provided the necessary and sufficient conditions for the origin and evolution of life on our planet, shows that they are concerned with the nature of the central star, the planetary system, and the specific life-bearing planet, as well as with the chemical and biological evolution processes involved. These constraints or universal requirements for life, are briefly described below.

1.   UNIVERSAL REQUIREMENTS FOR LIFE

1.1.   Stellar Constraints

1.1.1. Single central star. A central star is more favorable for life since binary and multiple star systems would prevent the existence of stable planetary orbits. However, recent studies indicate a possible compatibility of life with certain binary systems.

1.1.2. Planetary system. A solar-type planetary system requires the presence of several non-luminous companions with masses substantially lower than Jupiter. Although the interpretation of data on Barnard and Van Bisbroeck 8 presumed companions was apparently incorrect, there are interesting observations on Beta Pictoris (1), Epsilon Eridani (2) and other stars which deserve further study. The recent explanation of the Titius-Bode law (3) increases the prospects of its generalization to other planetary systems.

1.1.3. Elemental composition. The star should be of second or third generation. Such stars have the appropiate mixture of biogenic elements (H,C,N,O,S,P) necessary for the formation of biochemical compounds. A first generation star would be too rich in hydrogen and helium and too poor in higher atomic number elements.

1.1.4. Mass of the star. This should be approximately between 0.5 and 2 times the mass of our Sun for Main-Sequence stars. Stars heavier than 2 solar masses evolve too rapidly to permit the origin as well as the evolution of life to the level of a technological

G. Marx (ed.), Bioastronomy – The Next Steps, 161–165.
© *1988 by Kluwer Academic Publishers.*

civilization. On the other hand, stars which are less than half of a solar mass may not provide a sufficient flux of energy.

1.1.5. Lifetime. Based on the terrestrial life record, the lifetime of the star should be at least 1 billion years for unicellular life to emerge. Longer lifetimes (5 billion years or more) will increase the probability of emergence of intelligent life. Being less dependent on the lifetime of the star, microbial life should be more abundant in the Universe.

1.2. Planetary Constraints

1.2.1. Mass. The mass of the planet should be sufficient to retain the volatile molecules of the biogenic elements (H,C,N,O,S,P), but not too large, as in the case of Jupiter, to retain large amounts of hydrogen. In this case, even if some biochemical molecules were formed, they would have a short lifetime since they would be quickly transformed back to the hydrides.

1.2.2. Orbit. The orbit of the planet has to be at an optimum distance to the central gravitational body so that it may provide the necessary energy to maintain a reasonable temperature (0 to 100°C) on the surface of the planet. An almost circular orbit, of about 1AU, is sufficient to allow relatively fast rates of synthesis of biochemical compounds.

1.2.3. Atmosphere. The atmosphere is required to provide an environment for the synthesis of simple organic molecules from the biogenic elements, or their hydrides. The atmosphere must also help protect emerging biochemical molecules against ultraviolet light and other solar radiation.

1.2.4. Hydrosphere. One of the most important planetary constraints is the existence of a large body of liquid water. This provides a medium for the trapping of molecules and ions, and it constitutes the major interactive environment where the synthesis of biomonomers can occur. Aside from the Earth, only Mars and Europa (4), possibly Venus too, have had a substantial body of liquid water in the past.

1.2.5. Lithosphere. Solid surfaces are necessary in order to provide the boundaries for localized evaporation, and condensation of monomers to polymers. In contrast to the well defined phase separations in the Earth, Jupiter has no clear separation between the solid and other phases which would be required for the condensation of monomers to polymers.

1.3. Chemical Constraints

1.3.1. Solvents. Two types of solvents appear to be necessary: hydrophilic (liquid water) and hydrophobic (liquid hydrocarbons and lipids). Liquid water is the ideal solvent. Liquid ammonia (-77.7°C) and liquid methane (-185.5°C) (the latter present in Titan) are too cold for significant reaction rates of biochemical synthesis. Water is also the most stable and most abundant triatomic molecule in the Universe.

1.3.2. Composition and concentration. Compounds of carbon and other biogenic elements should exist in high concentration, so that biochemical compounds can be formed at significant reaction rates. Comets may have contributed high concentrations of carbon compounds to the Earth's surface (5). Evaporation of primitive Earth ponds was probably one of the best ways for the concentration of monomers and the synthesis of biopolymers.

1.3.3. Effective energy sources. Ultraviolet light is one of the most prevalent forms of energy. Electric discharges and shock waves (collisions) are more effective in the production of hydrogen cyanide, cyanoacetylene and other precursors of aminoacids, purines and pyrimidines.

1.3.4. Redox potential. The redox potential has to be between both extremes. A completely reducing environment (e.g. excess of hydrogen, Jupiter) would inhibit the formation of biochemical compounds. Alternatively, a completely oxidizing environment (e.g. Venus, Mars) would cause the conversion of all the carbon compounds to carbon dioxide.

1.3.5. pH range. The pH should be close to neutrality, so that the functional biochemical compounds, polypeptides and polynucleotides, are not hydrolyzed. These compounds would be stable at a pH range from 5 to 7.5.

1.4.   Protobiological Requirements

1.4.1. Replication of informational molecules. (Proto RNA-DNA) Linear homochiral polymeric molecules (protoRNA or protoDNA) are needed to store the information in a simple linear code capable of undergoing template replication with high fidelity. This is accomplished by hydrogen bonding between purine and pyrimidine base pairs. Prochiral analogs of RNA and DNA have been synthesized abiotically. The informational molecules should also be capable of being decoded and translated into catalytic molecules to enhance replication.

1.4.2. Stereospecific catalysis.(Protoenzyme). Catalytic molecules are needed to bring about the synthesis of biopolymers at a high reaction rate and with high specificity. They have to be linear homochiral aminoacid polymers with their sequences derived from the informational molecules. They should be able to catalyze the synthesis of both types of biopolymers by feedback autocatalysis.

1.4.3. Information transfer (Proto-AA-t-RNA). The process of information transfer from replicative to catalytic molecules requires a special recognition and translation mechanism. This is provided in living systems by a hybrid molecule (AA-t-RNA) which uses the purine-pyrimidine hydrogen bonding mechanism of recognition by means of base pair triplets. Prebiotic models of similar hybrid molecules have been developed.

1.4.4. Polymerizing assembly (Protoribosome). A high fidelity of translation is required which probably could not be accomplished in free solution. This can be accomplished by means of an organized assembly of polypeptides and polynucleotides (ribosome) which activates the aminoacids and brings about the synthesis of catalytic molecules by template directed polymerization.

1.4.5. Protocellular structures (Protomembrane). All the above macromolecules could have been formed on the prebiotic Earth, and yet they could not have originated an organism in the proper sense of the word capable of Darwinian evolution. This would have required the prior formation of protocellular structures (liposomes made of amphiphilic lipids), which allow the enclosure and cooperative interaction of all the above functional biopolymers. The increased reaction rates within the protocellular structure, or protocell, is what primarily differentiated the living system from the inanimate world.

1.5. Evolutionary Requirements

1.5.1. <u>Multiplication of genome content</u>. Darwinian evolution demands an increase in genomic content with time. From the simplest protocell to a prokaryotic cell the number of genes may be increased by a factor of 1000 and another 1000 from a prokaryote to the eukaryotic cells of man. This is the result of gene doubling, rearrangement and mutation followed by natural selection. The latter was facilitated by major changes in the environment.

1.5.2. <u>Biogeochemical cycles</u>. Evolution provided a mechanism for life to change in accordance with changes in the environment. But the converse is also true. For instance, photosynthetic organisms evolved an enzyme for the dissociation of water. This facilitated the reduction of carbon dioxide to monosaccharides, and generated oxygen which was released into the atmosphere. In turn, this change introduced the aerobic era, completed the carbon cycle, and made possible the appearance of eukaryotic organisms.

1.5.3. <u>Emergence of the eukaryotic cell</u>. The symbiotic association of two or more prokaryotes into a more complex organism (eukaryote) introduced a major leap forward in the evolutionary process. This occurred about 1.4 billion years ago, thanks to the progressive accumulation of high levels of oxygen in the Earth's atmosphere, which by increasing some 20 times the metabolic rate significantly enhanced the potential for the evolution and diversification of living systems.

1.5.4. <u>Emergence of multicellular organisms</u>. It is remarkable that for most of the Earth's history, life on our planet was unicellular. Presumably the increased calcium ion content and other environmental circumstances facilitated the appearance of multicellular organisms about 700 million years ago. Once this process was established, evolution of higher forms of life proceeded at a continuously accelerating pace until the advent of man.

1.5.5. <u>Emergence of man</u>. The five major attributes that significantly increased the survival of man's ancestors are: a strong grasp and flexibility of upper extremities; a medium size brain capable of development; ability to walk erect on the lower extremities; ability to emit modulated sounds; plus family and social bonds. Of these attributes, the cooperative interaction of the first two (hands and brain) are probably the ones most responsible not only for man's survival, but also for the increase and further development of his intelligence. In other words, development of encephalization per se without the manipulative translation of ideas into tools, instruments and writing can never lead to an advanced technological civilization.

2. DISCUSSION AND CONCLUSION

The predominantly qualitative nature of the above constraints, or requirements, emphasizes the difficulties in subjecting to quantitative analysis the different evolutionary processes considered to be involved in the origin of life and emergence of intelligence. On the other hand, the recognition that most of these processes obey known physical or chemical laws provides strong support for the concept that life and intelligence must be universal. For instance, when we realize that glycine and gama-aminobutyric acid

(GABA), two effective neurotransmitters present in meteorites, were probably synthesized before the formation of the solar system, we cannot escape the conclusion that the Universe was in a sense already prepared for the emergence of intelligence a long time ago.

Among the different evolutionary processes described above, the one that offers a major conceptual and experimental challenge is the one more intimately involved with the emergence of the living system (see 1.4 Protobiological Requirements). Indeed, we do not know yet how to bring about, or trigger, the cooperative interaction between replicative and catalytic molecules within a protocellular structure, so that it leads to an autopoietic system capable of replication, mutation and evolution. Continued research in this field is necessary to improve our understanding of this process.

From an astronomical point of view, the challenge is of a more technological nature. Namely, we need to sharpen our instrumental methods and technology in the search, first, of other planetary systems, so that we can confirm and extend the initial findings about Beta Pictoris, Epsilon Eridani and other stars, and at the same time improve the precision of some of the factors in Drake's equation. However, a more direct and bold approach, justified on the basis of chemical and biological evolution theory, is the search for extraterrestrial intelligence (SETI). It is hoped that with the application of the new microwave megachannel analyzers, with the help of the NASA SETI program (6), and with the implementation of other programs by the world astronomical community, we will move forward in our search for advanced technological civilizations. The discovery of an extraterrestrial civilization would be, without question, the greatest discovery of mankind.

ACKNOWLEDGEMENTS

Supported in part by NASA, Grant NGR-44-005-002. The author gratefully acknowledges the help received from Gil Armangué, and the information received from B. Campbell, A. Delsemme, G. Marx, S. Squyres, B. Smith, R. Terrile and J. Tarter. Further discussion and references may be found elsewhere (7).

REFERENCES

1. Smith, B. and Terrile, R. J. 'A Circumstellar Disk Around Beta Pictoris' Science **226**, 1421-1424 (1984).
2. Campbell, B. 'A Search for Planetary Mars Companions to Nearby Stars' This volume.
3. Llibre, J. and Pinol, C. 'A Gravitational Approach to the Titius-Bode Law' Astron. J. **93**, 1272-1279 (1987).
4. Squyres, S. 'Europa: Prospects for an Ocean' This volume.
5. Oro, J. 'Comets and the Formation of Biochemical Compounds on the Primitive Earth' Nature **190**, 389-390 (1961).
6. Oliver, B. 'The NASA SETI Program: An Overview' in Search for Extraterrestrial Intelligence, Proceedings of NRAO Workshop no. 11, Green Bank, West Virginia, May 1985 (Eds. K. I. Kellermann and G. A. Seielstad) 121-133 (1986).
7. Oro, J., Rewers, K. and Odom, D. " Criteria for the Emergence and Evolution of Life in the Solar System". Origins of Life **12**, 285-305 (1982).

COSMIC METABOLISM: THE ORIGIN OF MACROMOLECULES

C.N. Matthews
Department of Chemistry, University of Illinois
Chicago IL 60680 USA

Abstract.   Three possible routes for the synthesis of macromolecules--
geochemical, biochemical, astrochemical--are shown to be directed by
water, hydrogen cyanide and silicates, respectively.

The origin of volatiles and refractories.
The hydride hypothesis for the origin of molecules assumes that hydrides
are readily formed within our galaxy because of the dominating presence
of hydrogen compared with all other bonding elements.   One of these
hydrides, water, has a unique role because it reacts readily with metallic hydrides to give refractory materials such as silicates, or with
ionic hydrides to yield salts such as sodium hydroxide, but does not
react with covalent hydrides such as the volatile compounds dihydrogen,
methane, ammonia, hydrogen sulfide, and phosphine.   The existence of
this watershed producing reduced compounds together with those that are
oxidized has profound implications for the origin of stars, planets, and
life.

The origin of proteins and nucleic acids.
The cyanide model for the origin of proteins in the reducing environment
of primitive Earth maintains that polyamidines formed by base-catalyzed
polymerization of hydrogen cyanide in the atmosphere are readily converted by water in the oceans to polypeptides.   In the absence of water--on
land--these polyamidines could have been the original condensing agents
directing the synthesis of nucleosides and nucleotides from available
sugars, phosphates, and nitrogen bases.   Most significant would have
been the parallel synthesis of polypeptides and polynucleotides arising
from the dehydrating action of polyamidines on nucleotides.   On our
dynamic planet this polypeptide-polynucleotide symbiosis mediated by
polyamidines may have set the pattern for the evolution of protein-
nucleic acid systems controlled by enzymes, the mode characteristic of
life today.

The origin of stars and planets.
According to the planetary connection, disintegration of planets and
other satellites -- moons, asteroids, comets -- during the red giant
phase of stellar evolution yields circumstellar dust and molecules that
become interstellar following ejection by planetary nebula activity,
nova cataclysms or supernova catastrophes.   Further production of dust

G. Marx (ed.), Bioastronomy – The Next Steps, 167–178.
© 1988 by Kluwer Academic Publishers.

and molecules - circumstellar, interstellar and protostellar - is promoted by the dust acting as aggregating agent, coordinating matrix and radiation shield. As well as being possible abodes of life, planets play an essential role in bringing about stellar evolution in spiral galaxies.

Taken together, these preferred pathways suggest that in spiral galaxies planets are natural companions of stars, and that on Earth-like planets life is a universal phenomenon.

1. Introduction

Two of the most far-reaching generalizations of our time -- the unity of cosmochemistry and the unity of biochemistry -- are concerned with the behavior of molecules within the solar system and indeed throughout our galaxy. What is the origin of these molecular species large and small, some found in isolation or as dust particles in stellar and interstellar environments, others existing as components -- gases, liquids, solids -- of bodies inorganic and organic?

I discuss here the essential role of three ubiquitous compounds -- water, hydrogen cyanide and silica -- in bringing about the synthesis of these universal structures. In this light I then propose answers to questions arising, such as

Why is life based on the elements H, C, O, N, P, S?
Which came first, proteins or nucleic acids?
Where have all the planets gone?

2. The Origin of Volatiles and Refractories.

A striking fact regarding the distribution of molecules in our galaxy is that reduced and oxidized species are often found together, usually in the form of volatiles (ices, atmospheres, mantles) and refractories (grains, minerals, cores), respectively. This was first demonstrated by the classic Miller-Urey experiment simulating conditions on primitive Earth. On the grounds that hydrogen is by far the most abundant element distributed within our galaxy (see Figure 1), Urey[1] had

COSMIC ABUNDANCE OF ELEMENTS

H	86.7	He	13.1
O	0.08	Ne	0.05
C	0.03	A	0.0005
N	0.008	Mg	0.003
S	0.001	Si	0.003
(P	0.00003)	Fe	0.0005

Figure 1. Relative number of atoms of most abundant elements

reasoned that new-born planets must possess reducing atmospheres consisting mainly of molecular hydrogen, methane, ammonia and water. By subjecting mixtures of these gases to continuous electric discharges, Miller[2] obtained several kinds of organic compounds including, most dramatically, some of the α-amino acids known to be among the building blocks of proteins today. At one stroke this imaginative investigation linked biochemistry to cosmochemistry, showing why organic molecules are largely built up from the elements H,C,O,N. The essential presence of P and S in living systems then led to further studies in which PH_3 [3] and H_2S [4] were added to the original reducing mixture. In discussing the special fitness for life of these six elements Wald[5,6] emphasized the remarkable versatility of carbon in being able to form single and multiple bonds with itself and other elements. Phosphorus and sulfur are also able to form double bonds, unlike silicon which exists essentially in silicate structures possessing strong Si-O chains. Abundance and fitness then, have a direct bearing on the question of why the molecules of life are built largely from H,C,O,N,P,S. A further criterion --selection-- is proposed here to account for the origin of their ancestors.

Building on the original premise of Urey, the <u>hydride hypothesis</u>[7] assumes that the first molecules formed within our galaxy were necessarily hydrides because of the high cosmic abundance of hydrogen atoms compared with all other bonding elements. Although the reactivity of hydrides varies widely, it is known that many hydrides react violently with air or water.[8] Indeed water, next to dihydrogen the most abundant hydride, has a unique role because it

(a)   reacts readily with metallic hydrides to give refractory compounds such as silicates, or with ionic hydrides to yield salts such as sodium hydroxide, but

(b)   does not react with covalent hydrides such as the volatile compounds dihydrogen, methane, ammonia, hydrogen sulfide and phosphine.

Hydrides continue to be synthesized in environments rich in dust and gas such as interstellar clouds and stellar nebulae. As they become concentrated during the process of star formation their interaction and aggregation produces planetesimals and, eventually, planets consisting of refractory and volatile components. Reaction of water with abundant hydrides such as those of Si, Al, and Mg would give rise to the familiar oxy-compounds of geochemistry while the fundamental molecules of prebiotic chemistry would be formed from the volatile hydrides, stable to water, of the elements H, C, O, N, P and S. On Earth these would congregate in the atmosphere, undergoing selective photochemistry to produce the carbon-based structures of biochemistry (see Section 3).

In today's solar system[9] pristine hydrides have been observed in meteorites (H_2O), comets (H_2O, CH_4, NH_3), giant planets (H_2, CH_4, NH_3, H_2O, H_2S, PH_3, GeH_4 on Jupiter; H_2, CH_4, PH_3 on Saturn; H_2, CH_4 on Uranus and Neptune) and on a moon (H_2, CH_4 on Titan). Present in interstellar clouds[10] are H_2, CH_4, NH_3, H_2O and H_2S. Interstellar detection of the less abundant metal hydrides may yet become possible with improved techniques in the far infrared spectral region.[11]

3. The Origin of Proteins and Nucleic Acids.

It is widely assumed that the prebiotic formation of primitive proteins occurred in two steps: α-amino acid synthesis brought about by the action of natural high energy sources on the components of a reducing atmosphere, followed somehow by polycondensation of the accumulated monomers in the oceans or on land.[12] The pioneering demonstration by Miller and Urey[1,2] that α-amino acids are readily obtained from methane, ammonia and water subjected to electric discharges, taken with subsequent syntheses of peptides from amino acids under many sets of conditions,[12,35] seems to be in accord with this view. A more critical examination of the experimental evidence, however, raises fundamental questions concerning selection, concentration, purification and interaction of such products. How plausible are these attempted simulations as models of prebiotic chemistry? On primitive Earth the inherent thermodynamic barrier to spontaneous polymerization of α-amino acids might not have been so easily overcome.

An alternative route for the origin of proteins considers instead the direct synthesis of heteropolypeptides from hydrogen cyanide and water without the intervening formation of α-amino acids.[13-16] Following the initial production of hydrogen cyanide in the upper atmosphere by photolysis of methane and ammonia, a key step was the rapid vapor phase polymerization of clouds of HCN to polyaminomalononitrile (I). Subsequent reactions of hydrogen cyanide with the activated nitrile groups of I then yielded heteropolyamidines (II) which settled in the oceans and became converted to heteropolypeptides (III) after a series of hydrolysis and decarboxylation steps (see Figure 2).

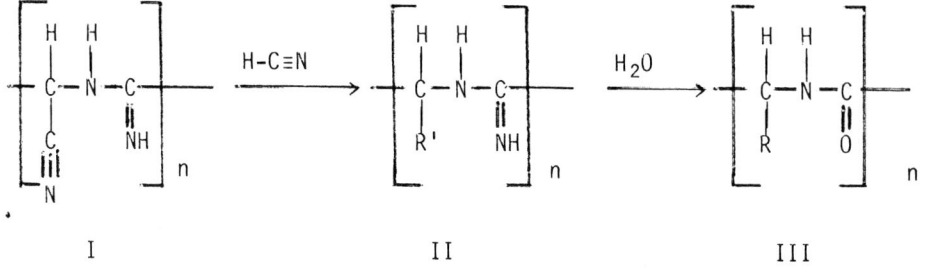

Figure 2. Conversion of polyamidines I and II--HCN polymers--to heteropolypeptides III.

Several kinds of experiments have provided results consistent with this route. In general, water-soluble, yellow-brown solids can be extracted from the black products of HCN polymerization or from electric discharge reactions yielding HCN from methane-ammonia mixtures. As predicted, acid hydrolysis of these yellow-brown polymers yields not just glycine but other α-amino acids as well, such as alanine, aspartic acid, glutamic acid, serine, and leucine.[13-16] Also as predicted, the glycine is perdeuterated when D_2O/DCl is used for hydrolysis instead of H_2O/HCl.[17] Non-destructive analysis of these polymer products became possible with the advent of cross-polarization magic-angle spinning solid state NMR spectroscopy (^{15}N, ^{13}C). In particular the unambiguous presence of secondary amide groups, as in peptides, has been established by double-cross-polarization studies on polymers synthesized from mixtures of H^{13}CN and HC^{15}N.[18]

Turning to extraterrestial studies,[19] most striking have been the direct images obtained by the Vega[20] and Giotto[21] spacecraft showing that the nucleus of comet Halley is covered with a surface layer as black as any body in the solar system. This non-volatile insulating crust of dark material may be largely composed of hydrogen cyanide polymers and related compounds,[22,23] since the expected predominance on cometary nuclei of frozen volatiles--methane, ammonia, and water-- subjected to high energy sources makes them ideal sites for the formation and condensed-phase polymerization of HCN. Dust emanating from the nucleus, contributing to the coma and tail, would also arise partly from the polymer. Results from the recent Halley missions support this view. Within the coma are hydrogen cyanide molecules[24] as well as cyanide radicals[25] derived from cometary grains, some of which consist of the elements H,C,O,N, others only of H,C,N.[26]

HCN polymer appears red in the photovisual spectral region (0.3-1.0μm)[27] and may account for the dark red color of D-type asteroids and other outer solar system bodies. VJHK photometry by Cruikshank et al[28] indicates that the bare nucleus of Halley and other comets resembles such asteroids.[29] Carbonaceous chondrites from the asteroid belt might therefore also possess HCN products. Studies of the Murchison meteorite have shown that free α-amino acids are present[30] together with acid-labile amino acid precursors of undefined structure.[31] Deuterolysis of these water-soluble yellow-brown solid extracts yields perdeuterated glycine as would be expected from peptide segments derived from HCN polymers.[17,32] Hydrogen cyanide polymerization could account, too, for much of the yellow-brown-orange coloration of Jupiter and Saturn, since HCN has been found in Jupiter's reducing atmosphere and in the atmosphere of Titan, the largest moon of Saturn. Most intriguing is an orange haze high in Titan's stratosphere that may consist of organic polymers,[33] possibly polycyanides formed directly from HCN.[34]

While life would not be expected to arise in such environments, the visible cyanide chemistry on these bodies is a continuing reminder that hydrides of the elements oxygen, carbon and nitrogen are a ready source of prebiotic molecules. In particular, the ubiquitous presence of HCN polymers invites the reexamination and possible reinterpretation of almost all studies concerned with the origin of proteins. In the Miller-Urey experiment, for example, it seems clear from our reinvestigations that the primary products were not α-amino acids, as claimed, but rather HCN polymers, the HCN being formed from methane and ammonia by electric discharge reactions and by elimination from intermediates such as aminoacetonitriles and HCN oligomers. The polymers then became hydrolyzed to amino acids either during reflux in the reaction flask, or later during the working-up procedure. The same conclusion applies to virtually all simulations of the chemistry of primitive atmospheres[35] as well as to studies of aqueous cyanide chemistry by Oró,[36] Ferris[37] and others. In our view, these investigations ostensibly yielding α-amino acids actually supply evidence for the abundant prebiotic existence of polymeric protein ancestors -- heteropolypeptides synthesized directly from hydrogen cyanide and water.

This ready conversion by water of polyamidines to polypeptides suggests that the polyamidines I and II might have played a further essential role in chemical evolution.[38] In the absence of water --on land-- they could have been the original condensing agents of prebiotic chemistry. Their reactive amidine groups, eager to become amides, would have brought about the stepwise formation of nucleosides, nucleotides and polynucleotides from available sugars, phosphates and nitrogen bases. Most significant would have been the parallel synthesis of polypeptides and polynucleotides arising from the dehydrating action of polyamidines on nucleotides:

polyamidines   A-A-A-A-A-A-A   P-P-P-P-P-P-P   polypeptides

-------->

nucleotides   N N N N N N N   N-N-N-N-N-N-N   polynucleotides

Optimum conditions might well have existed on a gradually warming Earth where prebiotic synthesis of organic molecules proceeded in the atmosphere in three overlapping stages defined by the relative volatility of methane, ammonia and water.[16,39] First, hydrocarbons were formed following the photolysis of methane to acetylene and the tautomeric C_3H_4 molecules methylacetylene and allene. Isoprene oligomers and polymers - the foundation for today's Isoprene Rule - were among the important products. During the second stage, with ammonia becoming more involved in atmosphere photochemistry, hydrogen cyanide and cyanoacetylene were major reactants. Polymeric peptide precursors I and II were formed in the stratosphere together with nitrogen heterocycles[35-37] possessing the basic skeletons of purines, pyrmidines, pyridines and porphyrins. Hydrogen sulphide also present took part in the modification of side-chains of the polymers. When most of the ammonia had been used up - its action as an ultraviolet screen thereby being considerably diminished - photolysis of water vapor that had been confined to lower levels become possible leading to the third stage when formaldehyde and sugars were synthesized as well as phosphates from phosphine. Meanwhile, as the macromolecules and heterocycles of the second stage gradually settled, contact with water introduced oxygen functionality mainly by hydrolysis of azomethines to carbonyl groups and of nitriles to amides and carboxylic acids.

As Earth's surface became covered with this organic shower, potential membrane material --carboxylic acids, carbohydrates, polypeptides-- accumulated in lakes and oceans, while on land the simultaneous synthesis of polypeptides and polynucleotides was promoted by polyamidines (I,II) --cyanide polymers-- perhaps assisted by clays. Potential metabolic machinery --hardware-- arising separately from the genetic apparatus --software--[40] soon interfaced to produce elementary replicating systems. Which came first, polypeptides or polynucleotides? The cyanide model suggests neither. Instead, they arose simultaneously from polyamidines. On our dynamic planet this polypeptide-polynucleo-

tide symbiosis mediated by polyamidines set the pattern for the evolution of protein-nucleic acid systems controlled by enzymes, the mode
characteristic of life today.

4. The Origin of Stars and Planets

In the tenuous interstellar medium of the Galaxy are cold, dark
clouds of gas and dust collapsing to form new stars and, presumably,
planets. This interstellar matter constitutes about 10% by mass of the
Milky Way and appears to be equally divided between atoms (mainly H and
He) and molecules (mostly H_2) with the dust amounting to about 1% of the
total gaseous mass. Most of the sixty or so interstellar molecules[10] so
far detected by radio and optical spectroscopy are organic compounds
ranging in size from methylidyne (CH) to cyanodecatetrayne ($HC_{11}N$),
among the most abundant being carbon monoxide (10^{-4} molecules relative
to H_2), water, ammonia, hydrogen cyanide (10^{-6}), acetylene (10^{-7}),
cyanoacetylene and formaldehyde (10^{-8}). Less certain are the identities
of the accompanying dust grains[41] of submicron dimensions which are
thought to have refractory cores and volatile mantles. Infrared and
ultraviolet spectral assignments point to silicates and graphite as
common core components. Mantle material might consist of frozen ices of
water, methane and ammonia or complex organic macromolecules.
    Controversy surrounds the origin of these diverse species, molecu-
lar and macromolecular. Can they be synthesized directly in interstellar space with its average density of only one atom per c.c.? Are
they, rather, placental remnants of star birth, expelled from developing
stellar nebulae? Or were they formed from atoms ejected from highly
evolved stars during continuous or catastrophic processes of mass loss?
    Most revealing in this respect has been the discovery of envelopes
of molecules and dust around cool stars in the red giant phase of
stellar evolution.[42] What could be the origin of this circumstellar
material? The widely held belief that synthesis occurs directly from
atoms in relatively cool red giant atmospheres seems plausible enough,
but only if condensation nuclei are first present.[43] These could be
supplied by the disintegration of solid bodies within the swollen giant.
Consider the following scenario for the evolving solar system, as yet
the only undisputed example of a star with planetary companions. In
about five billion years the sun will leave the main sequence, enter the
red giant phase while steadily losing mass and proceed through the
planetary nebula stage when a large fraction of its remaining mass will
be blown off. What happens to our planets during these traumatic
events? Incineration and/or vaporization is the customary answer,[44]
implying that some planets would revert to the atomic state while others
would continue to exist as bleak, charred wanderers in the sky. These
extreme situations by no means cover all eventualities. More likely the
engulfing of planets by the expanding sun would bring about extensive
synthesis of complex molecules through atmospheric and oceanic reactions. The formation of refractory grains by disruption of planetary
cores, crusts, and mantles would follow. Meanwhile, other interplanetary bodies -- moons, asteroids, meteorites, and especially comets in
the Oort cloud -- would also be contributing to the rich molecular mix.

Radiation pressure on this shell would accelerate the outflowing gas of the red giant leading to further dust-promoted synthesis of molecules.

How does such circumstellar matter become distributed in interstellar space? For solitary star-planet systems of solar order, the planetary nebula phenomenon most probably would be responsible. Molecules such as H_2 and CO have already been detected in such regions as well as carbon-rich dust, perhaps consisting of graphite grains and organic compounds of high molecular weight.[45] The prediction that extended oxygen-rich haloes might in some cases also be present has been confirmed by the detection of silicate spectra (10 μm) in several compact planetary nebulae.[46] In addition to the remnant red giant envelopes, new dust and molecules would be formed by condensation on nuclei after the transition to planetary nebulae.[47] Gradual dispersion of this gas and dust would then enrich the interstellar medium.

Would similar products also be expected from multiple stars that exhibit nova or supernova (Type I) behavior? Statistical studies[48] show that roughly half of all stars are not single, while computer simulations[49] indicate that a double star system can have stable planetary orbits, notably when the planets are close to one of the two binary stars, or when they are distant from both. Since most close binaries have a white dwarf component, disruption of planets during a preceding red giant phase could have produced circumstellar dust and molecules which would certainly have become involved in the mass exchange processes characteristic of such pairs. Eventually dust would be ejected during nova outbursts or during Type I supernova explosions brought about by the accumulation of material on the white dwarf.

What about the contribution of supernova explosions (Type II) arising from the violent death of massive stars? Again, although it is widely accepted that solid particles readily condense out during the expansion period it seems that such condensations must necessarily be preceded by dust formation arising from planets and other bodies.

Underlying this planetary connection are three propositions.

(i)   Planets form readily around single or multiple stars from refractory and volatile molecules within stellar nebulae.

(ii) Star-planet systems produce circumstellar shells of molecules and dust following planetary breakdown during the red giant stage of stellar evolution.

(iii) Circumstellar material becomes interstellar through planetary nebula activity, nova cataclysms, or supernova catastrophes.

In each situation--protostellar, circumstellar, and interstellar--further production of grains and molecules by way of hydrides is promoted by the dust acting as aggregating agent, coordinating matrix, and radiation shield. Interstellar molecules synthesized on dust grains would either be released --H_2 especially-- or retained as mantles. The mantles themselves could become a source of free organic molecules through photochemical degradation processes.[50] Within denser clouds, gas phase interactions[10,51] of atoms and ions, radicals and molecules might further yield compounds protected by the pervasive dust from the high energy flux of interstellar space. Most significantly, in these clouds dust-mediated processes such as sedimentation, accretion, and fragmentation would lead to the formation of protostars surrounded by newly-formed molecules, soon to become stars encircled by planets.

In a sense then, just as life begets life, so do planets beget planets, the ejection of dust from dying star-planet systems being the progenitive step in this cosmic metabolism (Gk meta, change; ballo, throw). But what was the origin of the first planets or of the primeval dust? It seems that galaxies began with a tremendous burst of activity.[52] The very first stars were formed only from the primordial elements H and He. These massive and short-lived stars exploding as supernovae supplied the residual amounts of heavy elements that are found in the oldest visible stars of the Galaxy. Thereafter, as heavy elements from supernova activity rapidly enriched the interstellar medium, formation of massive star-planet systems became possible even before there was dust, as long as the Galaxy was still in a state of gravitational collapse. Subsequently, with more settled conditions prevailing, a critical level[53] was reached marked by the accumulation of dust from planetary disintegration and of heavy elements to the extent of about 1% of present solar abundances. A new mode of star formation took over, dominated by the cooling effect of dust grains[53] and H_2 molecules.[54] This dramatically lowered the temperature of molecular clouds, causing repeated fragmentation till clouds finally became dense and opaque enough for protostar formation to begin.[53,54] Long-lived, less massive stars appeared with their accompanying planets and the Galaxy began to acquire its present stellar distribution with most stars in the 0.1 to 1.0 solar mass range.[53] The birth-death cycle for star-planet systems now became established (Figure 3) with dust acting both as instigator and end product of the overall process. In dust-free galaxies -- ellipticals -- most stars are visibly ancient, however, formed closer in time to the Big Bang. Only in dusty spirals are such fossil populations continuously augmented by the cyclic regeneration of stars and planets and of life that asks: How? Why? and Where have all the planets gone?

5. Conclusions

The hydride hypothesis accounts for volatiles and refractories being found together in protostellar, circumstellar and interstellar environ-ments. The existence of this watershed producing reduced compounds together with those that are oxidized has profound implications for the origin of stars, planets, and life. According to the cyanide model the volatile component readily yields hydrogen cyanide polymers, required structures for the emergence of life. The planetary connection shows how the refractory component -- silicate dust -- brings about stellar evolution, the birth-death cycle for star-planet systems.

Taken together, these preferred pathways suggest that in spiral galaxies planets are natural companions of stars, and that on Earth-like planets life is a universal phenomenon.

176

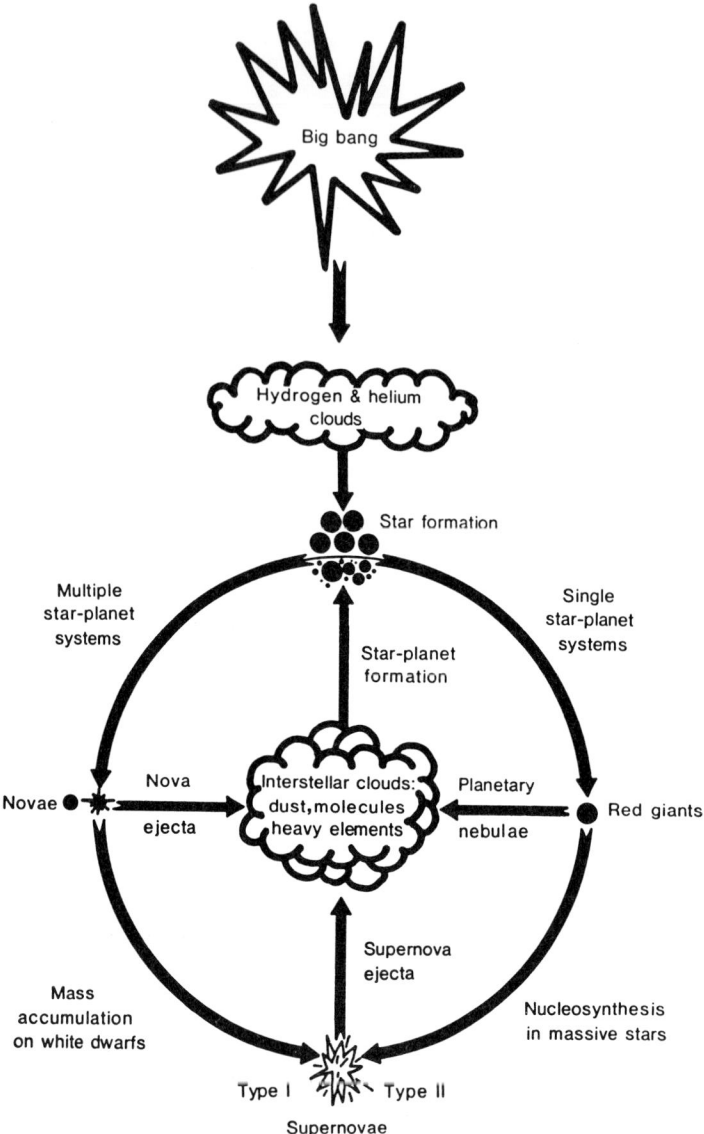

Figure 3. Evolutionary cycle for star-planet systems in spiral galaxies. The original stars were formed directly from hydrogen and helium clouds in the developing protogalaxy. Cyclic regeneration of star-planet systems now takes place within interstellar clouds of gas and dust concentrated in the spiral arms and center of the galaxy. (Adapted from Clark, D.: 1979, Superstars, Dent, London, England).

References

1. Urey, H. C.: 1952, Proc. Natl. Acad. Sci. U.S. **38**, 351.
2. Miller, S. L.: 1953, Science **10**, 528.
3. Palm, C. and Calvin, M.: 1962, J. Am. Chem. Soc. **84**, 2115.
4. Sagan, C. and Khare, B. N.: 1971, Nature **232**, 577.
5. Wald, G.: 1962, in Kasha, M. and Pullman, B. (eds.) Horizons in Biochemistry, Academic Press, New York, N.Y., p. 127.
6. Wald, G.: 1964, Proc. Natl. Acad. Sci. U.S. **52**, 595.
7. Matthews, C. N.: 1980, Abstracts, 6th Intl. Conf. Origins of Life, Jerusalem, Israel, p. 29.
8. MacKay, K. M.: 1966, Hydrogen Compounds of the Metallic Elements, E. and F. N. Spon Ltd., London, England.
9. Jones, B. W.: 1984, The Solar System, Pergamon Press, London, England.
10. Duley, W. W. and Williams, D. A.: 1984, Interstellar Chemistry, Academic Press, New York, N.Y.
11. Townes, C. H.: 1980, in Andrew, B. H., Interstellar Molecules, Reidel, Dordrecht, Holland, p. 644.
12. Miller, S. L.: 1984, in Nicolis, G. (ed.), Aspects of Chemical Evolution, Wiley, New York, p. 85.
13. Matthews, C. N. and Moser, R. E.: 1967, Nature **215**, 1230.
14. Matthews, C. N. and Moser, R. E.: 1966, Proc. Natl. Acad. Sci. U.S. **56**, 1087.
15. Matthews, C. N.: 1975, Origins of Life **6**, 155.
16. Matthews, C. N.: 1984, Proc. Roy. Inst. Gt. Britain **55**, 199.
17. Matthews, C. N., Nelson, J., Varma, P. and Minard, R. D.: 1977, Science **198**, 622.
18. Matthews, C. N., Ludicky, R. A., Schaefer, J., Stejskal, E. O. and McKay, R. A.: 1984, Origins of Life **14**, 243.
19. Matthews, C. N.: 1985, in Papagiannis, M. D. (ed.) The Search for Extraterrestrial Life: Recent Developments, Reidel, Dordrecht, Holland, p. 151.
20. Sagdeev, R. Z., Blamont, J., Galeev, A. A., Moroz, V. I., Shapiro, V. D., Shevchenko, V. I. and Szego, K.: 1986, Nature **321**, 259.
21. Reinhard, R.: 1986, Nature **321**, 313.
22. Matthews, C. N. and Ludicky, R. A.: 1986, Proc. 20th ESLAB Symposium on the Exploration of Halley's Comet, ESA S.P. 250, p. 273.
23. Matthews, C. N. and Ludicky, R. A.: 1987, Polymer Preprints **28**, No. 1, 104.
24. Schloerb, F. P., Kinzel, W. M., Swade, D. A. and Irvine, W. M.: 1986, Proc. 20th ESLAB Symposium on the Exploration of Halley's Comet, ESA SP-250, p. 577.
25. A'Hearn, M. F., Hoban, S., Birch, P. V., Bowers, C., Martin, R. and Klinglesmith, D. A.: 1986, Nature **324**, 649.
26. Clark, B. C., Mason, L. W. and Kissel, J.: 1986, Proc. 20th ESLAB Symposium on the Exploration of Halley's Comet, ESA SP-250, p. 353.
27. Cruikshank, D. P.: 1986, Advances in Space Research, in press.
28. Cruikshank, D. P., Hartmann, W. K. and Tholen, D. J.: 1985, Nature **315**, 122.

178

29. Hartmann, W. K., Cruikshank, D. P. and Degewij, J.: 1982, _Icarus_ **52**, 377.
30. Kvenvolden, K. A., Lawless, J. G. and Folsome, C. E.: 1973, _Scientific American_ **227**, June, 38.
31. Cronin, J. R.: 1976, _Origins of Life_ **7**, 337, 343.
32. Matthews, C. N., Nelson, J. E. and Minard, R. D.: 1980, Abstracts, 6th Intl. Conf. Origins of Life, Jerusalem, Israel, p. 100.
33. Owen, T.: 1982, _Scientific American_ **246**, February, 98.
34. Matthews, C. N.: 1982, _Origins of Life_ **12**, 281.
35. Ponnamperuma, C.: 1983, (ed.) _Cosmochemistry and the Origin of Life_, Reidel, Dordrecht, Holland, Ch. 1.
36. Oró, J. and Lazcano-Araujo, A.: 1980, in Vennesland, B., Conn, E. E., Knowles, C. J., Westley, J. and Wissing, F. (eds.) _Cyanide in Biology_ Academic Press, New York, N.Y. p. 517.
37. Ferris, J. P. and Hagan, W. J.: 1984, _Tetrahedron_ **40**, 1093.
38. Matthews, C. N.: 1986, _Origins of Life_ **16**, 500.
39. Kliss, R. M. and Matthews, C. N.: 1962, _Proc. Natl. Acad. Sci. U.S._ **48**, 1300.
40. Dyson, F.: 1985, _Origins of Life_, Cambridge University Press, Cambridge, England.
41. Knacke, R. F.: 1978, in Gehrels, T. (ed.), _Protostars and Planets_, University of Arizona Press, Tucson, Arizona, p. 112.
42. Zuckerman, B. A.: 1980, _Ann. Rev. Astron. Astrophys._ **18**, 263.
43. Matthews, C. N.: 1983, Abstracts, _7th Intl. Conf. Origins of Life_, Mainz, FRG, p. A1-10.
44. Motz, L.: 1975, _The Universe, Its Beginning and End_, Scribners, New York, N.Y. p. 283.
45. Field, G. B.: 1978, in Terzian, Y. (ed.) _Planetary Nebulae, Observations and Theory_, Reidel, Dordrecht, p. 367.
46. Aitken, D. K., Roche, P. F. and Spenser, P. M.: 1979, _Astrophys. J._ **233**, 925.
47. Sun Kwok: 1981, in Iben, I. and Renzini, A. (eds.) _Physical Processes in Red Giants_, Reidel, Dordrecht, p. 421.
48. Abt, H. A.: 1978, in Gehrels, T. (ed.), _Protostars and Planets_, University of Arizona Press, Tucson, Arizona, p. 338.
49. Harrington, R. S. and Harrington, B. M.: 1978, _Mercury_, 34.
50. Sagan, C. and Khare, B. N.: 1979, _Nature_ **277**, 102.
51. Watson, W. D.: 1978, in Gehrels, T. (ed.) _Protostars and Planets_, University of Arizona Press, Tucson, Arizona, p. 77.
52. Shklovskii, I. S.: 1978, _Stars: Their Birth, Life and Death_, W. H. Freeman, San Francisco, CA, Ch. 16.
53. Silk, J.: 1980, _The Big Bang_, W. H. Freeman, San Francisco CA.
54. Reddish, V. C.: 1978, _Stellar Formation_, Pergamon, Oxford, England.

EXOTIC CHEMICAL LIFE

Ronald D. Brown

Chemistry Department, Monash University

Clayton, 3168 Australia

For a long time now there has been an almost universal assumption that life in the universe has to be based on carbon chemistry. The only notable modern exception appears to be that of Cairns-Smith (1982, 1985) who has eloquently espoused the case for starting life with silicate clays, there being a subsequent genetic takeover in favour of carbon compounds. However in the past there have been speculations, albeit superficial by modern standards, of extra-terrestrial life based on other chemical elements. Let me attempt to review the situation through more modern eyes.

To make progress we must agree on a specification of living things. This has been a traditionally debatable point (see Fig. 1) but is not central to my present theme. Let me therefore assume without debate that we are considering an entity that is capable of existing for a considerable time in an environment less highly specialised than itself; that is capable of reproducing itself with very little error (so that some generations later the entity is still recognisable as generically related); and that can store and pass on to subsequent generations some coded information.

Terrestrial life separates itself from the environment by means of a membrane, based on lipids in the simplest cases. To contemplate life based on other elements we therefore have to envisage some membraneous material or, as the only apparently plausible alternative, we must contemplate a solid agglomerate such as a crystalline solid.

Crystallinity is not unknown for almost all elements of the periodic table. The formation of semi-permeable membranes is however a more demanding requirement, met by a rather wide range of high-molecular weight compounds based on carbon, a variety of compounds based on silicic acid and various oxy-acid materials in the gel state based on various other elements. The latter appear to require rather stringent environmental conditions of pH etc.. We should also note that here we are adopting the assumption of a liquid state based on water as the underlying liquid. Alternative liquids such as ammonia or hydrogen fluoride appear to be less accommodating, while higher-temperature molten-salt systems look even less inviting!

The Liquid Basis

We might here note that while it may be tempting to speculate on a variety of exotic liquids as the fluid base of living systems, to be cosmically realistic we must concentrate on simple molecules, essentially water, ammonia, hydrogen fluoride (if we are contemplating temperatures not too far from terrestrial temperatures) or perhaps lower temperature liquids like methane, ethylene, carbon dioxide and other small carbon-based compounds, ammonia and hydrazine, oxides of nitrogen, hydrogen sulfide, phosphene and the like. But here, as elsewhere, we must bear in mind cosmic abundances of the elements (see Table 1) which, *inter alia*, eliminates from serious consideration fluorine compounds and probably phosphine or other phosphorus compounds; and probably sulfur compounds.

At elevated temperatures we might consider molten SiO_2 or molten silicates for the liquid phase. Liquid silica however requires such a high temperature that it might not survive long enough before solidifying and thus not offer a sufficient time span for any primitive life system to emerge. Liquid silicates based

G. Marx (ed.), Bioastronomy – The Next Steps, 179–185.
© *1988 by Kluwer Academic Publishers.*

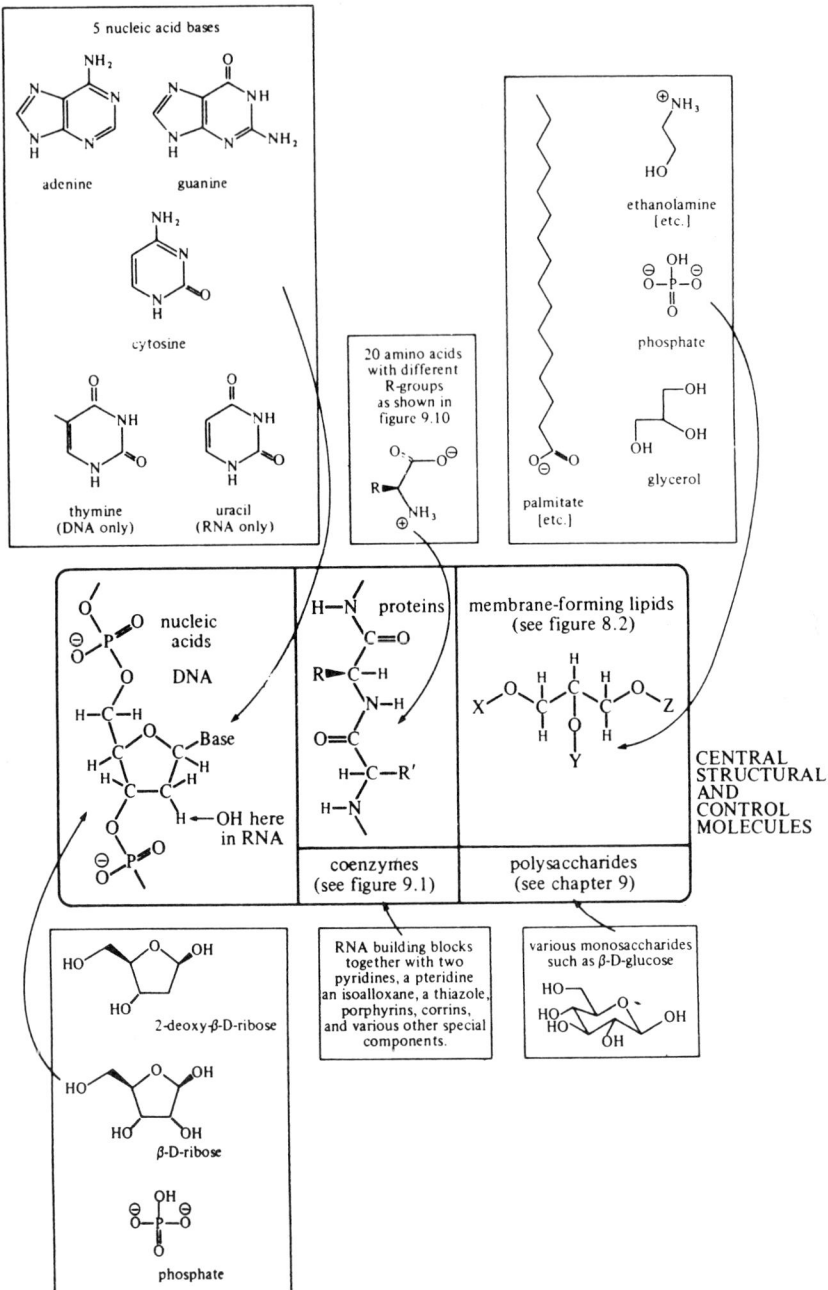

Fig. 2   The main groups of molecular "building blocks" involved in current terrestrial life. (From Cairns-Smith, 1982, with permission).

on the relatively abundant Mg, Na or Ca species survive at much lower temperatures and so just might be considered. However the laboratory chemical studies of such liquids as solvents does not seem to have included searches for a sufficient variety of the more complex solutes to offer us much basis for speculation about bizarre life systems for hot planets based on such liquids.

The role of the liquid basis within a living organism is to be solvent for a variety of nutrients and other substances involved in metabolism; to be a transport medium. This limits us to a very few substances that are both good solvents and consist of only cosmically abundant elements. I have not been able to think beyond ammonia (NH_3), carbon dioxide (CO_2), hydrogen cyanide (HCN) and perhaps hydrazine (N_2H_4); ruling out methane (CH_4) and other hydrocarbons on the ground of not being sufficiently versatile solvents. Other solvents familiar to chemistry, while being composed of the cosmically abundant elements (C, H, O, N) have more complex molecular structures, unlikely to be generated in sufficient abundance on a newly formed planet. Even for the select group of solvents just mentioned, it is apparent that water stands out as being a suitable liquid vehicle for transport of a very wide range of chemicals, including very many large species that we often refer to as biopolymers, peptides, enzymes, sugars and many other species, as well as offering environment control possibilities via variations in pH and being a good solvent for various small inorganic species such as NaCl. The possibilities with NH_3 or HCN seem far more limited, although we must always accept that we have not yet managed to encounter in the laboratory some new and versatile group of molecules that offer a new range of potentially biological features in ammonia, or hydrogen cyanide, etc.. For our present purposes however I shall now ignore such speculations.

Molecular Complexity and Variety

In deliberations about the origin of life on Earth there has been a long-running debate as to whether proteins or nucleic acids came first. There is now considerable support for starting with a crude form of RNA since it has recently been found that some RNAs can act as enzymes. But the real dilemma seems to be that even the simplest living system involves a number of genetic and metabolic requirements so that we have to contemplate a molecular make-up of some complexity and variety (see Fig. 2). Such complexity makes it difficult to discern just how it was achieved, given that we must start from quite simple molecules, presumably mixed with other diluting material that was not required for the evolution of a living system but which had to be separated in some manner from the desired constituents.

The problem of how the self-assembly occurred on the young Earth of complex molecules and systems of molecules is a continuing challenge. From the point of view of contemplating chemically exotic life it emphasises that if we are seeking alternative chemistries than the carbon-based system then we require chemistries as complex and varied as that of carbon. The short answer is that there are none. The next most complex would appear to be compounds based on silicon and oxygen but even they are much more limited in both variety and complexity. To contemplate such chemical bases requires life based on considerably less complex systems. Perhaps the silicate crystal proposition is a basis for a simpler life-system.

Information Storage

To provide a means of storing information or even just some substantial structural units for building up a complex entity, we must envisage molecules that are large in at least one dimension. Thus we ask which elements lend themselves to polymer structures, or to two-dimensional sheets or three-dimensional arrays. Many elements lead to compounds that are crystalline and so satisfy the requirement for three-dimensional arrays, but for molecules that are extensive in one or two dimensions we are back to seeking membrane materials or polymers. To go beyond carbon we have to contemplate

Figure 3

Codable Polymers

Siloxanes

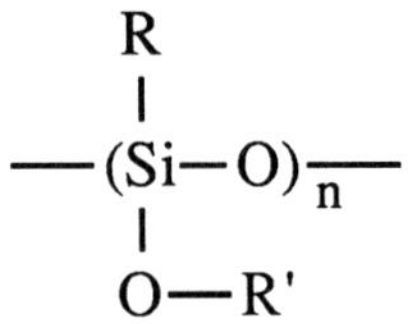

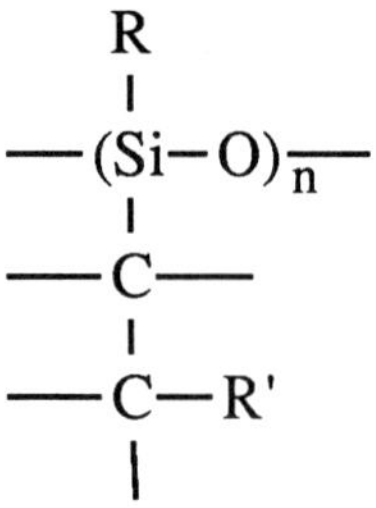

$$\begin{array}{c} R \\ | \\ -(Si-O)_{\overline{n}} \\ | \\ O-R' \end{array} \qquad\qquad \begin{array}{c} R \\ | \\ -(Si-O)_{\overline{n}} \\ | \\ -C- \\ | \\ -C-R' \\ | \end{array}$$

Polymer is usually water-soluble if $R' = -(CH_2CH_2O)_n-$.

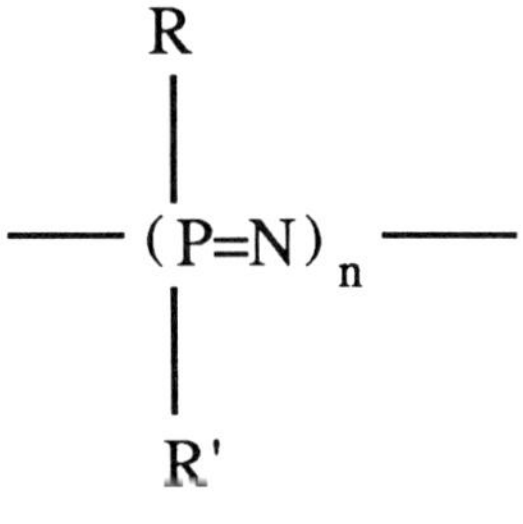

$$\begin{array}{c} R \\ | \\ -(P=N)_n- \\ | \\ R' \end{array}$$

Water-soluble if $R' = -O(CH_2CH_2O)_nCH_3$ $n = 2,3,....$

various oxy-acid systems, silicates or just possibly borates or some other oxyacid systems, molybdates, vanadates, niobates etc. Unfortunately, with the reasonable exception of silicates, the other elements central to such oxyacids are of negligible cosmic abundance and so can be discounted as plausible for the basis of life anywhere in the universe.

If we contemplate information storage at the molecular level then we require not only the feasibility of building up molecules that are very large in at least one dimension but whose structure includes features that can be varied without destroying the whole structure. This variation is of course required so that a varying pattern can be built up - a storage of data. But the data bank must have both "read" and "write" capabilities. Moreover if it is to be the basis of a genetic system then it must be copyable with very low error rate.

In the general chemical area I know of nothing remotely comparable to the nucleic acid systems, RNA and DNA, that have evolved as information systems on Earth. I can offer two very speculative examples of polymers with chains based on elements other than carbon, namely (Fig 3) siloxanes and polyphosphazines. However for coding information we would have to envisage side groups based on carbon both to give us a small "alphabet" for coding and also to confer water-solubility on the polymers. I can offer no useful suggestions of chemical schemes to generate the requisite patterns of side-chains, although attachment and detachment of these groups in these polymers are acid/base mediated processes that are fundamentally akin to the processes included in the coding of nucleic acids. And who would have thought up the rather complex mechanism involved in nucleic acid synthesis in advance to its being uncovered in the laboratory?

Again we are driven back inexorably to carbon-based compounds and a related water solvent-system. The only realistic alternative is to turn to an information system of a totally different kind, such as that which features in the Genetic Takeover system of Dr. Cairns-Smith, i.e. a pattern of dislocations on a crystal surface.

A considerable range of crystalline materials could be contemplated, even if we confine ourselves to compositions based solely on cosmically abundant elements. Terrestrial geology points towards salts of oxyacids - carbonates, silicates, sulfates or oxides. Cairns-Smith has developed an enchanting hypothesis (Fig. 4) focussed initially on silicates, these of course being the most commonly encountered minerals in the Earth's crust. The crystal habits of oxides, carbonates or sulfides may not seem so attractive, especially as dividing platelets are an important component of the concept of transmitting and replicating information in this hypothesis.

We should bear in mind that the universe is, by virtue of the dominant abundance of hydrogen, strongly reduced and we have been contemplating oxidized systems. Minerals in a reduced form, such as metallic solids are uncommon on Earth and are not usually encountered in crystalline form. In any case, metals tend to crystallise in cubic or hexagonal forms that do not readily flake. I find it hard to see how any dislocation pattern could plausibly be transmitted to new crystals of metal, elegant though it might be to fantasize on living crystals of gold!

We might however return to carbon and the classic layer lattice of graphite. It can certainly flake readily in a way vaguely reminiscent of cell division. Is there some possibility of having a coding of defects, or some chemical intercalations, that could be replicated? One is straining hard here to guess at possibilities and it seems pointless to pursue this further now. I conclude that crystal-based information systems, other than perhaps clays, do not offer any real hope for life elsewhere in the Universe.

184

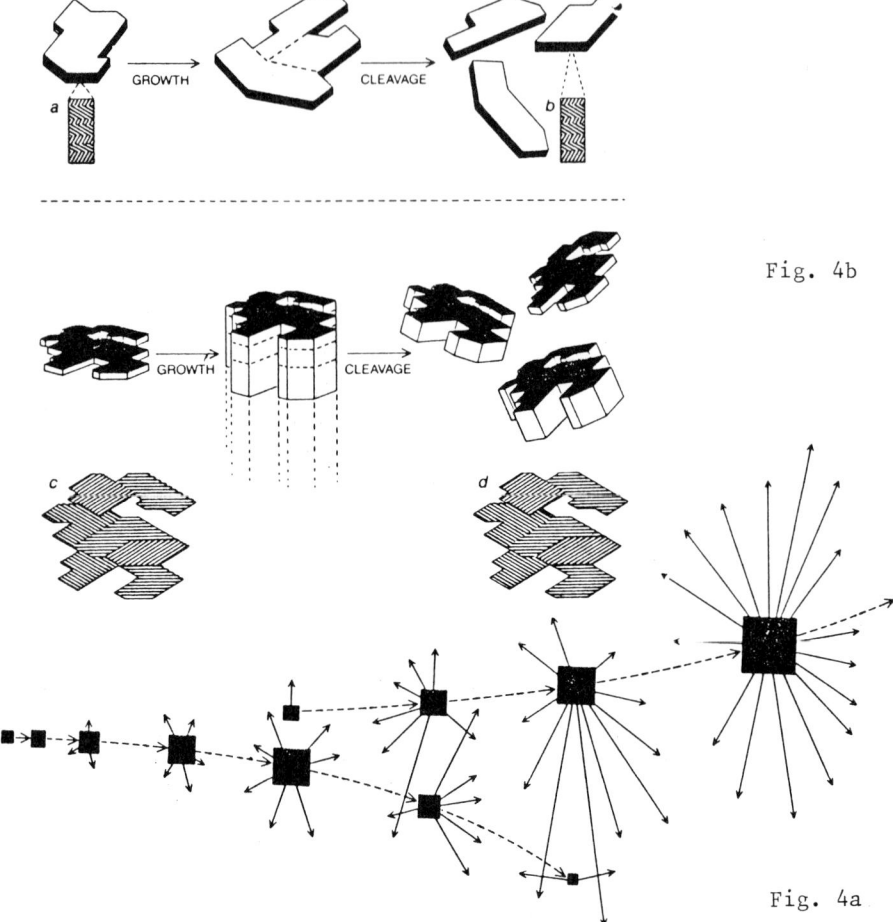

Fig. 4b

Fig. 4a

Fig. 4. (a)   Genetic takeover is seen as a key phase in early evolution. Originally there were naked genes of some unknown first genetic material (small black square on left) which evolved to control their immediate environment by specifying the production of increasingly elaborate surrounding phenotypes (the irregular outlines). A new kind of gene appeared (upper black squares) that could function only within a rather sophisticated phenotype but that proved to be more effective there than the original genes. The new genes gradually took over control of the phenotype, which came to be converted to their exclusive service. Eventually the original genes were dropped.

(b)   Crystal genes would need to display the right combination of structural growth and cleavage characteristics. Information might be stored in one or two dimensions in crystal genes. In a one-dimensional gene (top) information would be held in the detailed structure of a sequence of stacking layers, which remains constant (a,b) as the gene replicates. Growth takes place only on the edge faces and cleavage takes place only parallel to those faces.   The information-carrying layers could vary physically (there might be differently aligned crystal structures, for example) or in their chemical composition. In two-dimensional crystal genes (bottom) information would be held as a pattern (again either physical or chemical) on one face of the crystal; that pattern remains constant (c,d) as the gene replicates by growing on the shaded face and by being cleaved in a plane parallel to it.

(from Scientific American, June, 1985).

Conclusion

I conclude then that exotic chemical life, although I think it unlikely to resemble terrestrial life in full chemical exactitude, and almost certainly to be morphologically vastly different from life forms on Earth, is undoubtedly going to be based on carbon chemistry, the sole chemically plausible alternative being the montmorillionite clay system of Cairns-Smith. Both require water as the fluid support system and I am still putting my money on carbon to win!

References

Brown, R. D. and Rice, E. H. N., *Mon. Not. R. astron. Soc.*, 1986 **223**, 405.
Cairns-Smith, A. G., *Genetic Takeover and the Mineral Origins of Life*, C. U. P., 1982.
Cairns-Smith, A. G., *Sci. Amer.*, 1985 (June), **252**, 74.

The Most Cosmically Abundant Elements *			
Element	$\log_{10} N$†		
H	12.00		
He	10.93		
O	8.82	Ar	6.8
C	8.52	Al	6.39
N	7.96	Ca	6.30
Ne	7.92	Ni	6.30
Fe	7.60	Na	6.25
Si	7.52	Cr	5.85
Mg	7.42	Cl	5.6
S	7.20	P	5.52

* After C.W. Allen, 'Astrophysical Quantities" (1973).
† Logarithms of number abundances referred to H = 12.00.

I define life . . . as a whole that is pre-supposed by all its parts.

S. T. Coleridge (ca *1820*)

. . . organisms – that is to say, systems whose parts co-operate.

J. B. S. Haldane (*1929*)

We shall regard as alive any population of entities which has the properties of multiplication, heredity and variation.

J. Maynard Smith (*1975*)

I suggest that these three properties – mutability, self-duplication and heterocatalysis – comprise a necessary and sufficient definition of living matter.   *N. H. Horowitz* (*1959*)

. . . it might be claimed that the most important fact about them [living things] is that they take part in the long term processes of evolution.   *C. H. Waddington* (*1968*)

Fig. 1.   Some definitions of "life".

A SCEPTIC'S VIEW

R. Shapiro

Chemistry Deparment, New York University
4 Washington Place, New York 10003, USA

ABSTRACT. Two positions that concern the origin of life on earth and the possible existence of extraterrestrial life are compared. In one view, life on earth is considered to have started with the prebiotic synthesis of replicating RNA molecules. Life elsewhere, if it exists, is likely to have a biochemistry much like our own, and is therefore restricted in its origin to earthlike worlds. As such planets may be quite infrequent in their occurrence, life may also be quite rare in the universe. This position is often stated with conviction, but it goes far beyond the skimpy evidence available. It is therefore termed "the believers' view". The pessimism inherent in this position has served to quench enthusiasm for planetary exploration and efforts to detect extraterrestrial life.

In an alternative position, "the skeptic's view," the above questions are considered to be open. We do not understand how life began here. A variety of environments within our solar system are possible sites for the origin of life. Definite information can be obtained only by observation. A search for signs of life is therefore given the highest priority in future planetary missions.

1. INTRODUCTION

A common definition of the word skeptic is "one who questions the validity or authenticity of something purporting to be knowledge". As a contrast to this position, I will use the word "believer" to mean "one who completely and unquestioningly accepts something, even in the absence of proof". Skepticism is always appropriate in science, but particularly so in speculative areas such as the origin of life, and its distribution in the universe, where hard evidence is skimpy or totally lacking.

187

G. Marx (ed.), Bioastronomy – The Next Steps, 187–194.
© 1988 by Kluwer Academic Publishers.

I shall argue that many generalizations concerning these topics, while stated as truth in assertive terms, are in fact based on little or no evidence. These positions, which I term "the believer's view", will be contrasted with an alternative, "the skeptic's view".

2. THE BELIEVER'S VIEW

The believer's view holds that (1) much progress has been made in understanding the origin of life on earth; (2) life here began with the spontaneous assembly of the first replicating RNA molecule; (3) life in places other than earth must also be based on proteins and nucleic acids, or at least on a system of carbon chemistry carried out in water; (4) life elsewhere will therefore require conditions similar to those on earth; (5) the Viking experiments demonstrated that Mars (the most earthlike world in the solar system) has no life; and therefore, (6) life exists only on earth in this solar system.

A full exposition and discussion of the merits of this position would take us far beyond the scope of this article. The interested reader should consult two earlier books by this author (Feinberg and Shapiro, 1980; Shapiro, 1986). A few quotes from exponents of this set of positions will help us to appreciate the thrust of their arguments, however. The RNA origin of life has been supported by Gilbert (1986): "One can contemplate an RNA world containing only RNA molecules that serve to catalyze the synthesis of themselves. The first stage of evolution proceeds, then, by RNA molecules performing the catalytic activities necessary to assemble themselves from a nucleotide soup." Darnell and Doolittle (1986) wrote in a similar vein: "RNA chemistry is assumed to be the dominant successful polymer chemistry in prebiotic times."

George Wald has argued the extraterrestrial biochemistries will resemble our own, stating (1974): "I have become convinced that life everywhere must be based primarily on carbon, hydrogen, nitrogen and oxygen, upon an organic chemistry therefore much as on the earth, and that it can arise only in an environment rich in water". He further illustrated this point in 1973: "So I tell my students: learn your biochemistry here and you will be able to pass examinations on Arcturus." Norman Horowitz has supported this theme, stating (1976): "The capacity for generation, storing, replicating and utilizing large amounts of information implies an underlying molecular complexity that is known only among compounds of carbon."

This position is in accord with Horowitz's pessimism concerning prospects for Martian life (1986): "Viking found no life on Mars and, just as important, it found why

there can be no life. Mars lacks that extraordinary feature that dominates the environment of our own planet,
oceans of liquid water in full view of the sun". Other
supporters of this claim have included James Lovelock
(1986): "We know there is no life on Mars; indeed we knew
it long before the Viking spacecraft landed there," and
Lewis Thomas (1978): "Mars from the look we've had at it
thus far is a horrifying place. It is, by all appearance,
stone dead. It is surely the deadest place many of us have
seen, and it is hard to look without wincing. Come to
think of it, it is probably the only really dead place of
any size we've caught a glimpse of, and the near view is
incredibly sad".

Some of us may wonder how the moon came to be omitted
from Dr. Thomas' list of dead places. Other writers have
more than compensated for this omission, however, by vastly
extending the dead zone in their speculations. The Committee on Planetary Biology of the U.S. National Research
Council wrote, for example (1981): "We view the search for
present life in the solar system as completed: There is
strong evidence that neither the planets (other that earth)
nor their satellites provide conditions consistent with the
maintenance of life. The major limitations are the absence
of liquid water or temperatures incompatible with the survival of the organic compounds requisite for living systems." Norman Horowitz (1986) has provided a further esca-
lation: "Since Mars offered by far the most promising
habitat for extraterrestrial life in the solar system, it
is now virtually certain that earth is the only life-bearing planet in our region of the galaxy". Finally, Michael
Papgiannis (1986) wrote: "The complexity of these requirements might make planets like ours, that are able to retain
liquid water on their surfaces for billions of years, a
rare occurrence in the Cosmos, with obvious implications on
the frequency of life-harboring solar systems in the
galaxy".

Although the above opinions are stated with a conviction quite suitable for a believer's viewpoint, they are
still speculations that go far beyond the data supporting
them. The skeptic's view offers an alternative and more
conservative interpretation that is also consistent with
the evidence at hand.

3. THE SKEPTIC'S VIEW

The position of the skeptic is that (1) we know very little
about the origin of life (although much has been learned
about molecular biology and evolution); (2) it is highly
implausible, with our current knowledge of chemistry, that
life began with RNA; (3) we have no information concerning

other bases for life; no current knowledge excludes life
forms based on very different systems of matter (4) earth-
like environments are promising sites to search for life,
but very different ones may also sustain life; (5) the
Viking life detection experiments produced ambiguous and
confusing results; we do not know whether life exists on
Mars, even at the sites visited by Viking; and (6) life
may or may not exist elsewhere in our solar system.

4. THE UNKNOWN ORIGIN OF LIFE ON EARTH

The conclusion concerning RNA follows from chemical considerations: the spontaneous assembly of a molecule as complex
as RNA by prebiotic processes appears quite implausible
(Shapiro, 1984). The advocates of an RNA world have usually presumed the opposite: Dickerson (1978) wrote, for
example: "It is not difficult to account for the appearance of the bases and sugars of nucleic acids on the primitive earth...current knowledge of the chemistry by which
amino acids, bases, sugars and other monomers of life could
have been synthesized on the primitive earth is really
rather impressive." Similarly, Eigen and Schuster (1978)
stated: "Here we simply start from the assumption that
when self-organization began, all kinds of energy-rich material were ubiquitous, including in particular: amino
acids in varying degrees of abundance, nucleotides involving the four bases A, U, G, C, polymers of both preceding
classes...having more or less random sequences."
     Unfortunately, the literature cited in these works in
no way sustains the claims for the facile prebiotic synthesis of RNA or its building blocks. The sugar D-ribose, for
example, is a vital unique component of the RNA backbone.
The formose reaction (Mizuno and Weiss, 1974) has been the
only one cited as a source of prebiotic ribose. This reaction requires implausibly high (for prebiotic purposes)
formaldehyde concentrations, and produces ribose only as a
short lived component in a complex mixture (Shapiro, 1987).
The complexity of the furmose mixture is so great that it
has been termed "a carbohydrate analog of petroleum"
(Weiss, et al., 1970). One research group that has been
closely involved with this field concluded: "We do not believe that the formose reaction as we and others have carried it out is a reasonable model for the prebiotic accumulation of sugars" (Reid and Orgel (1967))
     Further difficulties arise when nucleoside synthesis
is considered. The formose reaction would be inhibited if
nitrogenous substances such as the RNA bases and their precursors were present. If they were absent however, ribose
destruction would take place without nucleoside formation
(Shapiro, 1987). To circumvent this difficulty, the advo-

cates of a prebiotic origin for RNA have proposed scenarios
that include a separate origin for ribose and the bases
(Schlesinger and Miller, 1973), and transport by glaciers
to bring them together (Schwartz, 1981, Schwartz, et al.,
1982).

The skeptic's alternative to these exercises in imagination is the presumption that RNA first arose later in
evolution, through biological catalysis. Joshua Lederberg
wrote in 1960: "There is some controversy over whether nucleic acids were the first genes, partly because they are
so complex, partly because their perfection hints at a period of chemical evolution rather than one master stroke."
Subsequent work did not alter this outlook. Freeman Dyson
(1985) said in summary a generation later: "The results of
thirty years of intensive chemical experimentation have
shown us that the pre-biotic synthesis of amino acids is
easy to simulate but the pre-biotic synthesis of nucleotides is not. We cannot say that the pre-biotic synthesis of nucleotides is impossible. We only know that if
it happened it happened by some process which none of our
chemists has been clever enough to reproduce." This conclusion had also been reached by Cairns-Smith (1982): "The
importance of this work lies, to my mind, not in demonstrating how nucleotides could have been formed on the
primitive Earth, but in precisely the opposite: these experiments allow us to see, in much greater detail than otherwise would have been possible, just why prevital nucleic
acids are highly implausible."

If we assume that nucleic acids arose during the
course of biological evolution, then some other chemical
substance preceded them as the genetic material, and was
subsequently replaced. Cairns-Smith (1982) has termed this
process "genetic takeover." Various suggestion have been
made for the prior genetic material, including proteins and
related substances (Fox and Dose, 1977), simpler RNA-like
materials (Joyce, et al., 1987) and clay minerals (Cairns-
Smith, 1982). The issue remains unresolved.

5. POSSIBILITIES FOR LIFE BEYOND EARTH

Our ignorance of the chemistry involved in the origin of
our own type of life should make us cautious in placing re-
strictions on this process if it should occur elsewhere.
In an analysis of possibilities for the start of life in a
variety of environments, physicist Gerald Feinberg and I
(1980) were able to deduce only three necessary conditions
for the process: (1) a flow of free energy, (2) a system
of matter capable of interacting with the energy and using
it to become ordered, and (3) enough time to build up the
complexity that we associate with life. In our analysis,

alternative life forms become worthy of consideration; for
example life operating in non-aqueous solvents such as ammonia or hydrocarbons, and life based on mineral systems,
rather than on organic chemistry.

Such suggestions are speculations, and can be reflected or confirmed only by experiment and observation.
From this viewpoint, however, a number of locations in our
solar system meet the above conditions and are worthy of
detailed investigation. One obvious candidate for further
study is Mars.

The three Viking life-detection experiments had been
designed to detect metabolic processes typical of organisms
on earth. With this limitation in mind, Sagan and Lederberg wrote (1976), in advance of the Viking results, that
"while Viking 1976 represents a significant first step in
the in situ search for a Martian biology, negative results
would exclude an important subset, but only a subset, of
the possible classes of Martian organisms." They recommended that subsequent missions "should have a wider range
of biology experiments, able to search for a broader spectrum of conceivable Martian organisms."

Criteria for negative results had been drawn up in advance for each of the life-detection experiments on Viking,
using lunar and sterilized earth soils. None of the three
experimental results matched these criteria. One gave a
positive result (it provided data similar to that given by
microbial soils on earth), while the other two afforded results that had not been anticipated in advance. The mass
spectrometer on Viking detected no organic compounds in the
Martian soil. This result eliminated the subset of microorganisms that live in soil rich in organic debris (a
situation common on earth).

Taken together, these results are ambiguous and far
from definitive. Possible, the results may be explained
entirely in the basis of inorganic chemistry. However, a
number of possibilities for Martian life remain open
(Feinberg and Shapiro, 1980): (1) Life existed in the samples tested by Viking, and was responsible for a portion of
the life-detection results. Either microorganisms based on
carbon existed in the sample, but were not accompanied by
organic debris in amounts sufficient for detection by the
mass spectrometer, or mineral life of the type suggested by
Cairns-Smith was present. (2) Life was absent in the
Viking samples, but was present nearby, for example within
the rocks or deep in the soil. (3) Life was absent at the
Viking sites, but was present elsewhere on Mars, in under-
ground oases of water, at the edge of the icecap, or other
specialized locations. Believers of various persuasions
have chosen to interpret the Viking results in terms of the
absence (most commonly) or presence (occasionally) of life
on Mars. The skeptic prefers to draw no definite conclu-

sion, but awaits the results of further biological investigation of Mars.

Mars of course is not the only site in the solar system worthy of exploration for life, nor is it necessarily the most promising one. Europa may contain a vast ocean beneath its icecap (Squyres, et al., 1983). If this feature were combined with the presence of volcanic vents on the ocean floor, then a possible environment for the generation of life would exist. Sites of this type have in fact been suggested for the origin of life on earth (Corliss, et al., 1981). Other suitable targets for biological exploration include Titan with its dense atmosphere and possible hydrocarbon ocean, and Jupiter. The latter planet may contain sites within its clouds, in which water is present at temperatures and pressures close to those on earth (Feinberg and Shapiro, 1980). It is possible to speculate, or make dogmatic declarations about may or may not be found in each of the above locations, but the most valid scientific course would be to send instruments or observers there, and look. Unfortunately, the possibilities for doing this have grown less favorable in recent years.

6. A SUGGESTED GOAL

Positions of faith are often coupled with optimism, and skepticism with pessimism, but the reverse is true in the present situation. The questions of life's origin and its place in the universe are among the most exciting in all of science. the believer's view outlined above has been publicized widely, and quenched enthusiasm for those possible goals of the planetary program that would draw the widest support and interest. As a result, this program has ground to a halt in the 1980's, amidst widespread public apathy. This situation is best illustrated by the progress of Project Galileo, the only planetary mission currently scheduled by the United States. The launch of the Galileo spacecraft, planned in the 1970's and first scheduled for 1982, has been delayed many times. The spacecraft will most likely depart in the 1990's and then it will proceed to Jupiter by an indirect route, much slower than the one originally selected.

A different climate could follow if the skeptic's view were generally adopted, and the possibility of life in the solar system were left open. An inspirational and unifying goal is badly needed to stimulate further planetary exploration, and the following one is suggested: the most significant scientific goal in the exploration of our solar system is the search for signs of life. Such signs could include existing life (Europa, Mars, Jupiter), extinct life

(Mars), chemical systems evolving in the direction of life (Titan) and artifacts of intelligent extraterrestrial life.

Such a program need not exclude more conventional science. If public interest were restored, more missions could be run, and much more learned in many areas than at present.

Cairns-Smith, A.G.: 1982, *Genetic Takeover*, Cambridge Univ. Press, Cambridge, England.

Committee on Planetary Biology and Chemical Evolution, Space Science Board, Assembly of Mathematical and Physical Sciences, National Research Council, 1981: *Origin and Evolution of Life-Implications for the Planets: A Scientific Strategy for the 1980's*, National Academy of Sciences, Washington D.C.

Corliss, J.B., Baross, J.A. and Hoffman, S.E.: 1981, *Oceanol. Acta*, **Special Issue**, 59.

Darnell, J.E. and Doolittle, W.F.: 1986, *Proc. Natl. Acad. Sci. USA*, **83**, 1211.

Dickerson, R.E.: 1978, *Scientific American* **239**, September, 70.

Dyson, F.: 1985, *Origins of Life*, Cambridge Univ. Press, Cambridge, England.

Eigen, M. and Schuster, P.: 1978, *Naturwissenschaften* **65**, 341.

Feinberg, G. and Shapiro, R.: 1980, *Life Beyond Earth; The Intelligent Earthling's Guide to Life in the Universe*, Morrow, New York.

Fox, S.W. and Dose, K.: 1977, *Molecular Evolution and the Origin of Life*, Revised Ed., Marcel Dekker, New York.

Gilbert, W.: 1986, *Nature* **319**, 618.

Horowitz, N.: 1976, *Acc. Chem. Res.* **9**, 1.

Horowitz, N.: 1986, *To Utopia and Back*, W.H. Freeman, New York.

Joyce, G.F., Schwartz, A.W., Miller, S.L. and Orgel, L.E.: 1987, *Proc. Natl. Acad. Sci. USA* **84**, 4398.

Lederberg, J.: 1960, *Science* **132**, 393.

Lovelock, J.: 1986, *Nature* **320**, 646.

Mizuno, T. and Weiss, A.H.: 1974, *Advan. Carbohyd. Chem. Biochem.* **29**, 173.

Papgiannis, M.: 1986, *Abstracts, the Fifth ISSOL Meeting and the Eighth International Conference on the Origin of Life*, Berkeley, California, 017.

Reid, C. and Orgel, L.E.: 1967, *Nature* **216**, 455.

Sagan, C. and Lederberg, J.: 1976, *Icarus* **28**, 291.

Schlesinger, G. and Miller, S.L.: 1973, *J. Amer. Chem. Soc.* **95**, 3729.

Schwartz, A.W.: 1981, in E.K. Duursma and R. Dawson (eds.), *Marine Organic Chemistry*, Elsevier, Amsterdam, pp. 7-30.

Schwartz, A.W., Joosten, H., and Voet, A.B.: 1982, *Biosystems* **15**, 191.

Shapiro, R.: 1984, *Origins of Life* **14**, 565.

Shapiro, R.: 1986, *Origins: A Skeptic's Guide to the Creation of Life on Earth*, Summit, New York.

Shapiro, R.: 1987, *Origins of Life*, **18**, in press.

Squyres, S.W., Reynolds, R.T. and Cassen, P.M.: 1983, *Nature* **301**, 225.

Thomas, L.: 1978, *New York Times*, July 2, 1978.

Wald, G.: 1973, in R. Berendzen (ed.) *Life Beyond Earth and the Mind of Man*, NASA Scientific and Technical Information Office, Washington D.C., p. 15.

Wald, G.: 1974, in J. Oro, S.L. Miller, C. Ponnamperuna and R.S. Young (eds) *Cosmochemical Evolution and the Origins of Life*, Vol. I, D. Reidel, Dordrecht, the Netherlands, pp. 15-27.

Weiss, A.H., LaPierre, R.B., and Shapiro, J.: 1970, *J. Catalysis* **16**, 332.

HYDROTHERMAL ENERGY FLOW OF PLANETARY BODIES AND THE CREATION OF LIVING SYSTEMS

J.B. Corliss

Department of Atomic Physics, Eötvös University
Puskin 5, Budapest 1088, Hungary
Department of Chemistry, Georgetown University
Washington D.C. 20057, USA

ABSTRACT. An understanding of the events leading to life on Earth, presently our one observable example of the process, will provide insight into the possibilities of life elsewhere in the Universe. Life on planet Earth is here characterized as an attractor emerging within the natural hierarchy of dissipative systems which transfer heat from the planetary interior. The key elements in this hierarchy, the submarine hot springs, are examples of the entropy bursts which punctuate the evolution of the universe and lead to the creation of organized structures and systems. The phase space trajectories which the fluid components are constrained by history to follow in these powerful and potent flow reactors can lead to the assembly of complex organic molecules and organized structures which are thermodynamically and kinetically inaccessible to near equilibrium systems or unconstrained chaotic dissipative systems. This model implies the possibility that life can emerge and be sustained in parts of planetary systems where the flow of solar radiation is negligible, such as the Galilean satellites of Jupiter where tidal heating produces significant heat flow.

1. ENERGY FLOWS IN THE UNIVERSE

The nearly ubiquitous flow of energy outward from centers of gravitational accretion in our universe is essential for creating and sustaining life. The concentration of energy into these accretionary objects - stars, planets, and moons - occurs through several mechanisms. The release of gravitational potential energy during accretion is common to all such bodies. In stars, it provides the high temperature which triggers the fusion reactions which lead to the flow of radiation into the surrounding space. It is this flow of radiant energy which sustains the only example of life we have observed, that on Earth. In planetary bodies with silicate mantles, the accretionary heating can be supplanted by the fission of radionuclides (isotopes of U, Th, K) inherited from the supernova which produced the dust cloud from which planetary systems accreted.

The spacecraft imagery of the Galilean satellites of Jupiter disclosed a spectacular example of another mode of planetary heating. These satellites, Io, Europa and Ganymede, are locked into an orbital resonance which distorts their orbits. The resulting tidal forces deform the solid planet generating enough heat to produce spectacular volcanic activity on Io, and perhaps enough to produce a liquid water ocean under the icy surface of Europa (Squyres, this volume).

If water is present in abundance on a planet which is hot inside, its properties will lead it to an important role in the transfer of this heat to the surface. On Earth, the flow of energy from the interior, derived both from accretionary energy and fission of inherited radionuclides, has created an organized pattern of thermal convection cells within the Earth's core and mantle which drive the plate tectonic evolution of the crust. The Earth cools itself by creating oceanic crust; sea water convection quenches and cools this crust at submarine spreading centers (Fig. 1). The relevance of this process for life in the Universe is based on its relevance to life on Earth.

G. Marx (ed.), Bioastronomy – The Next Steps, 195–200.
© *1988 by Kluwer Academic Publishers.*

2. BIOGENESIS IN SUBMARINE HYDROTHERMAL SYSTEMS ON EARTH

The accretion and early evolution of earth led to the close approach of two huge convecting, long time-constant thermal reservoirs, the oceans and the earths mantle, with a temperature contrast of over 1000°C, creating and driving a system of dissipative structures which transmit thermal energy between them. These periodic attractors, submarine hot springs, first appeared at the end of the Hadean era of the Earth's surface. At this time, about 3.9 billion years ago, gravitational accretion subsided (Taylor, 1979), the crust solidified, the atmosphere cooled, the oceans formed (Smith, 1981), and the cooling of the Earth became dominated by the creation of new oceanic crust (Nisbet, 1984, 1985). At the present time, 25% of the heat flowing from the mantle is transferred through submarine hot springs (Sclater, et al., 1981), and the total volume of the oceans cycles through these hot springs and is heated to ~350°C in ~6 to 20 million years (Jenkins, et al, 1978). Ocean crust formation accounts for over 90% of the Earth's total volcanism in recent Earth history, based on both volume of magma produced (Horai and Uyeda, 1969), and the heat released by cooling of this magma (Sclater, et al., 1981).

The model for biogenesis in these systems (Corliss, et al., 1981) can be outlined as follows:

(1) At the cracking front (Lister, 1974)(Figs. 1 and 2) sea water is heated rapidly to ~600°C, carbon is extracted from the rock, nitrogen form the atmosphere is provided by sea water, and reactions with ferrous iron in the rock produce a reduced fluid containing low concentrations of methane (Welhan and Craig, 1983), ammonia (Van Damm, et al., 1983) and hydrogen. This fluid reacts and equilibrates with the rock as it mixes with cooler sea water to about 350°C. This equilibrium fluid then rises, either (a) directly to the sea floor, emerging as "smoker" plumes, or (b) mixing in the rock with cold sea water to emerge at ~5-30°C, following a highly constrained mixing trajectory. In the "smoker" plumes, the dissipation of energy occurs in the chaotic mixing of the plume; the constraint imposed by the mixing trajectory in the rock is more interesting for biogenesis.

(2) It is suggested that high-energy organic monomers would be synthesized in the fluid at or near the cracking front, and then rapidly quenched by mixing, in analogy to the Miller-Urey reactions. The mixing trajectory followed by the fluids allows the "freezing" of thermal energy into the high energy bonds of these molecules.

(3) The convective flow solves the serious problem of short-lived intermediates (hydrogen cyanide, sugars, etc) in prebiotic synthesis. The rapid rise of the fluids allows very short-lived intermediates to participate in the hierarchical chain of synthesis reactions.

(4) Fractures in the upper parts of the hot spring flow reactors (Fig. 1) are the optimal geochemical environment for the synthesis of clay minerals (saponite, nontronite, ...). They can extract and accumulate organic matter, thus solving the problem in prebiotic synthesis of achieving high concentrations of organic matter from an initially very dilute aqueous solution.

(5) These fractures provide a space where the organic components can leave the mixing trajectory and enter a fluctuating steady-state. The catalytic surfaces of clay minerals can provide initial information to organize organic components in the fluids, creating high concentrations of various random polymers. The steady flow provides a constant supply of energy-rich molecules, including simple high-energy organic molecules as well as a disequilibrium assemblage of oxidized and reduced inorganic molecules and ions.

(6) Eigen (1971) has proposed that such a flow reactor is the site of the emergence of autocatalytic self-replicating cycles of interacting protein and nucleic acid fragments, as attractors in the phase space, which evolve to form "quasi-species" of information-bearing molecules. The flow constitutes a dilution flux, a necessary condition for competition and evolution of the "quasi-species".

(7) It is reasonable that lipid vesicles (Deamer, 1986)(or proteinoid protocells) might form in such an environment as stable equilibrium structures, attractors in the phase space. Such permeable, bounded volumes in a steady state flow reactor are examples of the classic reaction-diffusion systems which have been the focus of the physical and thermodynamic description of self-organizing behavior (Turing, 1952, Nicolis and Prigogine, 1977). Spherical organic structures are common in Archaean

Figure 1. The submarine hot spring flow reactor. High temperature reduced fluid components produced by sea water-magma reaction at the cracking front flow upward into cool clay-lined cracks which extract organic components. Primitive cells evolving in these fracture flow reactors are carried into the redox gradient of hydrothermal mounds at the interface with the surface environment on the sea floor. In shallow seas on the Archaean earth, they were exposed to a flux of UV-filtered solar radiation.

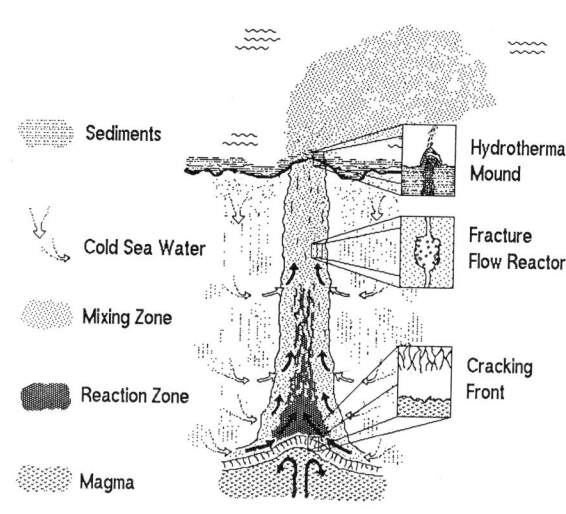

Sediments

Cold Sea Water

Mixing Zone

Reaction Zone

Magma

Hydrothermal Mound

Fracture Flow Reactor

Cracking Front

Figure 2. Trajectories of hot spring fluids in pressure-enthalpy space. The fluids descend at low temperature [1], are drawn into the cracking front [2], heated to high temperature [3], and either rise to emerge in smoker vents [4], or are quenched by mixing to emerge at low temperature [5]. The rapid heat-quench cycle freezes thermal energy into high energy bonds. The steady-state flow reactors [5] provide the competitive environment necessary for the evolution of imperfectly self-replicating primitive biological structures.

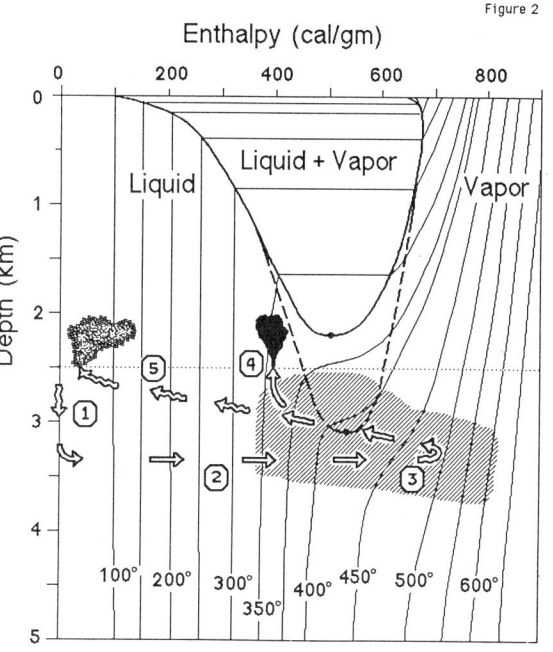

Enthalpy (cal/gm)

submarine hot spring chemical sediments (e.g. Knoll and Barghoorn, 1977).

(8) The growth and replication of such primitive cells in such a flow reactor would lead to selection and evolution of the most rapidly growing "quasi-species" of cells (Feistel, 1983); i.e. selection of those entities at the hierarchical level of cells which contain the most efficient self-replicating entities at the hierarchical level of molecules. The feedback cycle between between entities at these two levels, between code and structure, is fundamental to the hierarchical evolutionary ascent to present-day life on earth.

3. THE SEARCH FOR THE REMOTE DOMAIN OF LIFE

The hypothetical self-replicating cells of Archaean hot springs lie in a remote domain of the phase portrait which describes all of the possible microstates of that particular combination of elements and energy flows, which is in turn a remote domain of the phase portrait of all possible combinations of matter and energy in the universe. The question of the origin of life is the question of how a collection of matter in a natural system was led through a sequence of events into the limited domain in phase space where a living cell could appear as a stable system; as an attractor. The hot springs are powerful dissipative systems which export a significant quantity of entropy to their environment. Thus it is thermodynamically possible, within the constraints of the Second Law (Prigogine, 1955), for low entropy structures to emerge within them. In what sense is the domain of life remote? What are the qualities of the hot springs which could lead matter into this remote domain?

The remoteness of life can be stated (Wicken, 1987) in terms of the improbability of both: (1) the energetic state of the molecules in the cell, and (2) the informational state" of these molecules. Morowitz (1968) demonstrated the infinitesimal probability of the energetic state of the molecules in a living cell - the potential energies of their electrons - by comparing the energy of an equilibrium distribution of the relevant atoms of a cell (C,H,N,O,P,S) with the bond energies of those same elements as they occur in an *E. Coli* cell. He argued that living systems could not have emerged as a fluctuation in an equilibrium system, but only in a flow of energy which raised the average electron potential energy above equilibrium values. Yockey (1977) points out the similar infinitesimal probability that a biologically meaningful sequence of amino acids or nucleotides could have arisen as the product of a random process. Biological structures are far from equilibrium and biological information is far from random.

The key phrases are "equilibrium system" and "random process"; both of these concepts are mathematical idealizations which do not exist in the real world. Real systems are dissipative systems; real processes are "informed". As Wicken (1987) points out : "There is no a priori connection between dissipation and structuring. The reason the two tend to be coupled... is that the the forces of nature are for the most part associative ones. ...The potential energy wells into which natural processes tend to flow are correlated with the buildup of structure".

These "potential energy wells" are, in more general terms, "basins of attraction". Dissipative system subject to a flow of energy typically go through evolutionary events which lead from one stable state to another; in mathematical terms, they pass through a sequence of bifurcations which separate basins of attraction. The events which have the greatest potential for creating new structures are those in which a large amount of energy is rapidly dispersed; rapid heating followed by rapid cooling, providing a burst of entropy production and the emergence of high-energy structures. One example is the origin of the Universe itself in the Big Bang and the subsequent emergence of matter from energy; another is supernova explosions and the resultant accretion of the dispersed mass into new stars and solar system. In submarine hot springs, the trajectory of the fluids through the cracking front and upward through the mixing gradient provides such an evolutionary event, storing thermal energy in equilibrium structures, e.g. complex organic molecules. This is the trajectory of the Miller-Urey experiment (Miller, 1953). It leads matter close to the life attractor.

The next and final step in the search is involved with the emergence of dissipative systems within this highly constrained energy flow, specifically in the fluctuating steady-state fracture flow reactors in the cool upper parts of the hot springs (Fig. 1). Studies of the dynamics of replicating systems suggest that once a collection of self-replicating dissipative organic structures emerges, a vast new terrain of bifurcations and basins of attraction becomes accessible for exploration. The diversity of the members of such a population can lead to populations which can learn(Allen and McGlade, 1987), and thus the information barrier is surmounted, and the threshold from the non-living to the living is crossed.

4. THE EVOLUTION OF HOT SPRING LEARNING SYSTEMS

The learning required of primitive living systems can be characterized as a simulation by one hierarchical dissipative system of another with which it is in communication (J. Nicolis, 1986). The stream of information which enters the primitive cell in the fracture flow reactor (Fig. 1) consists of the fluctuations in concentrations, temperature, flow rate, etc. of the fluid, and the properties of the molecules which might penetrate the cell boundary; e.g. hydrogen sulfide and carbon dioxide. The collective soup within classes of replicating primitive cells evolves, in response to the competitive pressure of the dilution flux and the resultant selection at the cellular level, so as to increase the growth and replication rate of the cells in which it is confined. The information passed on at replication is carried in the conformational properties of the molecules of the evolving autocatalytic-metabolic networks in which the simulation of the external environment resides.

These primitive cells are the progenotes of Woese (1987); "an entity that, by definition, has a rudimentary linkage between its genotype and phenotype". Its proteins and nucleic acids (RNA) form classes with relatively large errors in replication (compared to those of modern organisms) which leads to diversity, precisely the requirement for learning (Eigen, 1971; Allen and McGlade, 1987). Given this model for the beginning of Darwinian evolution, we can search for the connections with research which follows the evolutionary tree from modern organisms back to its roots. The new revolutionary understanding of bacterial evolution, and evolution in general for that matter, provided by the work of Woese (1987) and his colleagues, clearly shows, I believe, such a connection.

The hot spring progenote could readily diffuse through the permeable ocean crust into the variety of environments provided by convective circulation and mixing. They would also be carried upward into the interface with the surface environment, where reducing hot spring fluids diffusing vertically through rock fractures or sediment interact with sea water (Fig. 1). Chemical deposits typically form hydrothermal mounds or chimneys at this interface. If the flux of photolytically produced oxygen extends into sea water, even at very low concentrations, as suggested by the oxidized nature of Archaean hot spring deposits, a redox gradient would be established within these deposits providing a variety of redox reactions suitable for mediation by micro-organisms. The flow of heat and fluid components and the dissipation of chemical energy - and radiant energy if the sea is shallow - in this space, establish a potent dissipative structure, the prototype of modern laminated microbial communities.

The hot spring affinities of the modern archaebacteria suggest that they are the descendents of the progenotes which adapted to various parts of the hot spring environment. The mixing of oxidized and reduced fluids provides all of the appropriate electron donors and acceptors necessary for the various chemosynthetic metabolic pathways of archaeabacteria. The halobacteria could evolve from methanogens in shallow hydrothermal mounds. "The most prominent phenotypic characteristics of the eubacteria tree are its metabolic diversity and the widespread distribution of anaerobic, photosynthetic and thermophilic phenotypes" (Woese, 1987) This versatility and diversity reflects well the environment of the hydrothermal mounds to be simulated by the progenote. These mounds would also be the sight of the emergence of the next hierarchical level of living systems, the ecosystem, in which a new collective behavior of organisms emerges. The symbiotic relationships established could lead to the formation of the eucaryotic cell (Margulis, 1981) in this setting.

5. CONCLUSION

The domain of life is remote only from systems near equilibrium; that is to say, in systems without a history. The particular sequence of trajectories which appeared in hot springs are necessary behaviors imposed by their history, constraints imposed by the history of the microscale entities within the hot springs, and the higher level systems of the hierarchy of which the hot springs are a part, and so on, backward in time, and upward in the hierarchy.

But entwined with this necessity is also the operation of chance. Even with the unique and seemingly ideal constraints on the Archaean hot springs imposed by the evolutionary history which preceded them, the synchronicity required to bring together all the necessary elements in a single hot spring seems intuitively improbable. But the pulsations of magma injection and sea water convection which were the sea water hot springs of the Hadean-Archaean oceans occurred along thousands of kilometers of mid-ocean ridge for millions of years. Each experiment was unique, and this diversity provided a "search" through the relevant phase space, and this search led to the remote attractors which provided for the creation of the living cell. And the improbable occurred.

In a similar sense, the countless planetary systems which presumably (Campbell, this volume) emerged as attractors in the evolution of cosmic dust clouds also represent such a search through a phase space. In our quest for life elsewhere in the Universe, we seek a second example of a system which located this domain of life in which we reside. This second example must necessarily have some of the same general features as that search which was successful on the Archaean Earth.

REFERENCES:

Corliss J. B., J. Baross, and S. E. Hoffman (1981) Proc. 26th International Geological Congress, Paris, 1980, Oceanol. Acta: Special Issue, p. 59-69.
Deamer, D. W. (1986). Origins of Life, **16.**
Eigen, M. (1971) Naturwiss. **58,** 465.
Feistel, R. (1983) Studia Biophys. **93,** 113.
Horai, K. and Uyeda, S. (1969). In: Hart, P. J. (ed.) The Earth's Crust and Upper Mantle. Geophys. Monogr. Ser., Vol. 13, Amer. Geophys. Un., Washington, D. C.
Jenkins, W. J., J. M. Edmond, and J. B. Corliss (1978) Nature, **272,** 156.
Knoll, A. H. and Barghoorn, E. S. (1977) Science, **198,** 396.
Lister, C. R. B. (1974) Geophys. J. R. Astron. Soc. **39,** 465.
Margulis, L. (1981) Symbiosis in Cell Evolution, W. H. Freeman, San Francisco.
Marx, G. (1984). Proc. Genl. Assem. European Phys. Soc., August, 1984, Prague.
Miller, S. L. (1953). Science, **117,** 528.
Morowitz, H. (1968) Energy flow in Biology, Ox Bow Press, Woodbridge, Connecticut.
Nicolis, G. and I. Prigogine (1977) Self-organization in Nonequilibrium Systems, Wiley.
Nicolis, J. S. (1986) Dynamics of Hierarchical Systems: An Evolutionary Approach, Springer.
Nisbet, E. G. (1985). J. Mol. Evol. **21,** 289.
Nisbet, E. G. (1984). Can. J. Earth Sci. **21,** 1426.
Prigogine, I. (1955) Introduction to the Thermodynamics of Irreversible Processes, Wiley.
Sclater, J. G., B. Parsons and C. Jaupert (1981). J. Geophys. Res. **86B,** 11,535.
Smith, J. V. (1981). Phil. Trans. R. Soc. Lond. **A301,** 401.
Taylor, S. R. (1979). In: McElhinny, M. W. (ed.), The Earth, Academic Press.
Turing, A. (1952) Phil. Trans. Roy. Soc. London, **237B,** 37.
Von Damm, K. L., B. Grant and J. M. Edmond (1983) In: Rona, P. A., et al. (eds.) Hydrothermal processes at sea floor spreading centers, NATO Conf. Series, Ser. IV: Mar. Sci. Vol 12. Plenum.
Welhan, J. A. and H. Craig (1983) In: Rona, P. A., et al., previous citation.
Wicken, J. S. (1987) Evolution, Information, and Thermodynamics, Oxford.
Woese, C. R. (1987) Microbiol. Rev., **51,** 221-271.
Yockey, H. (1977) J. Theor. Biol. **67,** 377-398.

SURVIVAL STRATEGIES FOR LIFE
IN HIGH UV, VERY LOW DENSITY ENVIRONMENT

G. Horneck
Institute of Aerospace Medicine DFVLR
P.O. Box 906058, Cologne 5000 F.R.of Germany

ABSTRACT. During the last 20 years, from laboratory studies and in-flight investigations on the response of microorganisms to space environment, it has been found, that desiccated systems can survive short-time exposure to free space, if shielded against solar UV radiation. Simultaneous treatment with solar UV radiation and space vacuum exerted a synergistic interaction in cell killing. From action spectroscopy for lethality and formation of photoproducts in the DNA the hypothesis has been supported that "vacuum-specific" less reparable photoproducts are causally involved in this synergistic response. These data shed some light on the likelihood of an interplanetary tranfer of life.

During its more than 3.5 Ga lasting history on Earth, life has evolved to the present day manifoldness and distribution. Today life has spread over the entire surface of the globe, it has conquered nearly every niche. Microorganisms are teeming throughout the biosphere, as a consequence of their ready dissemination by wind and water, of their short generation time which implies fast adaptation to a new environment by mutation and selection, and of the ability of some of them, to survive unfavorable conditions in a more resistant, interim dormant state, the spore.

The bacterial endospore can be considered as a model system of a cryptobiotic cell that is extraordinarily resistant to harsh environmental conditions, such as heat, radiation, or desiccation, respectively. Its resistance mechanisms towards environmental extremes appear to be based on changes in cell structure and composition, associated with dehydration and specific mineralization of the cell protoplast /1/. Bacterial spores are more resistant than vegetative bacteria to UV-radiation. The photoproducts that accumulate in spore DNA are different from the cyclobutyl pyrimidine dimers, the main photoproduct produced in vegetative cells. The spore photoproduct (5-thyminyl-5,6-dihydrothymine, TDHT) is preferentially formed in spore DNA after transition of the DNA from the wet state (of vegetative cells) to the dry state (of spores) /2/.

G. Marx (ed.), Bioastronomy – The Next Steps, 201–205.

The question arises, whether resistant microbial systems, such as bacterial spores, would also be capable to withstand the hostile environment of free space, if they are accidentally subjected to it. In order to cope with such a new and extreme environment, the "space travellers" would be requested to be endowed with a complete set of resistance mechanisms against the concerted action of the environmental factors of free space. The space environment represents a definit barrier for any active biological process, such as growth, metabolism and reproduction. This is attributed to the high vacuum yielding pressures down to 10^{-14} Pa, the solar electromagnetic radiation reaching from 2×10^{-12} to 10^2 m wavelengths – of special concern is the solar UV-radiation below 300 nm –, the solar corpuscular radiation, emitted in solar wind and during solar flares, the galactic cosmic radiation, comprising approximately 86% protons, 12.7% He-ions, 1% electrons and 1.3% heavy ions with charge $Z>2$, and the extreme temperatures of the body determined by the deep space temperature of 4K and by its position towards the sun. The effect of any one factor can rarely be considered in isolation, because there may be antagonistic or synergistic influences. Microorganisms which succeed in surviving in such an extreme environment will thus be those that are resistant to the complex interplay of all relevant factors determining that milieu.

With the development of spaceflight, the opportunity arose to investigate the problem of survival of resistant forms of microorganisms in space directly, either by attempting to collect microbial samples in higher altitudes or by sending bacterial and fungal spores into the upper atmosphere and into space and analyzing their response to the space environment or to certain related factors after recovery /3/. By using meteorological rockets expecially equipped with a microbe sampling device, viable microorganisms at altitudes up to 60-70km were collected /4/. They were predominantly black conidia and spores of fungi. It is assumed that pigmentation offers a selective advantage for the spores, because it protects them against the intense solar UV-radiation.

For microorganism, directly exposed to free space, solar UV-radiation especially of wavelengths in range around 260nm turned out to be the most deleterious factor of space. This spectral part is specifically absorbed by the DNA. Bacterial spores, that were simultaneously exposed to solar UV-radiation and space vacuum, responded with a significantly increased sensitivity against UV /4/. This phenomenon indicates a synergistic action of space vacuum and solar UV on bacterial spores. The UV-sensitivity to the full spectrum of solar UV (>170nm) as well as to selected wavelengths in the biologically effective range between 220nm and 280nm in space vacuum was increased by a factor of up to 10 (Fig. 1) /6/. This UV-supersensitivity of spores in space vacuum is due to the production of special photoproducts, which are less accessible to cellular repair pathways or completely unreparable. These are, in addition to the spore-photoproduct TDHT, trans-syn isomer of pyrimidine dimer and DNA-protein-crosslinking.

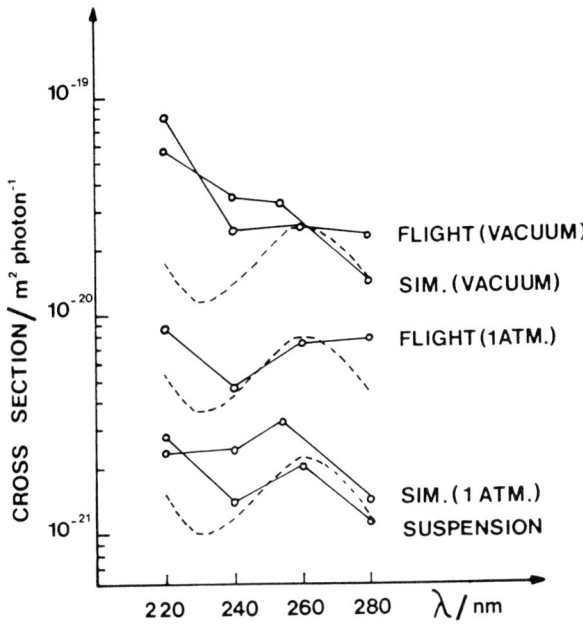

Figure 1. Action spectrum of UV-inactivation of spores of Bacillus subtilis, UV-irradiated in dry monolayers at atmospheric pressure or in vacuum; data from Spacelab 1 space experiment and simulation experiment on ground; DNA absorption spectrum (dashed line) (from Ref. 8).

These results indicate that even the most radiation resistant biological systems in free space will be completely inactivated within minutes by the synergistic action of solar UV and space vacuum. However, recent laboratory experiments have shown that at the temperature of 10K that prevails in interstellar dust, bacterial spores are extremely more resistant against UV-radiation that at room temperature /7/.

The chance for survival of microorganisms in free space is highly increased by an adequate shielding against solar UV-radiation. Practical shielding may be achieved by embedding the spores in dust or clay material or by shadowing. However, damage to genetic material is already produced by space vacuum alone, indicated by an increase in mutation frequency, in growth delay and DNA-protein-crosslinking /8/. Nevertheless, some bacterial spores can tolerate this vacuum-induced impairment, probably by DNA repair mechanisms starting during germination, which enables them to withstand extended vacuum desiccation without any remarkable loss in viability. B.subtilis spores survived vacuum treatment (app. 10^{-6} Pa) for up to 6 months.

The ultimate limit for survival of spores in space may be set by the heavy ions of cosmic radiation. Effective shielding is practically impossible. Due to lack of repair processes in dormant systems, primary radiation damage will accumulate with time. Assuming an average size of a spore of $1\mu m^3$ the maximum time that a spore in space could escape a hit from a cosmic iron particle, e.g., may be 10^5-10^6 years.

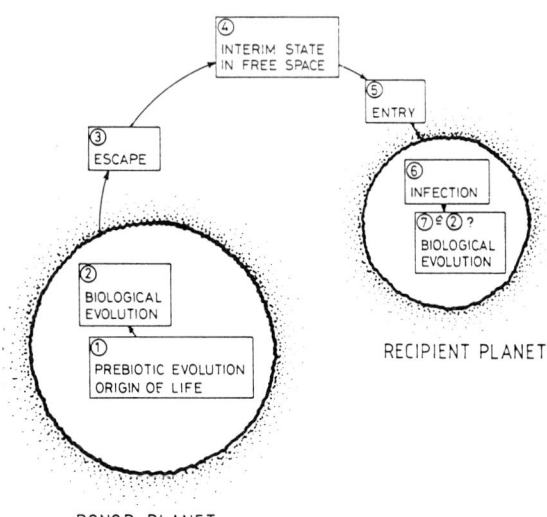

Figure 2. A model scenario for various steps of interplanetary transfer of microorganisms.

A model scenario can be imagined of a potential interplanetary transfer of resistant microorganisms, such as bacterial spores (Fig. 2). Under the aspects of current knowledge of comparative planetology, solar system chemistry and physics, abiotic organic chemistry, fossil findings and comparative studies on present-day microorganisms, a stepwise analysis allows to separate open questions accessible to experimental investigation from those of more speculative nature. So far, the Earth is the only planet known to support abundant life. Comparative planetology will increase our understanding, whether the chain of events towards the appearance and distribution of life on a planet is a universal phenomenon or a unique event.

To accelerate small particles of the size of a microorganism or small rock fragments to escape velocities on an Earth-like planet, regular dynamic forces, like meteorological drifts, thermal movements, magnetic or electric fields, or solar radiation pressure are insufficient /9/. However, meteorites of lunar and probably of Martian origin, collected from the Antarctica, indicate that matter can be exchanged by natural processes between bodies of our solar system. Potential mechanisms are vulcanic eruptions, fly-by meteorites or meteorite impacts /10/. Today, planetary missions provide man-made modes for such transport.

The limits for life in an anabiotic state to survive the unfavorable
conditions of space can be deduced from in situ experiments in space
and from simulation experiments on ground. However, one has to bear in
mind, that all these exposure experiments, since performed with contemporary organisms, cannot easily be deduced to a scenario several
10^6 or 10^9 years ago. For the subsequent steps requiring a safe
capture by another celestial body and the inition of a new chain of
biological evolution we are lacking of any proof.

This model scenario of interplanetary transfer of life, that is closely related to the hypothesis of panspermia, first formulated by S. Ar-
rhenius /11/, is still contingent on many ifs and buts. We do not
know, wether "panspermia" was a likely process in the history of our
solar system, nor whether it was feasible at all. However, space technology has offered a new tool to investigate some steps, especially
the impact of free space environment on the viability of resistant
microoganisms in a direct approach.

References

1) A.D. Russell, The destruction of bacterial spores, Academic
   Press, 1982
2) W. Harm, Biological effects of ultraviolet radiation, Cambridge
   Univ. Press, 1980
3) G. Horneck, Adv. Space Res. 1, #14, 39-48 (1981)
4) A.A. Imshenetzky, S.V. Lysenko, G.A.Kazakov, Appl. Environm.
   Microbiol. 35, 1 (1978)
5) H. Bücker, G. Horneck, H. Wollenhaupt, M. Schwager, G.R. Taylor,
   Life Sciences and Space Research, 12, 209-213 (1984)
6) G. Horneck, H. Bücker, G. Reitz, H. Requardt, K. Dose, K.D. Martens, H.D. Mennigmann, P. Weber, Science 225, 226-228 (1984)
7) P. Weber, J.M. Greenberg, Nature 316, 403-407 (1985)
8) G. Horneck, H. Bücker, K. Dose, K.D. Martens, A. Bieger, H.D.
   Mennigmann, G. Reitz, H. Requardt, P. Weber, Adv. Space Res. 4,
   #10, 19-27, (1984)
9) H. Bücker, G. Horneck, Life Science and Space Research, 7, 21-27,
   (1969)
10) P. Janle, Proc. 4. DFG Kolloquium über Planetenforschung,
    2.-5.10.1984, Schloß Ringberg, Tegernsee
11) S.Arrhenius, Die Umschau 7, 481 (1903)

ANTICRICK: THE NOTHING IS NOT REGULABLE

T. Gánti
Ecological & Modelling Research Group, Hungarian Academy
Kun Béla tér 2, Budapest 1083 Hungary

ABSTRACT. Regarding the living systems as programcontrolled
fluid automata which have at least three main subsystems,
i.e. the machinery subsystem, the controlling subsystem
and the boundary subsystem, the biogenesis, according to the
chemoton theory, seems to be a necessity, which happens
every where and everytime where and when the circumstances
are suitable for it.

The discovery of Watson and Crick, the double stranded
structure of DNA and the foundamental principles of the
storage and reproduction of the genetic information is
one of the greatest discovery of our century. A new
discipline, the molecular biology has grown up based on
their findings. They found the key of heredity, the golden
key of it.

The Watson-Crick model made possible the revealing of
the big secret: how a chemical compound can hidden in itself
the huge quantity of the hereditary information, how it can
be multiplicated this during the replication by the pairing
rules of the nucleotides.

This model was the one upon which the investigations
were based, that showed how the sequence of the nucleotides
determines the sequences of amino acids in proteins, and
how this sequence determines the tercier structure of the
polipeptid chains making them possible to fulfill their
wonderful regulating role.

This phenomenon is universal in the earth: the
genetic information is carried by DNA, in every living being
and the regulation happens to be done by proteins.

Naturally enough after the disclosing of the abiotic
formation of biologically important small molecules, i.e.
amino acids, organic acids, sugars, etc., the proteins and

207

G. Marx (ed.), Bioastronomy – The Next Steps, 207–211.
© *1988 by Kluwer Academic Publishers.*

nucleic acids became the centre of the research of
biogenesis. This happened about a quarter of a century ago.
These two and a half decades were enough to the birth a
lot of brilliant ideas about the origin of the genetic
information and the genetic code, to the realizing of
innumerable and successful experiments about the spontaneous
formation of proteins and nucleic acids. I stress: they gave
very important results concerning to the biogenesis. But I
also stress: these didn't solve the problem of the biogenesis
and they are not able to solve this problem for themselves.

Why can I say it? Because there is no inventor, who
could invent a key without a corresponding lock. No key can
function without its lock. Neither a golden key can.
Because information cannot function without a user system.
Neither a huge quantity of information can. Because
regulating system does not exist without regulated ones.
Neither a wonderful regulating system does. The nothing is
not regulable!

The living systems consist of three basically important
subsystems: the controlling-regulating subsystem, the
skeleton-boundary subsystem and the fluid machinery, which
is controlled by the genetic information, which is regulated
by the enzymic system and which is compartmentalized by the
boundary system. Within the investigations of the biogenesis,
most of the scientists investigate the formation of the
controlling-regulating subsystem alone, some /very few/
scientist the origin of the compartmentalization. But
nobody investigate the controlled and regulated part, i.e.
the fluid machinery. In other words, the investigations
concentrate to the side of the things fundamentalized by
the Watson-Crick model and nobody concentrate to the other
side, to the opposite side, the "anti-side", to the
machineries of life. The expression of "AntiCrick" refers
first of all to this fact.

If the controlling-regulating subsystem originated
alone it must have been a miracle. In this case, we are
either alone in the Universe, or the life was spread by
directed panspermia, as Crick supposed. But if the controlling regulating subsystem originated together with the
controled-regulated one, the spontaneous formation of
living systems was a necessity, which - according to the
laws of nature - happens everywhere and every time when
and where the circumstances are suitable for it, without
any kinds of panspermia. The "AntiCrick" expression
secondly refers to this fact.

What does "fluid machineries" mean? We are accustomed
to the machineries, which are assembled from hard constructing
parts, i.e. from cog wheels, vires etc. A clock e.g. consists
of several cog wheels, the strictly determined organization
of which makes possible the transfer of the energy from
its winded up spring to its hands, through the series of the

paths geometrically constrained by the copled cog-wheels.

And what is about the oscillating chemical systems which operate like a clock in pure liquid without any type of geometrical organization? And what about the cytoplasm of a wandering amoeba, in which the cytoplasm mixed up and whirl continously during the wandering and even operates not only regulated, but programcontrolled manner? What kinds of organization may exist in them?

There exists organization which is totally independent of the geometrical space. If we imagine an abstract chemical state space e.g. an abstract stoichiometric space, each compound takes its separated place in this abstract space. During the reaction the chemical energy moves from one point to the other of this abstract space. We can transfer the chemical energy from one point of this abstract space to an arbitrarily selected other point by the series of the reaction, i.e. by a selected reaction chain, in just similar manner as the electric energy is transferable from one point of the geometrical space to the other, by wires.

The chemical cycles - e.g. the Krebs cycle or the enzymic reactions - behave in this abstract space as wheels do. The operation of two chemical cycles may be connected by a stoichiometric coupling. In this case they not only behave as the coupled cogwheels do, but their functioning may be described by the same mathematical equation, only the number of the cogs must be changed up by the number of moles of the coupling compound (1,2).

Thinking in this manner, we are able to construct several machineries in the chemical space just in the same manner as the mechanical engineers can construct their mechanical machines in the real geometrical space. These are the so-called fluid machineries or fluid automata, the organization of which is independent of the geometrical space (2). The simplest living systems are pure fluid automata, the developed ones are mixed fluid and mechanical or mixed fluid-mechanical and electric automata.

It is totally impossible even to sketch the theory of fluid automata in some pages , but you can find the detailes in (2). This volume describes the fluid machineries step by step, from the simplest chemical reaction through the oscillating chemical systems to the most complicated, proliferating programcontrolled fluid automata, the so--called chemotons, which behave just as the living systems do. In each step, in each construction the detailed organization manner is given by the detailed topologies, by the exact stoichiometric descriptions and by the detailed kinetics. A newly deverlopped mathematical operation, the so-called cyclic stoichiometry (2,3,4,5) makes possible the exact stoichiometryc description of these machineries by algebraic-like equations.

The fluid automata are not only conceptions and not only theoretically designed and quantitatively described systems. Some of them, also quite complicated ones, have been realized as industrial technologies in the Reanal Factory of Fine Chemicals for producing several sugar phophates and adenozine-diphosphate (2,6,7). Some of these technologies have been patented too (8,9).

Theoretical considerations as well as detailed computer simulations have proved that the models of the proliferating programcontrolled fluid automata i.e. the chemoton model, may be regarded as the abstract models of the most primitive living systems. Having such a quantitatively treatable model of the simplest living systems in mind we can pose the question of the genesis of life in a concrete manner, in a really scientific manner. The second volume of this monograph investigate this question (13).

Making use the valuable experimental results of Miller, Oró, Orgel, Jukes, Fox, Ponnamperuma, Matthews, Moser, Buvet, Halmann and many other scientist of the chemical evolution it is possible, by the aid of the abstract chemoton model, to design concrete prebiotic chemoton models. Such a model is presented in (13). Its metabolic map contains about one hundred intermediates - mostly well known prebiotic compounds. It is totally selfreproductive, i.e. the quantity of each compounds of it doubles during one generation time. After the doubling of the quantity of its compounds it divides into two equal compartmentalized system (14,15). It is controlled genetically by RNA; not by the sequence however, but by the composition of its RNA. The hereditary controll by sequences developped during a next step of the evolution. It is regulated stoichiomet-rically and kinetically, but not catalytically. The catalytic i.e.enzymic regulation also developped during the next step of the evolution. The functioning of the system is also stoichiometriacally calculated in details exactly given how many formaldehyde, cyanide, cyanoacetylene, ammonia, phosphate and water molecules are needed as source materials for the doubling and how many CO_2 and urea molecules are formed as waste materials.

Having such a concrete prebiotic chemoton model we can investigate the possible evolutional pathways of such systems. These investigations showed that the formation of enzymes and genes were not accidental and the formation of sequences also were not random events. The book (13) outlines these evolutional processes u p to the appearance of prokariotes.

This volume is in print and sorry the first volume is printed only in Hungarian presently, with an English summary. But the basic principles of the chemoton theory - unfortunately without the results concerning to the prebiotic evolution, which have been developped later -

is obtainable also in English (16).

REFERENCES

1. T.GÁNTI (1980): 'On the Organizational Basis of the Evolution.' Acta biol.Acad.Sci.Hung. 31 449-459
2. T.GÁNTI (1982): The Chemoton Theory. Vol. I. The theoretical Bases of Fluid Machineries. OMIKK, Budapest (In Hungarian).
3. T.GÁNTI (1976): 'Chemical Systems and Supersystems I. Chemical Cycles.' Acta Chimica Acad. Sci. Hung. 91 357-368
4. T.GÁNTI (1976): 'Chemical Systems and Supersystems II. Stoichiometry of Self-reproducing Chemical Systems.' Acta Chimica Acad. Sci.Hung. 91 369-385
5. T.GÁNTI (1978): 'Chemical Systems and Supersystems III. Models of Self-reproducing Chemical Supersystems: the Chemotons.' Acta Chimica Acad.Sci.Hung. 98 265-283
6. T.GÁNTI (1975): 'Syntheses by enzyme systems of yeasts'. Szeszipar 23 57-62 (In Hungarian)
7. T.GÁNTI, A.CSOKA (1975): 'Preparation of Phospho-enolpiruvic Acid from 3-Phosphoglyceric Acid.' Magy. Kém.Folyóirat. 81 335-336
8. T.GÁNTI, L.PATAI, R.BODOKI, T.PALÁGYI (1970): Hung. Patent. No. 160 180
9. A.CSOKA, T.GÁNTI (1973): Hung. Patent No 166 473
10. T.GÁNTI (1979): A Theory of Biochemical Supersystems and its Application to the Natural and Artificial Biogenesis Akadémiai Kiadó, Budapest, and University Park Press, Baltimore
11. T.CSENDES (1984): 'A Simulation Study on the Chemoton' Kybernetes 13 79-85
12. T.GÁNTI (1978): 'On the Reality of Extraterrestrial Biogenesis' XXIX th IAF Kongress, Dubrovnik (Preprint No IAF-78-A-51)
13. T.GÁNTI (1987): The Chemoton Theory. Vol.II. Theory of Living Systems. OMIKK, Budapest, (In press)
14. T.GÁNTI (1975): 'Organization of Reactions into Dividing and Metabolizing Units: The Chemotons.' BioSystems 7 15-21
15. T.GÁNTI, Cs.GÁSPÁR (1978): 'The Conditions for a Growing Microsphere to Divide Acta Chim.Acad.Sci.Hung. 98 278-283
16. T.GÁNTI (1987): The Principle of Life. 6 th edition. OMIKK, Budapest

THE RIGHT-LEFT ASYMMETRY IN BIOLOGY

Ya. B. Zel'dovich

Space Research Institute, Adademy of Sciences
Profsojuznaja 84/32, Moscow 117810, USSR

I. SITUATION BEFORE PARITY VIOLATION DISCOVERY

The atoms and simple organic substances are optically inactive. They do not rotate the plane-polarized light. There are complicated organic molecules which are active. The simplest examples are built as a carbon atom with 4 valence bounds directed like lines from center of the tetrahedron to its angles. If all four groups $R_1 R_2 R_3 R_4$ bound to the central carbon atom are different, there are two modifications. (Fig.1A, 1B) To prove it, make the upper angle of tetrahedron be R_1 group! The $R_2 R_3 R_4$ are in the lowest plane. Seen from above $R_2 R_3 R_4$ can be located clockwise or anticlockwise (in the second case $R_2 R_3 R_4$ are clockwise). The two configurations – one of which is called R (right), the other L (left) – cannot be transformed one into other by any rotations in space. They are related by plane mirror reflection – but one cannot do it with the rotation of real molecules.

R and L rotate the plane of polarization in opposite directions This rotation can be interpreted otherwise: imagine the plane polarized wave as a superposition of two circular polarized waves of equal amplitudes with the momentary adjustment of R vectors like a right-hand or left-hand corksrewer. A given phase adjustment gives linear polarization.

As long as the refractive indicies for R and L-circular polarized lights are identical, their phase relation remains constant and thus after summing the R and L beams we obtain plane. The R organic substance have different refractive indices for R and L circular polarized light. The phase adjustment between them changes! This is the ultimate cause of the rotation of the plane of polarization. The same is true for L substance – with the sign of R-L refractive index and the rotation being reversed.

213

The other effect, confirmed experimentally, is the difference in photochemical reactions yield for creation of R substance (or decomposition of R) by R and L light. In all cases the effect is rather small, because the difference between circular and random plane polarization of light consists in spatial variation of the electric vector. Therefore the effect is proportional to a/λ where a is the dimension of molecule and λ is the wave length of light.

In the approximation declared above ("no parity violation") the energy of R and L molecules are identically equal. The $R \rightleftarrows L$ transformation does not violate energy or momentum conservation. Therefore in principle the process is allowed and possible. It is called racemisation. But practically in the organic substances the energy of inter-mediate stages with distorted angles between valence bounds is of the order or hundreds kilojoules per mole. The groups are heavy, therefore quantum tunnelling does not occur. At high temperatures the racemisation is induced by thermal motion, but usually it is accompanied by thermal destruction of the substance.

About the thermodynamical equilibrium between R and L: assume that the direct racemization is rapid enough or it goes through some low molecular inactive intermediates: $R \rightleftarrows A+B$, $L \rightleftarrows A+B$ and/or some catalysts are used. Obviously L=R is a solution of thermodynamic and kinetic equations. In 1938 I proved and published a rather obvious theorem: in a closed system with ideal components the equilibrum is unique. Therefore knowing the solution we may be sure that no equilibrium optically active solution exists. The word "ideal" is used above in the sense of ideal gas: no Van-der-Waals gas correction, no Debye-Huckel or other correction in liquid phase.

Now we are proceeding to analyze the violation of the conditions on which the theorem is based.

Let the system be concentrated enough and far from ideal. We shall take into account the physical, molecular interactions. Obviously the interaction of a pair RR is identically equal to the interaction of the pair LL. But the interaction RL between different molecules need not bed the same! Symbolically I(RR)=I(LL)=I(RL). (I for interaction.)

The famous experiment of Pasteur consisted of crystallization of a solution containing R-tartaric and L-tartaric salt in equal quantities. It turned out that the crystals were pure R and pure L. In contrast to the mixed solution, the crystalline phases were pure. The habitus of R crystals i.e. the details of their forms were slightly different from the to L-crystals. A small facet on the R-crystal was located just mirror-like to the same facet of L-crystal. Of course, the very precise view of a genius like Pasteur was needed to observe this small difference.

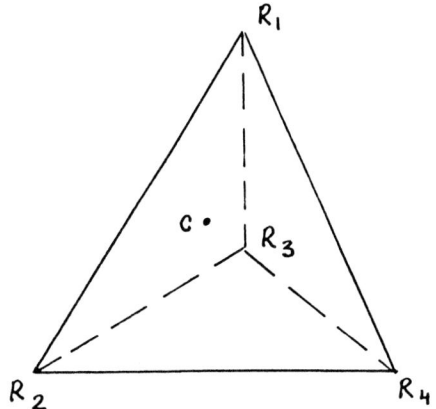

Fig. 1A

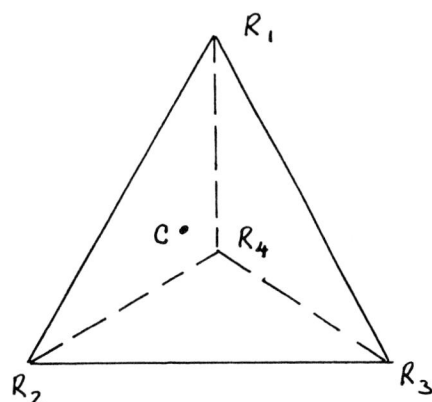

Fig. 1B

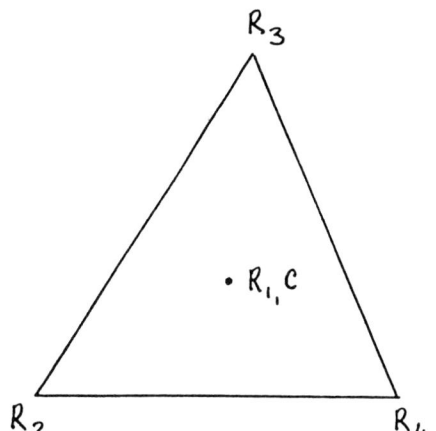

Fig. 1A - from above

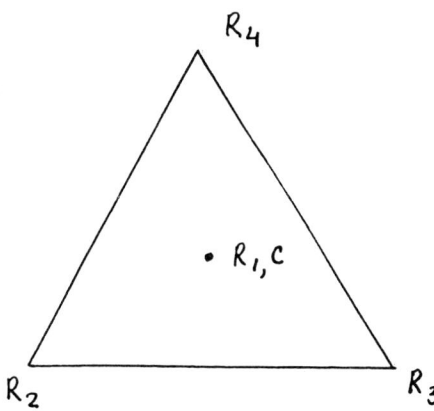

Fig. 1B - from above

But after grasping this, Pasteur divided the crystals, into two groups. Dissolving one of them, he obtained a pure R solution, rotating duly the polarization plane of light. The other group of crystals gave the L-solution. Of course the important point was the practical absence of racemization at room temperature of the active (asymmetric) tartaric acid ion.

Pasteur's experiment contained one more lesson: it has shown, that the RR and LL attractive forces are much stronger than the RL ones.

In principle one could imagine mixed RL crystal with, say, chesslike or layer over layer of R and L molecules. But Nature does not use this possibility, it does not like it usually. Obviously the build-up of pure R and pure L crystals is preferable from the energetical (thermo-dynamical) point of view.

I propose a Gedanken - experiment. Assume the race-mization in the solution (liquid) phase is not prohibited for example due to some catalyst. In the liquid phase it would lead to the 50% R; 50% L composition independently from the initial R-L ratio of the solution.

Now let us slowly evaporate water, growing the tartaric salt concentration. At some moment after the saturated concentration is achieved, the first crystal embryo is built. It will be an R or an L embryo, depending on pure chance. The formation of an embryo needs a definite supersaturation. Its growth is achieved at smaller supersaturation. If the embryo is R, its growth is diminishing the R concentration of the solution making it smaller than the underturbed L concentration.

Now, if there is the racemization catalyst, the L excess will be transformed into R, sustaining R=const and R will by crystallized further and further. After total evaporation of water an R monocrystal will be obtained. In the other case of L embryo, all substance is transformed into L monocrystal.

The lesson to be reached from these Gedanken-experiments is that depending on a parameter - the amount of water or concentration of solution, the system undergoes bifurcaton. At large water content (much larger than needed to dissolve all tartaric salt present) there is one sable equilibrium state: the racemic solution. After saturation concentration is overreached, there are two stable equilibriums (with R or L crystals) and one unstable (the supersaturated solution.)

A mixture of R and L crystals with some solution is not stable: due to surface energy of the crystals, the true equilibrium is always a single crystal, which necessarily must be R or L. Graphically the situation is shown in Fig.3. The full lines show stable equilibrium, the dotted line shows the unstable one. This type of dependence is now very

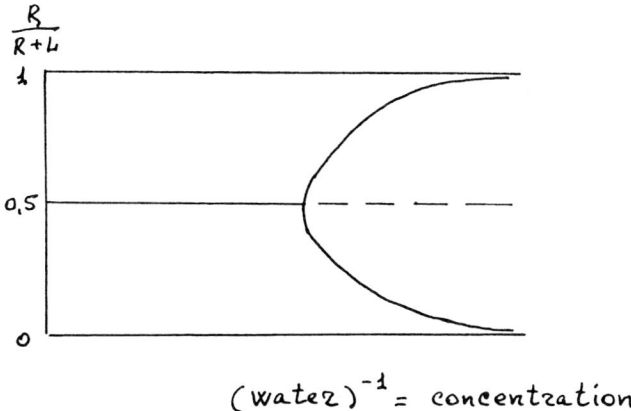

$(water)^{-1}$ = concentration

Fig. 2

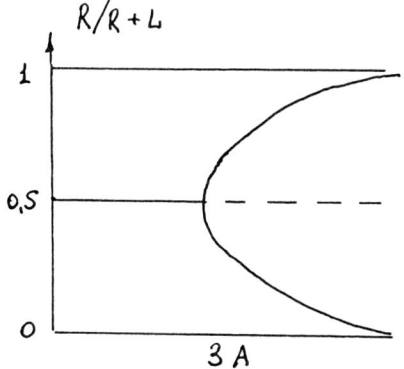

3 A

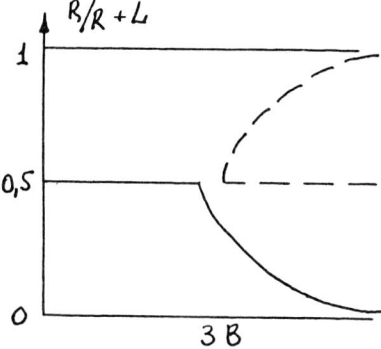

3 B

Fig. 3

popular in theoretical physics. It is called "spontaneous breaking of symmetry".

One could imagine the same situation without crystallizaton, due to specific interaction of L and R in liquid phase. Let the energy of the mixture be

$$E = -aR^2 - aL^2 - bRL = -a(R+L)^2 + c(RL); \quad c = 2a - b.$$

the term with c>0 reflecting that L and R do not like to be in contact — at least less than 2R or 2L.

The entropy of the mixture has a term

$$S = -\gamma(\ln L/(L+R) + \ln(R/(L+R))).$$

The equilibrium condition is the minimum of free energy F=E-TS. Take the total density L+R=const, denote R/(L+R)=X, L/(L+R)=1-x. One obtains F=const (x)+ δ.x(1-x)- εln a, c, t — ln x(1-x) with δ and ε depending on a,c,t. Now denote x(1-x)=y, $0 \leq y \leq 1/4$,

$$F = const + \delta y - \varepsilon \ln y, \quad dF/dy = \delta - \varepsilon /y. \text{ Here } dF/dx = 0$$

is the equilibrium condition. It gives dF/dy . dy\dx=0 The solutions are
(A) dF/dy=0, $y=y_0 = \varepsilon/\delta < 1/4$ or y=1/4, dy/dx=0, x=1/2.
(B) $y \geq 1/4$, dy/dx=0 at x=1/2.
There is one and only one symmetric solution: $y \geq 1/4$, x=1/2, R=L in the case (B). In the case (A) there are three solutions and it is easy to show that the symmetric one is unstable. Symmetric dependence of energy and other thermodynamic functions upon R and L does not guarantee the symmetry of equilibrium. A nonlinearity strong enough (bRL term) leads to a pair of stable nonsymmetric solutions of symmetric equations.

So we have demonstrated that molecular interaction really invalidate the uniqueness of the equilibrium theorem. The author has put forward the idea of pseudoscalar liquid crystals whose difference from normal liquid is the chirality, though they are not build from chiral asymmetric molecules. (JETP, 1971, v.67, pp 2357-2361, in Russian, English translation available.)

There is also another way of getting rid from the constraints of the theorem. It consists in considering a nonequilibrium situation i.e. an open system with inflow of some initial products ("food"), which is optically inactive i.e. R-L symmetric and some chemical reactions, leading to the formation of R and L. One imagines a well stirred vessel with an inflow of food and an outflow of the mixture obtained by all the reactions. The concept of a well stirred reactor was introduced at the beginning of the century by Bodenstein and Wohlgast. The result of chemical reactions is given by the solution of a system of algebraic equations; of course they are applicable to the stationary state of the stirred reactor only.

The time-dependent study of kinetics would lead to differential equations. But it turns out that the properties of stirred reactor equations are not those of equilibrium. The reactor equations are readily giving more than one solution, they can give nonsymmetric solutions even with symmetric food and symmetric other conditions that are needed is autocatalysis: reactions of the type n.R + food -> -> (n+1) R. The greater is the amount of R, the greater is its production. The concurrence about the use of food leads to damping of L. With equal probability the reversed situation occurs (L production, R damping) occurs.

The goal of all different examples considered above *) is to explain, how nonsymmetric biological structure could be built from symmetric abiotic organic products with symmetric thermodynamical and chemicalkinetic properties. The asymmetry of biological molecules (aminoacids, sugars) in vivo is absolute; there is a general belief that the explanation of the origin of life is impossible without simultaneous solution of the asymmetry puzzle.

It is possible that the scheme is more complicated than manipulating with concentrations of R and L molecules. Life is undoubtedly connected with rather long polymer chains: peptids built of aminoacid units, DNA and RNA built from many nucleic acid units etc. Therefore, as poited especially by Goldanski in USSR, it is the comparative ability to incorporate an alien R in the L-chain or a similar L what can be the most important.

The study of random polymerization of rather long polymer chains is now pursued. This gives a framework for the incorporation of RL. Larger probability of RR and LL as compared with RL will play the same role as autocatalysis, giving spontaneous symmetry breaking in problems with symmetrical underlying laws.

The general property of this approach is the randomness of symmetry breaking. One can obtain instability of the symmetric situation, but one has no indication on the direction of evolution of instability R and L remain equally probable!

II. PARITY VIOLATION

A new approach appeared in 1957, when Lee and Yang put forward the hypothesis of parity violation in "weak interaction" of elementary particles. Soon Wu (also in the

---

*) For detailed study see Ya. B. Zeldovich, A.S. Mikhailov "Uspekhi Phys. Nauk" 1987, Vol 153, p.p. 469-496, English translation will appear in 1988.

USA) has shown that polarized nuclei of radioactive cobalt are emitting electrons preferentially along one side of the rotation axis of the nucleus. Obviously this was a breakdown of mirror symmetry: the trajectory of electron along the axis remains by reflection, the direction of rotation axis changes by 180^δ: the direction of axis defined by the clockwise rotation observed from the point on the axis. The next discovery was the polarization of the particles emitted in beta decay themselves. In particular, the electrons are moving and rotating around their trajectories like left hand (anormal) corkscrewers, the positrons rotate in the inverse sense. Further came the idea of weak neutral currents also violating parity.

In contrast with parity violation by birth of electrons and positrons, neutral currents contribute to the scattering of electrons by nuclei and in general to the interaction of electrons with nuclei. The idea was put forward at phenomenological level by Bludman and myself, just as a complement to beta-decay interaction. Now the idea is a well-confirmed part of Salam-Weinberg theory. The neutral currents predict the rotation of plane polarized light by atoms.

After a difficult search the effect was confirmed by Barkov and Zolatdrev (Novosibirsk, USSR). A difference in energy between R and L molecules is predicted also. Their thermodynamic symmetry is now assumed to be only approximate, valid only as long as electromagnetic forces are considered, but "neutral currents" neglected.

At first glance it seems that now the puzzle of biological asymmetry is solved. Remember that the asymmetry seamed a miracle just on the background of exact symmetry of all physical and chemical properties. But such a statement about the puzzle solution would be very premature! The point is that beta decay has no direct relation to the origin of life. The neutral current effects in molecules are extremely small: of the order of 10^{-17} of thermal energy.

This ratio gives also the asymmetry created by neutral currents in equilibrium in a diluted ideal system. On the other hand, if the electrons from radioactive decay play important role in synthesis or destruction of asymmetric molecules, the asymmetry created would be of the order of magnitude $10^{-6} \approx$ (R/L-1).

These small numbers have nothing to do with the extreme chiral purity of biochemistry. The biochemical chirality cannot be explained directly by neutral currents and radioactive parity violation.

The autocatalytic mechanisms described in the first part of this article are inescapable. Parity violation effects can do no more, than trigger the choice of direction when the system is in its most sensitive state, at the bifurcation.

The diagrams of chirality dependence on the parameter inducing symmetry breaking are shown in two cases: without (Fig.3A) and with (Fig.3B) parity violation. With parity violation the curves are disjoint and when the parameter is changed we will always obtain one sign of chirality. In the example shown on the diagram, on the left side the upper and lower branches have the same probability. The right side diagram (with parity violation effects (Fig.3B) will always led from symmetry to the lower branch.

The physical conclusion is that in the case when radiochemical processes (with polarized electrons giving an initial asymmetry of the order $10^{-6}, 10^{-7}$) are important, they can determine the choice of chirality, fixed thereafter by autocatalysis. On the other hand the neutral current energetical effects of the order of 10^{-17} are too small even in the sensitive bifurcation situation.

The point is, that the simple theory of autocatalysis and bifurcation does not include random molecular fluctations.

In the simplest case the fluctuations are of the order of $\sqrt{N}$ when the number of molecules given by deterministic calculation is N. Therefore with N < 10 (corresponding to 10 gramms $= 10^6$ tonns!) in the region where embryos of life are born, the fluctuations are more important than neutral currents. The choice of R or L is random! We do not discuss here more difficult questions, like did the life begin in several places, with different chiralities and with a war for food between different domains? Much formal development is collected in the paper in Mikhailov and myself, mentioned above. But return to the picture in general! Too small is our knowledge concerning of the real chemistry of the birth of the life. The best one can do now is to enumerate various possible scenarios. The ultimate solution of the problem of the origin of life and of biochirality will be given hopefully in the year 2000. This is my prognosis and nobody will be more glad, than myself if this prognosis will be wrong and the answer will be given earlier.

Yakov Borisovich Zeldovich
(1914-1987)

Ya. B. Zeldovich was one of the most original physicist and astrophysicist of the modern era.

He started his scientific career working in physical chemistry at the age of 17. He created a model for the propagation of burning waves and explosions. Later he obtained fundamental results in the theory of shock wave, propagation, which was published in a book with Rayzer.

He worked for several years on nuclear physics, proposed the utilization of super-cold neutrons. Then published a series of papers in particle physics, on the theory of weak interactions, lepton charge conservation, prediction of neutral heavy mesons, partially conserved axiol vector currents etc. He returned repeatedly to the macroscopic consequences of the weak chiral assymmetry.

He was the first to point out that accretion onto black holes and neutron stars would lead to emission of X-rays. He had several other important results concerning the physics of black holes and neutron stars.

His work in physical cosmology helped enormously to turn it into a real science. He made fundamental connections between particle physics in the early universe and the subsequent stages of evolution. The Zeldovich-Harrison spectrum of fluctuations became especially important in the '80-s, after inflation was discovered. He was one of the first to discuss cosmic strings as a new possibility. His work on pancakes generated an enormous advance in our understanding of the large scale structure, his elegant approximate theory leading into the nonlinear regime. His vision of the topology of the universe has guided cosmology for several decades.

His work with Sunyaev on the Compton distortions of the microwave background predicted new types of observations, stimulated entire generations of astronomers to pursue these ideas. Combined with X-ray observations this provides a potential way to measure the Hubble constant in distant clusters of galaxies.

He has been a wonderful teacher, sparkling with ideas, and igniting everyone around him, to work day and night on the frontiers of physics and astronomy. He founded the school of physical cosmology in the Soviet Union. He had a rare career, living through one of the most exciting times in physics and astrophysics ever. His interests spanned a wide range of science and he left a significant contribution on every subject on which he worked. His papers also reflect his wonderful, warm personality. We enjoyed his lecture at our Symposium, this was his last one. He left us in a cool November night.

His students, his friends, the whole physics community, all miss him. Those of us fortunate enough to have had contact with him will never forget his exuberant personality and intellectual vigour.

A. S. Szalay
(Budapest-Baltimore)

CHEMICAL PRODUCTION OF OPTICALLY PURE SYSTEMS

G. Spach
Faculté des Sciences de Rouen
B.P. 67, Mont Saint Aignan 76130 France
A. Brack
Centre de Biophysique Moléculaire du C.N.R.S.
Cedex 2, Orleans 45071 France

ABSTRACT. The significance of optically active molecules and of the
related biopolymer chiral structures is underlined in present day
terrestrial living systems. These are dealing with the chemistry of the
tetravalent carbon atom, but chirality may be relevant to other kind, if
any, of living matter. Chiral fields that were possibly acting during
evolution may give a clue to the origin of optical activity arising from
a given handedness, but the opposite handedness is most probably not
biologically forbidden. In any case, enrichment in enantiomer requires
amplification mechanisms. Different examples are shortly displayed,
based on recent experimental results dealing with the ordered structures
of biopolymers. As they are generally restricted to a single class of
biomolecules, a tentative coherent approach to the problem of chirality
in the course of evolution is developed in the form of an integrated
model which takes into account as a whole the main homochiral molecular
families (aminoacids, sugars, lipids ...) and suggests the design of new
simulation experiments of prebiotic meaning.

INTRODUCTION

Present day terrestrial life makes use of and metabolizes a set of basic
optically active chiral molecules which can be sorted out into large
homochiral families : L aminoacids, D sugars and 3-sn derivatives of
phosphatidic acid (figure). They are found either in a free form or
combined into more complex chemical structures, which in turn govern
their biological activities : proteins, nucleic acids, polysaccharides,
i.e. mainly linear polymers which possess themselves chiral structures
(right and left handed helices) or lipid vesicles, i.e. colloids (1, 2).
    Now, during abiotic times, chiral molecules of prebiotic interest
were synthesized, among non chiral molecules, as a mixture of enantiomers of both handednesses in an equivalent (i.e. racemic mixture) or

223

Figure. Left : L-serine; middle : desoxy-β-D-ribofuranose; right : 3-sn-(or L-)phosphatidic acid (R = R' = H : sn-glycerol-3-phosphoric acid).

very nearly equivalent ratio on a time and space average. What are then the advantages for Nature to use optically active molecules instead of achiral molecules or racemic mixtures? Were there physical or biological reasons to select among the set of homochiral families a particular set rather than its mirror image or any other set? At which step of the evolution did that choice occur?

We will consider these questions keeping in mind the Colloquium subtitle "Alternate Biologies, Alternate Pathways?". In this presentation we consider that life is based on the chemistry of tetravalent carbon atom, although our developments should be valid for a life, if any, using another element. Special attention will be paid on aminoacids and peptides or proteinoids as they form an important class of biomolecules whose prebiotic synthetic pathways are presently best understood.

SIGNIFICANCE OF OPTICALLY PURE SYSTEMS

More than hundred years ago, Pasteur (3) established for the first time as a rule that optical activity was the demarcation line between animate and inanimate kingdoms and he expressed the feeling that life is dominated by dissymmetric actions whose existence is cosmic and wrapping.

Simplification, stability, necessity

In the existing complexity of life, biopolymers are fundamental tools, either nucleic acids for genetic instruction storage and its spreading or proteins as structure materials or catalysts. It is well known, especially for proteins, that optical uniformity of the building blocks confers to their ordered structures larger stability and resistance to chemical (hydrolysis) or physical (heat denaturation...) agents. For instance a polypeptide α helix built up with L and D aminoacid residues in the chain is less stable than a purely L polypeptide. Besides, disubstitution on the α-carbon atom of aminoacid residues results in a conformationnally restricted chain as the α helix is almost the only allowed structure for that kind of aminoacid.

Another advantage of the selection of homochiral enantiomers is illustrated by the enormous simplification of genetic and enzymatic systems, as compared to what would be needed to manage racemic mixtures of substrates.

The use of chiral building blocks is not only advantageous for stability and versatility of ordered structures, but it is also a must for the construction of the chain. Let us indeed consider again a simple polymer such as a polypeptide. The $C\alpha$ carbon atom of each residue requires four different substituting groups : one is the simplest atom, hydrogen; two others, an amino and a carboxylate, are condensed to built up the polymer backbone. These three substitutions are common to all "natural" aminoacids and their spatial arrangement defines the homochiral family. Now, the fourth group, varying from one aminoacid to the other, bears the chemical function that controls the properties of the proteins (interactions, reactivity, ...).

Although nucleic acids appear at first sight more complicated due to the presence of many asymmetric centers in the ribose or desoxyribose moiety, a closer examination reveals that the asymmetry at the level of the 4' carbon atom (figure) is relevant of the same analysis. Anyhow, because of the tight intertwining of all biologically important elements and especially the biopolymers, once optical activity is admitted at a level it becomes a necessity everywhere in the whole organization. A racemic life is highly improbable, and chiral symmetry breaking leading to the emergence of optically pure systems probably arised before the first closest ancestor of our present living matter appeared. This does not mean that in its very earlier stages, life was fully optically pure, as for instance D aminoacids or achiral disubstituted $C\alpha$ aminoacids are found in some peptides of present day microorganisms, such as gramicidin A, cell walls, alamethicin ..., although it is not known whether there are vestiges or improvements.

Thus, although the pathways from racemic (inanimate) matter to optically active (animate) matter are still a challenging and badly understood problem in the course of evolution to life, there is no doubt that use of chiral molecules gathered together in optically pure systems is not only an advantage but also a must to reach structural complexity with still chemical simplicity.

Abundance of chiral molecules

At this stage one should ask oneself if the abundance of chiral molecules in a sort of prebiotic reactor was high enough to allow their emergence out of the primordial soup. A statistical evaluation of the occurrence of chiral molecules in a chemically evoluting system can give a rough though indicative estimation of the actual situation. Obviously chemical reactions at work, relative concentrations of reagents, and so on, may modify the statistics.

As a general case let us consider one asymmetric center per molecule, supposed to be substitutable by n chemical groupings (R_1, R_2, ...R_n) differing by the functions they bear. It has i types of identifiable sites on which the R groups can be located, each type at the number of σ_1, σ_2, ...σ_i with $\sum_1^i \sigma_i$ = s. If in addition a number n_i of groupings among the n groups have no affinity for the σ_i sites of type i, then the probability p, as a function of n, n_i and σ_i with $(n - n_i) \geq \sigma_i$, for chiral molecules to be formed, as a racemic mixture, among the total number T of chemically different molecular species that may occur is given by :

$$p = \frac{\Pi_1^i(n - n_i)!}{\Pi_1^i(n - n_i)^{\sigma_i} \times \Pi_1^i(n - n_i - \sigma_i)!}$$

the number N of chiral racemic species is :

$$N = \frac{\Pi_1^i(n - n_i)!}{\Pi_1^i \sigma_i! \times \Pi_1^i(n - n_i - \sigma_i)!}$$

and the total number T of chemically distinguishable molecules is :

$$T = \frac{\Pi_1^i(n - n_i + \sigma_i - 1)!}{\Pi_1^i \sigma_i! \times \Pi_1^i(n - n_i - 1)!}$$

As expected, when n increases indefinitely, p and the ratio N/T tend to unity. Let us consider two simple cases :
a) A carbon atom substituted by n groups on s identical sites (σ_1 = s, $\sigma_{i \neq 1}$ = 0, n_i = 0). Let s = 4 and n = 4, then p = 0.094, N = 1 [$CR_1R_2R_3R_4$] and T = 35. Let s = 3 and n = 3, then p = 0.22, N = 1 [$CHR_1R_2R_3$] and T = 10. This corresponds to group redistribution as could occur in thermo or photocraking reactions. This situation may be relevant to the atmosphere of Titan, a planet-size moon of Saturn, where an active photochemistry of methane, ethane and nitrogen is taking place. We have elsewhere (4, 5) make out a list of the most simple, i.e. of molecular weight less than 100, chiral compounds that can be formed with carbon, hydrogen and nitrogen. It includes some fifteen derivatives that can be ranked into families (hydrocarbons, amino, nitriles ...). Thus the expected order of magnitude for the occurrence of these compounds may be roughly 0.1 mole.
b) A carbon atom having two types of sites ($\sigma_1 + \sigma_2$ = s) and, for the sake of example, a preference for n_2 groups to occupy site 1. Let s = 3, σ_1 = 2, σ_2 = 1, n_1 = 0, n_2 = 1, n = 2, then p = 0.5, N = 1 and T = 2. This situation is reminiscent of the esterification of glycerol, a prochiral molecule, with phosphoric and carboxylic acids [$CH_2(OPO_3H^-)$-$CH(OCOR_n)$-$CH_2(OCOR_n)$]. This example looks particularly interesting since appreciably different physical properties (solubility, micellization...) are expected for the various combinations of phosphoric and carboxylic glycerol esters, reducing their physical dispersion (triphosphate esters are for instance much more water soluble than tricarboxylic esters, etc.). In this respect the length of the fatty moiety of the carboxylic acids is also to be taken into consideration. Interestingly, the polar head of these glycerol derivatives lead to a homochiral family of products. The probability here is rather high but of course depends on the relative abundance and accessibility of phosphoric and carboxylic

acids as well as on their relative affinities towards the hydroxyle sites.

A more realistic example is found in the mixture of various aminoacids formed in the course of laboratory reactions simulating their prebiotic synthesis, or extracted from meteorites. In both instances, the mole ratio of chiral α aminoacids is about 20 per cent.

Thus the abundance of chiral compounds is far from negligible even in very simple systems and increases with the chemical complexity. Nature had full possibility to utilize these derivatives and they were selected for their advantages. It remains to know if simple mechanisms can be suggested that can extract from a crude mixture optically active molecules functionnally and chirally useful for further evolution.

CHIRAL FIELDS FOR OPTICAL ACTIVITY PRODUCTION

Chiral fields are provided by an axial force combined with a polar one, such as collinear magnetic and electric fields. Their action may, if interaction conditions with matter are realized, promote the synthesis of a favourite enantiomer from an achiral substrate, or the selective degradation of one enantiomer in a racemic mixture.

Parity violation and other fields

Most poeple agree that the classical electromagnetic interactions such as circularly polarized light or other fields that can be imagined acting on Earth would probably never meet a very high yield of optically pure compounds. They would also probably cancel on a time and space average. Apart from thinking that such a situation may arise somewhere else in the Univers from where optically pure molecules were spread on Earth, one has to consider chiral fields acting permanently and proceeding from an universal asymmetry.

This requirement is fulfilled by parity violation weak interactions as described in recent reviews (6, 7). It results that L α aminoacids in aqueous solution or engaged in α or β conformations, as well as hydrated D glyceraldehyde are theoretically more stable than their parent enantiomer (8). However the energy difference is so small that significant enrichment in enantiomer needs amplification mechanisms that can be sorted out into two non exclusive categories : catastrophic or cumulative.

ENANTIOMER ENRICHMENT THROUGH CHIRAL SYMMETRY BREAKING PROCESSES

In a rather simple kinetic model proposed in 1953 by Franck (9), an open flow reactor run in far from equilibrium conditions is fed by achiral compounds forming reversibly, directly and autocatalycally, two enantiomers which in turn react into an irreversible combination flowing out the reactor. When certain conditions of fluxes, concentrations ,... are reached, the racemic production may become metastable and the system bifurcates permanently to the production of either one or the other enantiomer, depending on a small excess of an enantiomer. In absence of any chiral field, both enantiomer productions are equally probable. It has been claimed (10) recently that within determined conditions, the parity violation effects may be large enough to drive significantly, up to 98 %, the production of the advantaged enantiomer, although this possibility was disputed (11, see also 12). It must be noticed that presently no experimental set of biologically meaningful reactions was suggested, so the model remains speculative, and it may well be that even considering the greater theoretical stability of L aminoacids and D sugars, their presence in Earth life is adventitious. Moreover, as the energy difference is so tiny, there are certainly no biological reasons to use a set of enantiomers rather than the opposite set. The discovery of extraterrestrial living systems may raise the doubt.

Although it is outside the scope of any biological evolution, the already mentioned surface of Titan might offer a test for a chiral symmetry breaking reactor. We suggested to take the opportunity of the Cassini mission to look for any optical activity in the ethane ocean (4), but due to the absence of data on the photochemistry and the very scarce known physical properties of the possible chiral molecules, we were unable to propose a realistic set of reactions leading to chiral symmetry breaking (5). However, search for optical activity on extraterrestrial bodies should be put on the items of exploration missions.

CUMULATIVE ENRICHMENT PROCESSES

Many processes for cumulative amplification of asymmetry, connected or not with parity violation, have been proposed and some were tested experimentally. They involve a large number of molecules in crystallization, surface adsorption, polymerization ... mechanisms.

Enrichment by use of ordered polypeptide structures

We have ourselves developed successful experiments (13-17) for enantiomer enrichment of aminoacids, starting with a slight departure from the racemic composition and implementing the α or β ordered structures of polypeptide backbone. Both structures offer chiral surfaces limited by chiral fringes or borders which can be used as templates or matrices for stereoadsoption or polymerization of substrates.

We have shown that the chirality of α helical poly benzyl DL glutamate is higher than the dissymmetry existing in the primary structure, i.e. the ratio of right to left handed α helix is larger than could be expeted from a linear correspondance with the enantiomeric monomer ratio. Thus a single screw sense helical poly DL peptide, with L>D initiates the preferential polymerization of the L monomer (13), as does a purely poly L peptide (14). However the enrichment is limited to the first stages of the polymerization reaction. Stereoselection by a growing β structure should be equally possible.

High enrichment was reached in a repetitive process including a polymerization of a nearly racemic mixture of aminoacid derivatives favoring the emergence of β stuctures, followed by the partial hydrolysis of the random coil parts of the polymer chain (15-17). After a few cycles of reactions it is possible to obtain full optical activity, at the cost of the yield which becomes smaller and smaller as the initial ratio of D and L monomers is nearer unity.

In this respect, the selection-extraction process of monosubstituted α aminoacids among a mixture of mono and disubstituted α aminoacids and ω aminoacids through the formation and the in situ polymerization of N-carboxyanhydrides of α aminoacids (18, 19) appears prebiologically more plausible than the thermal polymerization of aminoacids. Indeed, in addition to the selection mechanism of proteinous aminoacids, the former reactions have the advantage, contrary to thermal polymerization, to preserve throughout the optical activity of the starting aminoacid monomers.

An integrated model

The enrichment experiments just described as well as others found in the literature, although very significant, are generally restricted to a single type of compounds, mainly aminoacid derivatives. That is why we have recently (2, 20) elaborate a model that tends to take into account the whole of biomolecules and which extends the ideas expressed by Wald in 1957 (21). Following this author, living organisms acquired optical activity through molecular selection processes to form structures of higher order and complexity with gain of stability; perhaps at one time two populations of organisms, L- and D-, or precursors of organisms, were for a time surviving together, until one won the fight. There has been no "racemic" life based on polymers having in the chain an equal number of D and L monomeric units.

The model implies that functionnal, structural and chiral evolution should be examined altogether. It suggests the design of new simulation experiments. It is grounded on simple postulates, some intuitive, others derived from observation of present day facts :
1. Life exists and probably was born in small volumes isolated from the outside milieu by some membranes built up of optically active molecules. Spontaneous bidimentionnal resolution of chiral amphiphilic molecules might have been a key factor for the emergence of small enantiomorph vesicles (22, 23). Permstereoselectivity of such membranes might happen by the incorporation of chiral pores. In this respect, the peptide β barrel structures as found in the family of membrane proteins called the porins, appear good candidates that can be mimicked in the laboratory.
2. The choice made by Nature of a given set of homochiral families, L aminoacids, D sugars, 3-sn lipids and perhaps other molecules with isoprenoid fragments is probably not adventitious. As already mentioned, the choice of the mirror set is certainly not biologically forbidden, whereas other combinations might lead to mismatching pathways. All kind of homo and hetero stereointeractions leading to the selection of the "good" set should be looked for and examined. At this stage, stereocatalytic properties are probably a key step.
3. What is prebiotic must be simple. Now biologically important

molecules comprise n asymmetric centers which increases dramatically the number (2^n) of possible stereoisomers and renders more hazardous the choice of the "good" isomer. Speculations on chirally simplified analogues of biomolecules are wellcomed. For instance, a candidate analogue for ribose is glycerol (17), although it will introduce much flexibility into the polymer chain. Glycerol is found today into the phosphoric acid polyester backbone of natural teichoic acids, grafted with aminoacids or sugars. Acyclic and cyclic prochiral glycerol phosphate derivatives substituted by nucleic bases are currently studied as monomers in different laboratories. Cyclic monomers might become as important for nucleic acid analogues as are the cyclic N-carboxyanhydrides for proteinoids (24).

Some guide-lines of the model are summarized in the following scheme which should be taken as incentive rather than conclusive :

amphipatic chiral molecules		chiral α aminoacids		chiral hydrophilic biomolecules	
↓		↓		↓	
right and left handed optically active vesicles	+	homochiral oligopeptides in β barrel	→	stereoselective pores and vesicles	→

→ enantiomer enrichment of the inner part of the vesicles into homochiral optically active families   → stereospecific catalysts

CONCLUSION

Alternate biologies? alternate pathways? Terrestrial life based on the tetravalent carbon chemistry could (abundance of chiral molecules) and should (polymer functionalisation, structure stability, genetic and enzymatic simplicity) make use of homochiral families of optically pure compounds. This should also hold for a very unprobable life utilizing another element, as life requires highly elaborated structures and not just small molecules.

Except to imagine an extraterrestrial origin of the life in a location where for instance chiral fields were active enough to promote the synthesis of optically pure molecules, a terrestrial mechanism must be envisaged through which evolution started from a racemic or nearly racemic matter to an optically active animate matter.

Chiral symmetry breaking, still including possibly parity violation, should be looked for at a rather high level of organization, late after cumulative enrichment processes, involving the main biomolecules altogether, were at work. In any case an enantiomeric, in the chemist sense, biological world is fully conceivable.

Finally we would like to make a remark on the archaebacteria that develop on marine vents at high depth and temperature up to several hundred of degrees centigrade. They raise the question of aminoacid chiral stability as it is known that α aminoacids racemize at temperature above hundred degrees with a half time life of a few days. It would be very important to have more informations on the defense that these microorganisms developed against spontaneous racemization reactions.

REFERENCES

1. Ulbricht, T.L.V.:1981, *Origins of Life* 11, 55-70.
2. Spach, G. and Brack, A.:1983, *Structure, Dynamics, Interactions and Evolution of Biological Macromolecules*, Hélène C., edit., D. Reidel Publish. Co., Dordrecht, 383-394.
3. Pasteur, L.:1884, *Rev.Scient.* 7, 2-6.
4. Brack, A. and Spach, G.:1985, *The Atmospheres of Saturn and Titan*, *Proc. Int. Workshop*, Alpbach, Austria, **ESA SP-241** (dec.1985), 189-191.
5. Brack, A. and Spach, G.:1987, *BioSystems* 20, 95-98.
6. Mason, S.F.:1984, *Nature* 311, 19-23.
7. Mason, S.F.:1986, *New J. of Chem.* 10, 739-747.
8. Tranter, G.E.:1986, *J.Chem.Soc., Chem.Commun.*, 60-61.
9. Frank, F.C.:1953, *Biochim.Biophys.Acta* 11, 459-463.
10. Kondepudi, D.K. and Nelson, G.W.:1985, *Nature* 314, 438-441.
11. Omnes, R.:1987, discussion at the 5ème Colloque Européen du C.I.E.E.I.S.T. *La Terre et l'Origine de la Vie*, Université de Paris-Sud, 20-21 mai 1987.
12. Keszthelyi, L.:1984, *Origins of Life* 14, 375-382.
13. Spach, G.:1974, *Int. Symp. on Generation and Amplification of Asymmetry in Chemical Systems*, **Jül-Conf-13**, Zentralbibliothek KFA Jülich, 259-271.
14. Brack, A. and Spach, G.:1971, *Nature Phys.Sci.* **229**, 124-125.
15. Brack, A. and Spach, G.:1979, *J.Mol.Evol.* 13, 35-46.
16. Spach, G. and Brack, A.:1979, *J.Mol.Evol.* 13, 47-56.
17. Brack, A. and Spach, G.:1980, *J.Mol.Evol.* 15, 231-238.
18. Brack, A.:1982, *BioSystems* 15, 201-207.
19. Brack, A.:1984, *Origins of Life* 14, 229-236.
20. Spach, G.:1984, *Origins of Life* 14, 433-437.
21. Wald, G.:1957, *Ann.N.Y.Acad.Sci.* **69**, 352-368.
22. Arnett, E.M. and Thompson, O:1981, *J.Am.Chem.Soc.* 103, 968-970.
23. Arnett, E.M., Chao, J., Kinzig, B.J., Stewart, M.V., Thompson, O. and Verbiar, R.J.:1982, *J.Am.Chem.Soc.* 104, 389-400.
24. Spach, G.:1986, communication at the European Meeting on *Chiral Symmetry Breakings in Physics, Chemistry and Biology*, University of Rouen, june 17-19.

Session 5

IS INTELLIGENCE AN
INEVITABLE EVOLUTIONARY TRAIT?

During the history of terrestrial life it might be claimed several times by successful-looking creatures: I AM THE GREATEST!

- the blue-green alga developed photosynthesis when everyone was starving around them, and their number even now exceeds the population number of almost all other species;

- the scorpion is a genetically very stable structure, which resist even ionizing radiation, they have survived hundreds of millions years;

- the shark is surely not a genius but it is the most powerful creature in the oceans covering the larger part of our planet; even man gladly avoids meeting them;

- the Tyrannosaurus Rex exceeding everyone in muscular power;

- the Cro-Magnon man being proud of the brain; he succeeded to kill all the kins; now (s)he calls him(her)self modestly the Homo sapiens, the Crown of Creation;

- the Cray computer exceeds man in speed of thinking million times, artificial intelligent machines are emerging;

- the Galactic Club, what sci-fi writers dream about, may realize the slogan: E.T.-S OF THE UNIVERSE, UNITE!

Each creature looked glorious when it emerged. A few of them became milestones on the main highway of evolution, others were dead ends. Earlier time man considered himself to be the final terminal of the cosmic evolution, nowadays we are not so sure of it. Is intelligence of advantage or disandvantage for a species on a long run? Is it a convergent evolutionary trait (making communication among aliens possible)? These questions must be faced if we are about trying SETI and CETI, they have been discussed at Session 5:

- Neurons indicate convergence! - S. Finkelstein *proved.*

- Are the smart ones the better? - W. Calvin *asked.*

- Can we communicate with aliens at all? - Csányi *and* Kampis *questioned.*

- Communication is characteristics of intelligence! - Diana Reiss *(studying dolphins) claimed.*

- Are physical scientists themselves intelligent enough to be able to communicate? - *so* John Coffey.

George Marx

THE INTERFEROMETER PARADIGM
FOR MAMMALIAN BINAURAL HEARING

N. Cohen

Metropolitan College, Boston University

Boston MA 02115 USA

Binaural (stereo) hearing in man and other mammals may be modelled
by an interferometric array paradigm. The well known use of
amplitude differences at the two ears is but one of several relevant observables, which also include phase delay and group delay
differences, as well as their respective rates. This two element
interferometer appears to be nonlinear and spectral. Beam synthesis occurs with the decoding of harmonic information and the
generation of pseudo-UV coverage through cochlear overtone production. Diffraction limiting may be overcome through such overtone
generation. The precision or 'localization' of the synthesized
beam may be achieved through rapid (of order microseconds) beam
nodding, with acuity limited by the delay resolution of the
interferometric processing--this has been measured to be less than
10 microseconds in man. Interferometric rejection is used to
distinguish particular sounds in a noisy environment (the 'cock-
tail party effect').

Phase coherence of the interferometer is limited to frequencies
between 100-5000 Hz in man. However, as with radio VLBI, the
loss of phase information dilutes but does not prevent the use
of the interferometer--amplitude visibilities remain useful. There
is a strong correlation between communicative vocalization frequencies and the 'phase coherence bandpass' of the binaural
interferometer, a logical choice if one is to define a communication channel which is maximized for information transferral in
an often noisy environment. The binaural interferometric decoding
may also be useful in bat and odontocete echolocation, where it
has been hypothesized that the generation of fringe patterns,
reflected by the target, is indicative of holographic processing
(Cohen and Cummins,1987). Binaural hearing, like vision, is a
highly optimized system which takes advantage of sustantial
processing to accomodate often modest anatomical structure .

Cohen,N.,and Cummins,D.,1987,Cetus,7(1),32.

G. Marx (ed.), Bioastronomy – The Next Steps, 235.
© 1988 by Kluwer Academic Publishers.

FAST TRACKS TO INTELLIGENCE
(CONSIDERATIONS FROM NEUROBIOLOGY
AND EVOLUTIONARY BIOLOGY)

W.H. Calvin

Biology Program NJ-15, University of Washington
NJ-15 Kincaid 208, Seattle WA 98195, USA

ABSTRACT. It is often assumed that the evolution of intelligence is
inevitable, given the self-organizing seen in dissapative systems and
the gradual shaping-up of Darwinism. While compound-interest reasoning
suggests that small advantages will eventually triumph, eventually may
be a very long time: there are few examples of rapid brain growth,
suggesting that "smarter-is-better" is not a potent force for
evolution. Fast tracks to complexity may be much more important than
the slow-but-predictable paths (especially given the "windows of
opportunity" in the life cycle of a planet).
   Fast tracks are associated with 1) cycles of selection too
frequent for gradual evolution to track, giving advantages to variants
which can survive in either of several climates; and 2) novel
functionality that emerges from combinations of mechanisms, each
previously under natural selection for other uses. Thus frequent waves
of natural selection (due to low tides, droughts, ice ages, 28 Myr mass
extinctions, etc.) may play an essential role in compounding mechanisms
from which new emergent properties can arise.
   Intelligence in particular may involve a secondary use of neural
machinery for rapid movements such as throwing; the same machinery can
constitute a "Darwin Machine" handy for constructing candidate
scenarios that tend to explain the past and forecast the future, from
which one can select the best-seeming candidate. Since this Darwinian-
like simulation of real-world properties can operate in milliseconds
rather than millennia, using innocuous remembered environments rather
than real-time noxious ones, it greatly augments what we call language
and consciousness, key elements of the kinds of intelligence which can
lead to a technology able to export knowledge.

UPWARDS TO COMPLEXITY

The standard Darwinian evolutionary path is via the better-and-better
body styles (particularly brains) achieved through variants thrown up
by mutations and permutations of the genes, continuously edited by
various environments. Compound-interest reasoning suggests that body
styles with even slight advantages will exponentially come to dominate
within hundreds of generations, just as an investment made by Galileo
at one percent compounded yearly would be worth 32-fold more today.

G. Marx (ed.), Bioastronomy – The Next Steps, 237–245.
© 1988 by Kluwer Academic Publishers.

And thus, in this caricature of evolutionary thought, progress towards
intelligence seems inevitable (and indeed a soothing counter to the
inevitable disorder predicted by entropy) as long as variations keep
exploring the possibilities.

Closer acquaintance with biology and evolutionary theory shows
that the real world is full of dead ends, not just extinctions but also
stabilities: the so-called "living fossils" exhibit surprising
stability in the face of repeated challenges.

Shuffling the Genes

Since evolution is often very slow, we sometimes postulate a need for
more mutations -- but permutations are more the name of the game.
Bacterial evolution was apparently sluggish with just mutations and an
occasional conjugation; following 2.5 billion years of nothing but
bacteria and such, the eukaryotes institutionalized an additional
source of randomness about one billion years ago which we call _sex_,
shuffling the DNA deck during crossing-over while manufacturing sperm
and ova, and the great pre-Cambrian diversification of complex life
forms followed shortly thereafter.

Even if variation were somehow speeded up via more mutations or
more gene shuffling, that might actually slow evolution down. That is
because a major stimulus toward more elaborate organisms has been the
fluctuating climate: if evolution were fast enough to track it, we'd
likely see body styles fluctuating back and forth along the same path
that the weather takes -- for little net change.

Complexity via Multienvironmentality

But evolution is often too slow to track the Earth's climate, and so
those variants which happen along, capable of surviving both extremes
of climate, will have an advantage over the one-climate-at-a-time
efficient trackers. The very slowness of evolution serves as a drive
toward more complex organisms with the machinery for handling both
kinds of environment. And complexity is the overall trait which
underlies intelligence, primarily because new capabilities emerge from
combinations of mechanisms: rather than compound interest, we have
compounded mechanisms.

We track the seasons by varying the clothing we wear; when we
travel to Hungary in the summer, for example, we have the problem of
guessing whether or not we will need warm clothing. Those who always
carry both winter and summer outfits will be safest; those who carry
only enough for one climate at a time will be less burdened, and may
get the only available taxicabs while the cautious await their checked
baggage. If the weather was so unpredictably variable that tracking it
was impossible, then everyone would have to carry along both winter and
summer clothing -- but so long as climate fluctuations occur slowly,
the more efficient packers may out-reproduce the multi-climate packers.
But sometimes new properties arise from having both sets of clothing
available at the same time (perhaps a winter coat could be used as a
sail for wind-surfing?). Compounded mechanisms sometimes confer new
"emergent" properties, quite unlike anything existing. They are true
innovations, not just more-or-less predictable improvements.

This means that capabilities occasionally arrive unheralded by gradual predecessors: when feathers for thermal insulation on dinosaur forelimbs became numerous enough to reach the threshold for flight (flying isn't what got feathers started, as you need many before you can fly even a little), some protobirds probably discovered that they could glide. The combination of both forelimbs and feathers yielded something novel. Natural selection for insulation shaped forelimb feathers up to the threshold for flight, natural selection for a better airfoil shaped feathers thereafter, but the switch-over from one track to another was presumably a surprise, leaving the protobirds to explore their newfound abilities rather as we might try to figure out a holiday gift that arrived without an instruction manual.

Climatic Pumping of Complexity

Since selection is the flip side of the evolutionary coin from variation, we've learned to look for how evolutionary rates are stimulated by recurrent cycles of environmental change, particularly the ones too fast for evolution to track with adaptations.

    ** Thanks to the moon, there were regular cycles of extreme low tides selecting for intertidal plants and animals that could survive in a second kind of environment for longer and longer intervals until becoming land-dwellers. For the transition, they required both salt-water and air-dwelling mechanisms compounded.

    ** Thanks to axial tilt and land surface in the temperate zone, we have had yearly cycles of selection for species able to survive both summer and winter weather (most species simply stick to the tropics).

    ** Thanks to variations in tilt and drift in perihelion, we've had 100,000 year major climatic cycles (plus ice age stades) to shrink and expand the temperate zone populations.

    ** Thanks to whatever happens every 28 million years, we've had regular opportunities for new species to fill vacated niches after the mass extinctions.

Because there are usually "windows of opportunity" (due to the finite lifetime of one's sun if nothing else), evolutionary rates are all-important; it may be very important that a planet have a moon and an axial tilt, for example, just to keep nudging evolution along at a sufficient rate. However, if evolution were fast enough to track the climate fluctuations, we would lose an important advantage: <u>with evolutionary processes somewhat sluggish, one selects for the rare variant that can survive both extremes of the climate</u>. And hence one has a drift towards more and more complex organisms, thanks to slow evolution but fast environmental fluctuations.

IS INTELLIGENCE AN ADAPTATION?

The usual corollary to compound interest reasoning is the notion that for every feature, there was a corresponding selection pressure, e.g., adaptationist thinking considers the big toe as shaped up by walking. Adaptations are the way in which functionality becomes streamlined, made more efficient <u>once invented</u>. Adaptationist views of intelligence simply assume that smarter is better, that we pulled ourselves up by

our own mental bootstraps, from dull to bright, mostly in the last
2 million years when our brains enlarged 3-fold over the pint-sized
australopithecine brain.

So, is intelligence a convergent evolutionary trait? Like eyes
which have evolved independently some 40 times or more[1], will
intelligence strike repeatedly? All you have to do is look around at
the big brains (relative to body size; a crude index, but it suffices
for the present purpose) -- if an ever-bigger brain is so useful, why
haven't other mammalian lineages experienced some rapid brain spurts?
It doesn't require other examples of our 200% increase in a mere
2.5 million years -- I'd settle for a few examples of 50% in 5 million
years, but the fossil record just doesn't exhibit the alleged bigger-
is-better principle for us over and over, leaving us to fall back on
the singular and ever-so-slow enlargement of the brain from fish to
apes. Adaptationist-type evolution is typically too slow, too easily
closed off by dead ends (the "birth canal bottleneck" in the case of
big heads[2]) or slowed down by good-enough solutions that remove the
trait from regular exposure to natural selection. Adaptations are
usually concerned with improvements, rather than the initial invention
of a new functionality. For the functionality of looking-to-the-future
intelligence (as opposed to the usual animal cleverness in problem-
solving), we must try looking elsewhere.

SIDESTEPS: Unpredictable but Faster

The other evolutionary path emphasized by Darwin, but forgotten by most
who followed, is the sidestep ("functional change in anatomical
continuity"): **new functions often arise because a combination of
existing structures turns out to have an additional use** beyond those
which evolved the component parts. Something novel emerges from the
merger, as when those forelimb feathers became secondarily useful for
bird flight after enough natural selection for keeping warm.

And since brains are better at new-uses-for-old-structures than
any other organ, one must always consider non-adaptationist paths of
evolution as well as the obvious adaptationist schemes, especially
where intelligence is concerned. The stakes are not minor: rather
than our planning-for-the-future consciousness arising gradually
through its own selective advantages, there is the possibility that it
emerged nearly full-blown as a secondary use of neural machinery
originally under selection for a quite different purpose. In the case
of humans, as I shall presently argue, it might have involved another
forelimb function, a conversion even more revolutionary than flight
from thermal underwear.

While compounding percentages isn't as promising and inevitable as
it initially seems, the emergent properties that arise from compounding
mechanisms do hold the promise of providing fast tracks toward
organized complexity. Though unpredictability (compared to the ever-
greater-efficiencies of adaptations) is the hallmark of sidestepping
"conversions of function", intelligence is more likely to arise from
multifunctionality simply because of the general-purpose versatility of
what we call intelligence. Just as our general-purpose computers arose
out of a background of special-purpose devices (from the Jaccard loom
to calculating machines to the anti-aircraft gun's controller), so our

brain likely learned to "co-opt" (a very useful slang word meaning to divert from the primary purpose to a secondary use, to recruit help informally outside official channels via "co-operation") special-purpose neural machinery in various novel combinations during the off-hours. Certainly natural selection operates on the new use, once invented, but adaptationist reasoning usually is misleading -- or at least ineffectual, simply because of the unpredictability of the new functionality that arises from compounding mechanisms.

Brain Circuitry's Big Sidestep

A detailed candidate mechanism for how ape brains developed the neural machinery associated with human language and consciousness has recently emerged; while neither established nor comprehensive, it nonetheless serves in the present context as an example of how human-like intelligence could evolve, and so provides some lessons for contemplating fast tracks to extraterrestrial intelligence. And as an ecologically-minded graduate student pointed out to me, just in case we decide that terrestrial intelligence hasn't yet evolved, perhaps its planning-for-the-future aspects can suggest how we too could become intelligent.

Planning is all very useful, once you have the neural machinery, but most of the animal world operates on goal-plus-feedback -- even humans, most of the time. When I pick up a coffee cup, I am getting constant progress reports from the receptors embedded in my muscles and joints; with enough corrections, the cup arrives at my mouth. When I am hungry, I wander around the kitchen, exploring the resources. I could plan those actions in exquisite detail, and perhaps even execute some without any feedback (as when chewing food before the dentist's local anesthetic has worn off), but detailed planning is seldom seen in the animal world (the carnivores come closest, when carrying out an ambush).

But for some movements, a detailed plan is essential -- and so they are of considerable interest in looking at the evolution of human planning abilities. Movement command buffers, essential for planning ballistic movements (so fast that sensory feedback arrives too late to effect corrections), were surely under selection for throwing. For organisms that need to be both large (meters of conduction distance) and fast, one often needs the neural equivalent of an old-fashioned roll for a player piano. We carefully plan during "get set" to order to act without feedback. If those buffers are capable of sequencing other things when not needed for throwing-hammering-clubbing-kicking, then one might expect augmentation of such sequential abilities as stringing words together into sentences, or concepts into scenarios.

Sequencing tends to involve much of the left cerebral hemisphere in mammals[3]. One's left premotor cortex tends to program linked movements (as when we unlock, then push open, a door) for not only the right hand and arm, but the left hand and arm as well[4]. The left hemisphere (at least in typical right-handers, and many left-handers as well) is best at deciphering rapid sound sequences. It is from this general mammalian background of sequencing specialization that the human left hemisphere apparently became a natural home for many language-related abilities. Unlike the several dozen standard interpretations of the several dozen characteristic vocalizations of primate species, humans embed special meaning in word sequence -- "Dick hit Jane" means something different than "Jane hit Dick." The core of human language cortex (involving both inferior frontal and the usual cortical areas bordering on the Sylvian fissure) is a sequencing area for both incoming sounds and outgoing movements[5].

Massively-serial Circuitry for Throwing

Of all the ballistic movements, throwing makes the most demands on our nervous system. That is because it has a special timing requirement halfway through the movement: to let loose of the projectile at just the right split-second. Release too early and it lobs too high and goes too far, release too late and it hits the ground in front of the target. There is a "launch window" and it can be surprisingly tight.

In addition to the important techniques of group ambush tactics shared with the chimps and carnivores, humans hunt with projectiles. We are not entirely unique in doing this; apes throw as threats and both scare off leopards and gain social dominance over other chimps via throwing. Even if one narrows the definition down to predation, we see that archer fish spit, and that frogs throw their tongues for body-length distances. However, faster and farther throws are always better, provided accuracy can be maintained.

And that gets us into some interesting territory. I earlier proposed a biophysical model for throwing[6] that shows that typical human throwing requires sub-millisecond timing precision, far in excess of what one can expect from noisy neurons[7] in a single command buffer. This suggested that the precision of the release of a projectile must arise from the Law of Large Numbers (the same rationale as why, in order to halve a standard deviation, one averages four times as much data).

Thus there must be many sequencers which, at least temporarily, can be ganged in parallel. To reliably hit a rabbit-sized target from twice the distance requires that the jitter in rock release time be narrowed by a factor of 8, and the only known way of accomplishing this feat is, as one gets set to throw each time, to assign 64 times as many noisy neurons to the task and then average their recommendations for the release time.

The Virtues of Randomness

Noise is usually treated as an irritant by technology. Darwinism, however, views noise as a means of exploring new avenues; we see it here as a stimulus to evolve machinery that is not merely compounded but massively redundant.

But it is the secondary uses which may be revolutionary. I have even suggested that there may have been a "noise window" in hominid evolution: lacking sufficient neuron noise to overcome, ice-age hominids might have become proficient projectile predators without the massively-serial scheme[8]. Note that hunting via throwing via timing precision is the argument for why so many parallel planning tracks were evolved in the first place, but that the really interesting things are the possible spare-time uses -- if those extra buffers are capable of randomly sequencing other things when not needed for throwing-hammering-clubbing muscle commands.

The brain's construction of chained memories and actions suggests the metaphor of the candelabra-shaped railroad marshalling yard, with words for cars: imagine that many trains are randomly constructed on the parallel tracks, but that only the best is selected to be let loose on the "main track" of consciousness and speech. Presumably "best" is determined by memories of the fate of somewhat similar sequences in the past, and one presumes a series of selection steps[*] that shape up candidates into increasingly more realistic sequences. This selection among stochastic sequences is more analogous to the ways of Darwinian evolutionary biology than to the "von Neumann machine" serial computer. I like to call it a <u>Darwin Machine</u>[10,11]: it shapes up new thoughts in milliseconds rather than new species in millennia, and uses innocuous remembered environments rather than the noxious real-life ones, but otherwise it's strikingly similar to what Charles Darwin envisaged in 1838 as the route to those fancy eyes and organs which initially seem so nonrandom.

By providing many candidate queues, such massively-serial neural circuits might foster stringing words together into more sophisticated sentences, or schemata into more credible scenarios. Instead of our productive language and planning-for-the-future consciousness arising gradually through their own selective advantages, they perhaps emerged as novel spare-time uses of such neural machinery that was originally under selection for more mundane forelimb movements.

THE ESSENTIALS FOR EVOLVING INTELLIGENCE

This example of how intelligence might have evolved, whether or not it proves correct for the human case, does suggest how general-purpose computing machinery might evolve without retracing our long bacteria-to-apes route. While there is likely some natural selection for (or perhaps against) intelligence once it has evolved, such selection-for-smarts need not have been the source of innovation.

And such indirect routes are probably fastest, via conversion of function from some other combination of neural-like mechanisms. Since a niche once occupied will tend to resist new entries (as Darwin realized, "wedging" in via displacing another species is far harder than occupying an empty niche), it may be that the indirect routes will preempt the slower direct ones of the "smarter-is-better" variety. The meandering zig-zag route taken by successive sidesteps to novel functions may be hard for us to generalize about, but it does work, given enough cycles of faster-than-tracking-ability climate changes to compound mechanisms.

The throwing example suggests other routes besides whatever our particular route turns out to be. The other ballistic motions, such as hammering and clubbing, also require serial-order buffers for detailed planning of the movement during "get set." Inefficiencies such as slow conduction speeds in nerves may prove important; 150 meters/sec is about the top speed in mammals (though animals that have not invented myelin-separated repeater-stations are limited to a burning-fuse-type conduction that seems to yield no more than 1 m/sec). It might be that inefficiencies in conduction speed will lead to long loop times and force "get set" strategies with serial-order buffers. Slow conduction speed is also a characteristic of colonial animals (coral reefs) and

societies such as ant colonies; should they be exposed to selection
pressures that have a premium on a one-shot, quicker-than-feedback
response, they too might invent serial-order buffers for planning.

The Law of Large Numbers, however, is the really significant
principle which might often lead to the evolution of a great many extra
computing elements, essential only on occasion, and perhaps available
for other uses in between such tests. While the throwing problem
involved a precision problem in timing, any precision problem might
lead to the same result. For example, the precision depth
discrimination needed for hunting can also be solved by ganging
together many noisy depth discriminating neurons and averaging their
recommendations. Precision color discrimination (monkeys spotting
ripening fruit high in the trees amidst a background of breeze-blown
leaves), precision sound frequency determination (bats locating their
food amidst sound clutter), precision shape analysis -- all can use
averaging tactics. The best are tasks like throwing which have a one-
shot aspect forcing ensemble-averaging rather than time-averaging (in
echo-location, many repeats of the waveform can be time-averaged
together). Many such arrays may turn out to be too dedicated to their
single task to become more generally useful for other tasks -- I doubt
that the greatly enlarged inferior colliculus of the sonar-using
mammals is re-assignable in quite the way that cerebral cortex is; it
is widespread connections that should sometimes lead to "co-opting" for
secondary functions.

The present analysis has also emphasized a one-dimensional task
analogous to assembling a train of many cars; it seems likely that the
brain also engages in a two-dimensional version insofar as there are
many topographically-organized maps of the sensory worlds in the brain;
the primary visual cortex is perhaps the best-known "map" of the
individual's sensory world but, even in monkeys, there are more than
two dozen additional cortical areas with a complete visual field
representation. Can these be simultaneously assigned to a precision
visual discrimination (say, spotting fruit), but then unganged to do
several dozen separate tasks on less-demanding occasions? No one yet
knows, but there are already suggestions that selective attention to a
sensory task serves to "assign" more multifunctional neurons to that
task[12], the neurons shifting assignment for other tasks. And I have
shown that, certainly for the case of precision depth discrimination
but likely applying more generally to most precision discrimination
tasks, the same Law-of-Large-Numbers approach suffices[13].

Thus another route to massive numbers of general-purpose computing
elements might involve the Law of Large Numbers but a two- or three-
dimensional array rather than the one-dimensional serial order so
important in human evolution. Note, however, that one can handle
three-dimensional representations with lower-order buffers (for
example, a three-dimensional house can be represented by stacks of two-
dimensional blueprints, each of which can be sent cross-country by a
fax machine as a one-dimensional bit string, reassembled to duplicate
blueprints, which then serve to construct a duplicate house) -- but
true higher-order representations might allow quick comparisons of
shapes that would take forever by the dimensional-translation route.

Organisms that evolved under selection for quick precision shape
recognition might have secondary uses of their computing machinery
which are quite different from our scenario-spinning, sentence-

constructing, music-loving activities. It is difficult to imagine what
their "consciousness" would involve instead. But it seems likely that
any extra-terrestrial intelligence we encounter would turn out to have
some aspects of our compulsion to simulate the future, plan ahead,
criticize those plans before (rather than simply after) construction --
or they would never manage to evolve a technological society capable of
communicating with us.

REFERENCES

1. L. Salvini-Plawen and E. Mayr (1977). On the evolution of
photoreceptors and eyes. **Evolutionary Biology** 10:207-263.

2. W. H. Calvin (1987). Of fast teeth and big heads. **Nature** (in press).

3. J. L. Bradshaw and N. C. Nettleton (1981). The nature of hemispheric specialization in man. **Behavioral and Brain Sciences** 4:51-91.

4. D. Kimura (1982). Left-hemisphere control of oral and brachial
movements and their relation to communication. **Philosophical
Transactions of the Royal Society of London, Series B.** 292:135-149 (1982).

5. G. A. Ojemann (1983). Brain organization for language from the
perspective of electrical stimulation mapping. **Behavioral and Brain
Sciences** 6:189-230.

6. W. H. Calvin (1983). A stone's throw and its launch window:
timing precision and its implications for language and hominid brains.
Journal of Theoretical Biology 104:121-135.

7. W. H. Calvin and C. F. Stevens (1968). Synaptic noise and other
sources of randomness in motoneuron interspike intervals.
Journal of Neurophysiology 31:574-587.

8. W. H. Calvin (1986). **The River That Flows Uphill: A Journey from
the Big Bang to the Big Brain** (Macmillan, New York), p.407.

9. R. Dawkins (1986). **The Blind Watchmaker** (Longman, London).

10. W. H. Calvin (1987). Bootstrapping thought: Is consciousness a
Darwinian sidestep? **Whole Earth Review** 55:22-28.

11. W. H. Calvin (1987). The brain as a Darwin machine. **Nature** (in
press).

12. S. Hochstein and J. H. R. Maunsell (1985). Dimensional attention
effects in the responses of V4 neurons of the macaque monkey.
Society for Neuroscience Abstracts 11:364.6

13. W. H. Calvin (1984). Fine discrimination as an emergent property
of parallel neural circuits. **Society for Neuroscience Abstracts**
12:218.11 (1984).

FORM AND FUNCTION IN NEUROBIOLOGY

S. Firestein
Graduate Group in Neurobiology, University of California
Berkeley CA 94720 USA

ABSTRACT. A common strategy in modern neurobiology is to study aspects of nervous function in simpler animals whose nervous systems are less complex than our own and are thus more approachable. The unequivocal success of this strategy suggests that nervous tissue is more alike than it is dissimilar across the phyla. In particular the building blocks of nervous systems are seen to be remarkably similar and the functional differences from simple to complex nervous systems are the result of elaborations on the basics rather than of fundamental differences. Neural tissue possesses a kind of equipotentiality and may be capable of undergoing "elaborations" towards more complex forms without requiring qualitative changes. While this line of reasoning does not demonstrate that highly developed mental abilities may arise in forms other then the human it does suggest that it is not unreasonable to consider such development as possible.

The nervous system is commonly thought to be the most complex biological system. Even at a molecular level single nerve cells, or neurons, display a greater complexity in their function and a greater diversity in their form than virtually any other single cell type. Reflecting this diversity, the discipline of neurobiology is concerned with events which occur on the order of milliseconds (the action potential or the opening of single ion channels in the membrane) to many years (the maintenance of long term associations and memories) and with entities which range from only a few angstroms in size

247

G. Marx (ed.), Bioastronomy – The Next Steps, 247–252.
© 1988 by Kluwer Academic Publishers.

248

(proteins such as acetylcholine receptors or ion
channels) to nerve cells ranging from a few microns
to millimeters (one neuron in the spinal cord has an
axon which stretches over a meter to end in your big
toe), to brains which total 24 x 10^{12} cells and
occupy 1600 cubic centimeters·

Neurobiologists over the past 4 or 5 decades have
developed a strategy for dealing with this daunting
complexity, and as a result have been quite
successful in elucidating many of the mechanisms by
which nerve cells and nervous systems work. In
brief, the strategy consists of locating simpler
systems in which some dominant feature of nervous
tissue is both expressed and is accessible to
available experimental techniques.

Thus the pioneering work of Hodgkin and Huxley (1)
in describing the mechanism by which an action
potential, the primary electrical signal which
carries information in nervous systems, is produced
was the result of an elegant series of experiments in
the squid, an invertebrate which happily possess the
largest diameter axon on the planet (1 mm) and could
thus be studied by the newly developed method known
as "voltage clamping".

The most fertile work being done currently into the
molecular mechanisms of learning is being
accomplished by Eric Kandel at Columbia University,
and many others, working on the California sea slug,
Aplysia californica, another invertebrate (2,3).

Other invertebrate preparations which are in common
use today, and which are yielding important results
in the neurosciences, include the torpedo fish
(acetylcholine receptors (4)), the leech (development
of the nervous system (5)), the crayfish and lobster
(synapses and neural networks (6)), insects of
various types (development, neurogenetics, sensory
systems, assemblies of small dedicated neural systems
such as flight circuits (7)). This list is intended
to be only indicative of the wealth of studies
occurring in invertebrates alone and to suggest the
breadth of neurobiological questions which can be
successfully addressed in these simple animal forms.
It is not by no means exhaustive.

When the questions are such that a vertebrate
nervous system must be studied, the strategy is still
to find the simplest one available. Thus a
considerable portion of the research involving the
retina is carried out on amphibians (8). For
neuromuscular studies the frog is still the animal of
choice for most work. Even when it comes to higher
order learning, the rat, mouse and pigeon are

preferred over so-called higher animals. Indeed virtually the only instances in which higher animals are used in neurobiological research is in the development of clinical procedures such as surgical interventions for neurological disease or trauma.

In addition to the wealth of factual information being produced it is becoming clear that nervous tissue throughout the phyla is more similar than different. That is, a nerve cell is a nerve cell is a nerve cell whether it be functioning in the cerebral ganglion of a locust or in the hippocampus of a primate. Of course there are differences, but the commonality of underlying mechanisms in neural function is striking.

Let us consider, for an example, the basic functioning of the single neuron. In addition to maintaining virtually all the normal cellular functions (i.e. energy metabolism, protein synthesis, respiration, repair) nerve cells have the property of being transiently excitable. They accomplish this by maintaining an ionic gradient across their membranes resulting in an electrical potential, with the inside of the cell being negative with respect to the outside. This polarity can be changed transiently by the opening and closing of ion selective channels in the cell membrane. This transient depolarization and subsequent repolarization is commonly known as the action potential and it is virtually the same in all animals which possess nervous tissue. The action potential generated in the squid giant axon is phenomenologically and mechanistically the same as an action potential generated by a salamander olfactory cell or a cat retinal ganglion cell or vitually any other nerve cell in any other animal. And it is the action potential which is the communicative coin of all nervous systems.

Information in nervous systems is encoded in action potentials in two ways. Since the action potential itself is an electrical event of invariant amplitude and time course (that is each action potential is exactly like all others in its electrical form) the system is essentially binary in nature, either there is or is not an action potential. Except in a few specialized instances there are no graded responses.

The two main features of action potentials which encode all the information carried by the nervous system are the frequency with which they are generated and the pathway along which they travel. The frequency, which is generally in the range of 0 to 100 hz, encodes intensity; the higher the frequency the more intense the message. For example, in sensory cells the greater the frequency of action potentials the brighter the light or the louder the sound, etc.

The pathway encodes quality.   That is, visual information is visual because it comes from retinal cells whose axons all find their targets in the cerebral areas which process visual information. It is at least theoretically true that if one could re-route retinal axons to the aural portions of the brain then one would "hear" light.

All of the above description of the action potential, a fundamental unit of neural functioning, is true for all known nervous systems on the planet earth.

Indeed even beyond the production of the action potential by single neurons the method by which one neuron communicates with others is remarkably consistent throughout the phyla. It is accomplished through specialized structures known as synapses. The actual mechanism involves the release of certain molecules, neurotransmitters, from one cell into the synaptic region. This release is in response to the electrical depolarization of an action potential reaching the synaptic terminal of a nerve cell.   The neuro-transmitters, once released, diffuse across the small synaptic space between the two cells and are picked up by specialized receptors on the surface of the target cell.   When the receptors bind the neurotransmitter they cause an action potential to be produced in the target cell and thus the message is relayed from cell to cell. This description of neuron to neuron communication is virtually universal (at least in so far this biosphere is concerned).

The neurotransmitters themselves belong to a few classes of chemical compounds and number no more than a few dozen, although new ones are still being discovered.   They are all low molecular weight molecules or small polypeptides. Many also serve as hormones whose locus of action may be in the nervous system or elsewhere (i.e. the gut, liver, heart, kidneys, etc.).

The ubiquitous occurrence of these substances in all animals is cause enough for interest, but even further many of them are closely related to, or even identical with plant compounds; thus

we find that many neuroactive drugs are plant derived. Primitive colonial plant cells may have used them in a rudimentary intercellular communication system. Apparently these chemicals share a very distant evolutionary connection even before nervous systems per se arose (9).

Of course there are also differences, some subtle some quite obvious. The human brain is larger than most, especially when it is considered in relation to the size of the body within which it resides. It is also among the most complex at the level of cellular architecture; the density of neurons is high and the number of synapses between the cells is extremely large (there are some 24 trillion cells in the human brain and some of them are known make as many 100,000 synapses with other cells). Certain structures are enlarged while others are quite rudimentary; the forebrain, presumed site of abstract thinking, is quite considerable while the olfactory bulbs are shrunken. But for the most part these differences are at levels of development which might be called elaborations, the fundamentals are held in common with the rest of the fauna.

Perhaps the only really substantial improvement introduced in vertebrate, and particularly mammalian, nervous systems is that many axons are ensheathed in an insulating material known as myelin. This has the important effect of increasing the speed and to some extent the integrity of the conducted electrical signal, especially over long distances.

Otherwise the differences are ones of elaboration, even though some of the elaborations are substantial. However since the underlying mechanisms are so similar we might be forced to consider that neural material, when it arises, is virtually equipotential no matter in what form it develops initially. Whether it be the spider, the squid or the squab; the honeybee, the hippopotamus or the human, complex nervous systems can develop from simple material, and there is no neurobiological reason to suppose that only the human form is capable of developing an inventing, warring, painting, researching kind of nervous system.

NOTES

1. Hodgkin, A.L. and Huxley, A.F. (1952). A quantitative description of membrane current and its application to conduction and excitation in nerve. **J. Physiol. 117**:500- 544. This classic contribution to the neurobiological literature is thoroughly reviewed in virtually every advanced undergraduate and graduate textbook in the field.
Particularly recommended is **Ionic Channels of Excitable Membranes**, Bertil Hille (1986), Sinhauer, MA (pp 23-58).

2. Kandel, E.R. (1976) **Cellular Basis of Behavior.** Freeman, San Francisco.

3. Carew, T.J.and Sahley, C.L. (1986) Invertebrate learning and memory: from behavior to molecules. **Ann. Rev. Neurosci. 9:** 435-488.

4. Heidmann, T. and Changeux, J.-P. (1978) Structural and functional properties of the acetylcholine receptor protein in its purified and membrane bound state. **Ann. Rev. Biochem 47:** 317-357.

5. Stent, G.S. and Weisblat, D. (1982) The development of a simple nervous system. **Sci Am 246**:136-146.
A sizeable review which suggests the breadth and depth of neurobiological studies confined to the leech is **Neurobiology of the Leech**, K.J. Muller, J.G. Nichols and G. S. Stent, eds. Cold Spring Harbor Laboratory, Cold Spring Harbor, NY (1981).

6. Fentress, J.C. (ed.) (1976). **Simpler Networks and Behavior.** Sinhauer, Sunderland, MA.

7. Dumont, J.P.C. and Robertson, R.M. (1986) Neuronal circuits: an evolutionary perspective. **Science 233:** 849-853.

8. Dowling, J.F. **The Vertebrate Retina: An Approachable Part of the Brain** (in press).

9. An interesting non-technical discussion of the implications of flora-fauna common biochemistry as it relates to the brain appeared recently as: Roth, J. and LeRoith, D. (1987) Chemical Cross Talk: Why human cells understand the molecular messages of plants. **The Sciences (NY Acad Sci) 27,3:** 51-54.

CAN WE COMMUNICATE
WITH OTHER SPECIES ON THIS PLANET?
(Pragmatics of Communication between Humanoid and Non-humanoid species)

D. Reiss

Marine World Foundation, San Francisco State University
Vallejo CA 94589, USA

ABSTRACT. Preliminary discussions of the possibility of communication with extraterrestrials requires a careful examination of '(1) our own species specific or anthropocentric biases regarding the subject, (2) what we actually know about our own forms of human communication, (3) what we know about the communicative and cognitive abilities of other species, and (4) the possibilities and evidence for communication with other species on this planet. Comparative psychologists and cognitive ethologists are discovering surprizing complexity and plasticity in the communication, orientation, and navigation systems of many species. Ethological and interspecies communication research has provided evidence that diverse species either use or can learn to use, to different degrees, symbolic or referential communication for intraspecific or interspecific exchanges. This suggests that there may be a convergence or continuity in the communication and cognitive abilities in animals from different evolutionary paths.

1. INTRODUCTION

I suspect that I was invited to this conference because my research is involved with studying and communicating with a non-terrestrial, non-humanoid species, the bottlenose dolphin (Tursiops truncatus). My research investigates how dolphins communicate within their social group and aspects of their problem solving and learning abilities. In general I am interested in questions concerning intelligence and the communication systems and dynamics in other species and our epistemological and methodological approach in this area of research. In considering the possibility of detecting and communicating with extraterrestrial life forms, technology, or intelligence, certain viewpoints have been expressed suggesting that it is improbable that we could communicate with a non-humanoid intelligence (Tipler, 1980; Dobzhensky, 1972; Simpson, 1964; Mayr, 1978). This idea seems to be based on an underlying yet unsupported assumption that we humans possess a type of intelligence that has developed due to our physical and cultural evolution and that we should not expect to find anything

253

G. Marx (ed.), Bioastronomy – The Next Steps, 253–264.
© *1988 by Kluwer Academic Publishers.*

similar to it in type in other life forms. Implicit in this argument is the view that we cannot communicate with other forms of intelligence or with other species. Well, we really have no way of predicting at this time what an extraterrestrial being or intelligence might be like if it did exist and arguments regarding the probability of the existence and evolution of other humanoid or non-humanoid extraterrestrial life are based on insufficient data or subjective probability statements (Fine, 1973). In reviewing past discussions and different viewpoints regarding the probability and possibility of extraterrestrial intelligence and communication (Sagan, 1973; Regis, 1985) it was apparent that astrophysicists, biologists, and psychologists encounter some similar problems in exploring non-humanoid intelligence and communication on this planet as well as elsewhere in the universe. These problems involve (1) signal detection and pattern recognition problems when we are uncertain as to the nature or even existence of a signal, (2) how we decode and attempt to communicate with another species, and (3) how we look for and recognize 'intelligence' in other life forms.

The view that man is alone or unique as an intelligent life form is also heard in discussions concerning the relationship of homo sapians to the rest of the animal world. Throughout history humans have been considered to be the only species capable of symbolic or referential communication and tool use. Increased observations and improved methodologies for studying the behavior and communication patterns in other species have provided evidence that we are not alone in these abilities (see review in Griffin, 1981, 1982). Therefore, it would be advantageous in theorectical discussions of the possibility of communication with extraterrestrials to carefully examine (1) our anthropocentric biases regarding the subject of other life forms either on this planet or elsewhere in the universe possessing intelligence and the capability for interspecies exchanges, (2) what we actually know about our own form of human communication, (3) what we know about the communication or cognitive capabilities of other species, and (4) the possibilities and evidence for communication with other species on this planet.

2. SEARCHING FOR INTELLIGENCE

With the diversity of species on this planet we have a natural laboraory in which to investigate questions concerning the intelligence and communication of other species. But what are we looking for in searching for 'intelligence' and how do we, who supposedly possess it, look for it in other species? To date, there is not a coherent or satisfactory definition of intelligence for either human or non-human species. Intelligence means different things to different people. Although the nature-nurture question still fuels many a debate, the growing view is that, intelligent behavior, like other forms of behavior, is due to an epigenetic process, involving both the genes and environment (Barlow, 1983). A review of the current literature on the psychology, biology, and sociology of intelligence in humans and animals reveals a real confusion that must stem in part from a lack of definition of its nature. However, as Barlow has suggested, our theoretical understanding of information processing that must underlie 'intelligence' has influenced our approach to investigating it and we are now asking more about what it does for us, its function, rather than what it is.

Ashby (1956) has described intelligent behavior in man, other animals, or machines as the power of appropriate selection. There is a general consensus that intelligence is related to the way information is processed and used and involves sorting and pattern recognition of diverse stimuli (perception and encoding), storage and recall of information (memory), development of connections and integration of information (associationistic learning), the use and application of previously acquired information to new situations (learning) and the ability to appropriately select the correct information without trial and error procedures (power of appropriate selection) in similar, changing or novel social and environmental conditions (plasticity of behavior). The power of approriate selection or appropriate responding in a variety of situations and plasticity or flexibility and versatility in the acquisition and use of information seems to be a hallmark of what we perceive or recognize as intelligent behavior in humans or other animals. In discussing the ability of vertebrate species to appropriately respond to diverse and changing stimuli in their environment, Barlow (1983) states that "stimuli get associated with other stimuli and the basic problem is that the number of possible associations rises nearly as the square of the number of possible events for paired associations. This means that the task for looking for associations among the enormous number of signals that are used in a vertebrate brain is that of finding a path through a "combinatorial jungle" of quite unimaginable complexity". Barlow adds that in glimpsing this jungle one realizes that finding useful paths through it must be in part what intelligence is about. Macphail (1982) suggests that representational capacity or the role of language serves two functions. One is the communicatory aspect and the other is for encoding, categorizing, and storing stimuli. This representational scheme, he suggests, is what enables us to find useful paths through the "combinatorial jungle". Comparative neurophysiology suggests that the basic characteristics of neurons, synapes, and neuroendocrine mechanisms are similar (see Firestein , this volume for review). But does continuity in neurophysiology suggest continuity in the processing of information or cognitive abilities? To begin answering these questions we must find better ways of observing, analyzing and assessing the behavior of other species. However, in estimating our ability to assess intelligence and other cognitive abilities in humans and and other species we require information about sensory capabilities and detection thresholds of the species, signals used and perceived by the species, the past history of an individual in terms of what experience it has had with different stimuli in different conditions, and the nature of the social and physical environment of that individual. Ethological and laboratory studies conducted by comparative psychologists and ethologists are providing evidence that there is surprizing complexity and plasticity in the behavior, communication, orientation and

navigation systems in diverse species. Field and laboratory settings
have also demonstrated that some other species either use or can learn
to to use symbolic or referential codes (codes that refer to or stand
for other things) to communicate in intraspecific or interspecific
exchanges (reviewed in Griffin, 1981).

3. ADAPTIVE VALUE OF INTELLIGENCE

If we view intelligence as a capacity for information processing and the
power of appropriate selection as discussed previously, then it could be
argued that there is an adaptive value or economy of such an information
processing system for organisms whose goal is survival. This capacity
would be particularly important for social species which exist in ever
changing social and physical environments. Humphrey (1976) and Jolly
(1966) have suggested that conscious awareness in primates had to
develop as individuals had to attend to and use information about the
emotional states and intentions of other members of their social group.
In highly social species it is important for group members to be able to
"read" the signals, both overt and subtle, of group members and in some
cases of other species. This ability to interpret and act appropriately
to multitudes of sequential or concurrent signals from various and
changing sources may require an organism to process and use information
in flexible and economic ways. Therefore, we might consider it to be of
selective value for a social species to (1) have the ability to react to
new social and physical stimuli (new situations) in appropriate ways for
survival, (2) apply past experience to new situations, (3) form concepts
and generalizations in order to integrate and use old and new informatin
in a changing environment, and (4) respond appropriately to new
situations by applying past experience (information) without the time
consuming and often fatal process of trial and error.

4. PLASTICITY IN BEHAVIOR AND COMMUNICATION

Communication is a ubiquitous phenomenon found throughout the animal
world. As mentioned previously research in field and laboratory
settings have demonstrated surprizing versatility and plasticity in the
behavior and communication in species from different evolutionary paths.
The pioneering experiments of von Frisch (1967) and subsequent work by
Lindauer (1971) and Gould (1975,1979,1982) have revealed that the
social insect, the honeybee, symbolically communicates information to
other members of its hive about the direction, distance, and
desirability of food sources through a "waggle dance". The bees'
ability to communicate information about a food source which it visited
but is not present in the immediate environment when the information is
transmitted, suggests that the communication shows the characteristic of
displacement. That is, that information can be exchanged about things
not present in time or space. This research has also revealed the
capacity for bees to learn. In an experiment a food source was
presented to bees and then its location location and distance from the

hive was changed by increased increments over time by the experimenter. Researchers reported that the bees would fly beyond the last location of the food source as though anticipating the new location based on past experience or learning (see Griffin for review, 1984). This type of behavior shows the appropriate use of information acquired through past experience, or learning, and a degree of plasticity of response to a changing environment.

For the past twenty years it was believed that song birds could only learn their species specific songs and not the songs of other bird species. The learning of their own species song was shown by Marler (1970) to be be governed by a sensitive phase or time-dependent period for song acquisition, and due to a "neural template". This research was well accepted as it easily `fit' the ecological picture of different species having different songs or song dialects to serve as isolating mechansims between different groups as well as a bonding mechanisms within a group. However, Marler`s experiments were conducted by playing model songs to his subjects in the laboratory using tape recordings of bird song. More recent work using live and interacting birds as `tutors' in laboratory conditions have shown much more plasticity in the song acquistion of various species of birds (Baptista, 1972, 1983; Kroodsma and Pickert, 1980). By allowing their subject to socially interact (see, hear, and respond to) birds of their own species or different species, it was found that not only can birds learn songs beyond the proposed sensitive period but that they can even learn songs or elements of songs of other species of birds. This suggests that social interactions and social modelling of behavior may play a critical role in acquiring and developing communication in birds (see Pepperberg for review, 1985).

The critical value and importance of social interaction has been stressed by virtually all researchers investigating interspecies communication between humans and other species (Gardner and Gardner, 1969; Premack, 1976; Fouts, 1973; Rumbaugh, 1977; Patterson and Linden, 1981; Pepperberg, 1981, 1985, Reiss, 1981). Species from diverse phyla have shown the ability to learn, to varying degrees, symbolic or referential codes and to use these codes in interspecfic exchanges with humans.

Two factors that appear to be highly significant in developing and maintaining these interspecies exchanges are the use of an appropriate communication code for the interacting species and the social interaction between the participants. The initial success in teaching a non-human a symbolic code was with a young chimpanzee, Washoe (Gardner and Gardner, 1969). Prior to this pioneering work were several unsucessful
attempts to teach chimpanzees spoken English, an anthropocentric and inappropriate modality for non-human primates. The success therefore, in the Gardners approach was in selecting the appropriate modality, American Sign Language for the Deaf for their communication code with the chimpanzee. This decision was based on observations of chimpanzee natural behavior and communication in which these primates use a wide variety of gestures in their own exchanges. While chimpanzee vocalizations were not easily modified, their non-vocal behavior was

much more plastic.

Pepperberg (1981, 1983) has demonstated that by using a method called the "model-rival" technique she was able to teach an African Grey parrot, named Alex, to use functional English vocalizations to identify, request, classify, and vocally label numbers of objects. While vocal English was inappropriate for non-human primates, it was highly appropriate for the parrot whose vocal ability is highly plastic as evidenced by their well known mimetic abilities. The model-rival technique is a social situation in which the experimenter uses another person to <u>model</u> the desired vocal behavior for the bird and also functions as a <u>rival</u> for the bird for the experimenter's attention. Alex was shown how to request, obtain, and identify various items and activities by using this technique and giving him the actual referents themselves as the reinforcer for his correct vocal behavior rather than using an extrinsic food reinforcement.

While there is still the semantic issue as to whether other species can learn language or communication codes, most critics agree that these studies do demonstrate that some other species can learn referential and symbolic associations and can use them to communicate in rudimentary exchanges.

What do the results of these initial interspecies studies suggest about other communication systems? Should we expect to find a convergence in plasticity and referential forms of communication and intelligence in other species? It is difficult to predict how widespread these abilities might be through the animal world but several studies have indicated the use of referential communication in the natural communication systems of different species.

Historically the vocalizations of other animals have been thought to be emotional rather than referential in nature. This notion was based on a lack of evidence to show otherwise. Again, it was generally assumed that man was unique as a symbol or language using species.

Field studies have indicated that members of a social group of free ranging vervet monkeys use different alarm calls that appear to refer to different predators in their environment (Seyfarth, Cheney, and Marler, 1980). These apparent referential calls refer to specific predators such as leopards, martial eagles, and pythons. Different escape strategies are used by group members after the calls are produced. Infant vervets seem to generalize these calls to a wider range of animals in the environment and have to learn predator calls through experience. A similiar use of referential predator alarm calls has been reported in the California ground squirrel (Owings and Leger, 1980). In both the case of the vervets and the ground squirrels critics have argued that the calls might refer to different escape strategies. In either case the nature of the calls is referential rather than just emotional responses to environmental stimuli as was once thought. Ethologists are extending these types of investigations to other species in order to understand how extensive or widespread such referential communication may be in other social species.

4. THE DOLPHIN: AN INTERESTING MODEL FOR STUDYING A NON-HUMANOID INTELLIGENCE

4.1. Dolphin Communication: The Problem of Signal Detection

Cetacean communication and intelligence has been the subject of much interest and speculation in the scientific and popular literature. This interest is due in part to past speculation by some investigators (Lilly, 1961; Dreher, 1961; Dreher and Evans, 1962) that this large brained (approximately 1600 grams) highly gregarious mammal might possess a linguistic form of vocal communication. It was demonstrated that bottlenose dolphins do exchange vocal signals and exhibit turn taking behavior inthese exchanges (Lilly and Miller, 1961; Dreher, 1966). Later studies reported that while vocal exchanges do occur when physically separated but linked acoustically, there was not evidence from their acoustic repertoire and patterns to support the theory that they use a linguistic form of communication (Lang and Smith, 1965; Bastian, 1966; Gish, 1979).

The bottlenose dolphin is a mammal which has evolved to a completely marine existence. The extent of its adaptations to this environment is striking, including gross morphological, physiological, and perceptuo-motor variations ofthe normal mammalian patterns. Certainly one of the least well understood of these adaptations is the communication system of this highly social species. Our lack of understanding of dolphin communication stems in part, I believe, from our use of an inadequate framework or model of what we expect their communication signals to be like. Field and captive studies have indicated that these mammals, like many other mammals, use several modalities for communication including visual, tactile, acoustic, and perhaps gustatory channels (Tavolga, 1983). However, research on the multimodal communication and how these diverse signals are integrated and used in their communication has not been the approach generally taken. Rather, there has been an experimental bias towards focusing on the acoustic channel and their vocal communication. It is easy to understand why the acoustic mode would be priorily singled out as the primary signalling system. First, the hypothesis that the acoustic mode would be the optimal channel for communication in an aquatic environment focused attention towards vocal behavior and away from visual/kinesthetic behavior. Second, the size and structural complexity of the dolphin brain led to the speculation regarding linguistic abilities in this species. Thirdly, most of the typical mammalian appendages and morphological structures such as ear flaps, tails, erectile hairs, etc., that normally can be used to convey information between other animals and also to the human observer, have been lost in the dolphin's adaptation to a marine existence. This radical steamlining coupled with the idea that acoustic signals would be the optimal means of communication in the aquatic environment has inhibited systematic inquiry as to the integrated communication process.

Gish (1979) has noted and I have also observed in my research of dolphin communication, that there are relatively few vocal exchanges between dolphins when they are physically together in the same tank and

able to use other signals for communication. In my research conducted
in captivity with a social group of four bottlenose dolphins at Marine
World Africa USA in Vallejo, California, I have observed the important
role postural, spatial, tactile,and other visual signals play in
conjunction with vocal signalling.

The reason I am stressing this point and explaining the history of
how we have approached studying the dolphin is to emphasize that we must
stay open to the varieties of possible sources of signals when we try to
decode and communicate with other non-human life forms. Biologists as
well as astrophysicists are first faced with the problem of signal
detection, what is the signal and what is noise. Out of all the
behavior (noise) what are the intentional signals that the animals use
to communicate? As a member of the human species we are initially blind
and deaf to the signalling systems of these mammals. However, with
increased experience and the requisite instrumentation for recording the
signals we can begin the process of pattern recognition and correlations
between signals (both vocal and non-vocal) and changes in behavior.
Once we can observe the signals and patterns of signalling we can begin
to try to understand the "meaning" or function in social interactions.

4.2. Plasticity in dolphin behavior

Why do we think the dolphin is intelligent? Most of our observations
come from captive situations or a limited number of field observations.
Yet, most people, both laymen and scientist, recognizes something in the
behavior of this species as 'intelligent behavior". I would like to
suggest that it is the plasticity of behavior that these animals
demonstrate, that provokes this feeling in us. For example, the ability
for behavioral and vocal mimicry has been reported in the literature (
reviewed in Herman, 1980). For example, Tayler and Saayman (1973)
reported that a captive bottlenose dolphin would imitate the behavior
and postures of other species in its tank. This dolphin was also
observed to imitate a diver cleaning algae from its tank. The dolphin,
after observing the diver for some time picked up a piece of broken tile
from the tank, and holding it in its mouth proceeded to scrape algae
from its tank as the diver was doing. At the same time it also emitted
a bubble steam and produced a series of sounds that resembled the sound
of the air bubbles escaping from the diver's regulator. This account
also suggests that this non-handed species is capable of tool use.

The observation of this animal's behavior is a good
characterization of what the intelligence of these animals is like.
After studying these creatures for seven years I would also add that
they often seem like experimental psychologists, as they frequently test
the contingencies of their behavior. They are extremely quick to learn
new patterns of behavior in our formal and informal interactions with
them. In fact, I when I first began my research at Marine World, I was
repeatedly warned by an experienced dolphin trainer, not to adhere to
any set pattern with the dolphins because they would quickly learn it
and exploit it. For example, the conventional way that trainers
communicate to a dolphin that it has done the wrong thing repeatedly is
to give it a 'time out', by simply walking away and discontinuing the

activity and interaction for a period of time. Upon returning to the
dolphin, it usually gets the idea it did something wrong. While feeding
our dolphins, if they swim away we generally give them a time out, and
we made a rule that they get two time outs and if they continue to swim
away we end the feed. Our dolphins as we had been warned,quickly
learned the rules of the game and pushed them to the limit. They would
still occassionally swim away during a feed but would not leave a third
time. We set patterns and they quickly learned them. But, this is
essentially what many animals do. Our house pets quickly learn our
behavior patterns and synchronize their patterns with ours. I believe
this is very basic to communication and in social species we see this
tendancy for different species to adapt to the patterns of others in
their environment.

The dolphin, however, still shows a much higher degree of mimetic
vocal behavior than has been reported in other species outside of birds
and humans. Richards, et.al., (1984) has demonstrated the ability of
the dolphin to learn to mimic sample sounds generated by a computer. In
this experiment, the dolphin was trained using fish reinforcement to
mimic a variety of acoustic stimuli. In our research program at Marine
World, we provide our social group of dolphins with an underwater
keyboard. The keyboard consists of a 3 x 3 key matrix. The keyboard is
interfaced with an Apple computer which generates specific whistles into
the tank when the dolphin presses a particular key displaying one of a
set of three-dimensional visual forms. These different shaped forms are
intended to represent different objects and activities which the dolphin
can request. Locations of the different visual forms are randomized on
the keyboard. Therefore, if the dolphin selects a particular form or
visual symbol, it hears a computer generated whistle in its tank, and it
is given a particular object or activity, such as a ball, a ring, or a
rub. In this way we give the dolphin the opportunity to explore the
contingencies of keyboard use. The animals are feed three times a day
and the object and the activities the dolphins can obtain are the only
systematic rewards provided. This system was designed to offer the
dolphins stimuli in different modalities to give them a choice of
modality. While all the results of the experiment cannot be presented
within the scope of this paper, the most significant finding was that
the dolphins quickly began to spontaneously mimic the computer whistles.
IN the first weeks of the experiment they not only imitated the whistles
but but produced them before using the matching visual key. In the
following months they began to use some of the whistles appropriately in
other situations such as whistling the 'ball' whistle when playing with
a ball during a dolphin to dolphin exchange of a ball.

Another striking example of the plasticity of dolphin behavior are
observations of our dolphins at Marine World in both our research pools
and separate performance pools, creating their own "toys" and then
interacting with them. On numerous occassions we have observed the
dolphins assume a horizontal position near the tank floor and jerk their
heads upward slightly, releasing what looks like a silver ring of air.
The dolphins then watch the ring, follow it to the water surface, and
often touch it or swim through if the ring diameter is wide enough
(Fig.1).

6. CONCLUSION

In our exploration of the varieties and nature of communication and
intelligence in other life forms on this planet we are finding a higher
degree of complexity and versatilty than was expected. Ethological and
interspecies research is suggesting that some species from diverse
evolutionary paths show sufficient plasticity in their learning and
behavior for them to be able to acquire new types or codes of
communication for at least rudimentary interspecies exchanges. However,
much more research is needed to determine the extent of interspecies
possibilities or limitations.

If we consider intelligence to be closely related to information
processing, the power of appropriate selection, and behavioral
plasticity, then we might view intelligence as having adaptive
significance and survival value, at least in social species, and expect
to find a convergence toward intelligent behavior throughout the social
species on this planet. We can only speculate about other worlds.

Fig. 1. A dolphin watching and following an air ring which it has just
produced.

REFERENCES

Ashby, R. 1956. **An Introduction to Cybernetics.** New York: Chep and Hall Ltd.

Baptista, L.L. 1972. Wild house finch sings white-crowned sparrow song. **Z. Tierpsychol.** 30:266-270

Barlow, H.B. 1983. Intelligence, guesswork, language. **Nature.** Vol.304

Bastian, J. 1966. The transmission of arbitary environmental information between bottlenosed dolphins. **Final Report. Contracts.** No. 60530-10046 and No.60530-11450.

Dobzhansky, T. 1973. **Genetic Diversity and Human Equality.** Basic Books. New York.

Dreher, J.J. 1966. Cetacean communication: small group experiment. In: **Whales, Dolphins and Porpoises.** K.S.Norris, ed. Calif. Univ. of Calif.Berkeley Press.

Dreher, J.J. and Evans, W.E. 1962. Linguistic consideration of cetacean sound production. **Lockheed Calif. Co. Report.** No. 16175.

Fine, T. 1973. Nature of probability statements in discussions ofthe prevalence of extraterrestrial life. In: **Communication with Extraterrestrial Intelligence CETI.** C.Sagan. ed. Cambridge, Mass. MIT Press.

Frisch, K. von. 1967. **The Dance Language and Orientation of Bees.** Cambridge, Mass. Harvard Univ. Press.

Gardner, R.A. and Gardner, B.T. 1969. Teaching sign language to a chimpanzee. **Science,** 165:664-672.

Gish, S.L. 1979. A quantitative description of two-way acoustic communication between captive Atlantic bottlenosed dolphins. Ph.D. Dissertation. U.C. Santa Cruz.

Gould, J.L. 1975. Honeybee communication: the dance-language controversy. **Science,** 189:685-693.

_____. 1979. Do honeybees know what they are doing? **Nat. Hist.** 88. 66-75.

_____. 1982. **Ethology, the Mechanisms and Evolution of Behavior.** New York. Norton.

Griffin, D.R. 1981. **The Question of Animal Awareness.** New York.Rockefeller Press.

_____. ed. 1982. **Animal Mind-Human Mind.** New York. Springer-Verlag.

_____. 1984. **Animal Thinking.** Cambridge, Mass. Harvard Univ. Press.

Herman, L.M. ed. 1980. **Cetacean Behavior: Mechanisms and Functions.** New York. Wiley Press.

Humphrey, N.K. 1976. The social function of intellect. In: **Growing Points in Ethology,** P.P.G. Bateson and R.A.Hinde, eds., Cambridge,Mass. Cambridge univ. Press.

Jolly, A., 1966. Lemur social behavior and primate intelligence. **Science,** 153: 501-06.

Kroodsma, D.E. and Pickert, R. 1980. Environmentally dependent sensitive periods foravian vocal learning. **Nature.** 288:477-479.

Lang, T.G. and Smith, H.A. 1965. Commuication between dolphins in separate tanks by way of an electronic acoustic link. **Science,** 150. 3705.

Lilly, J.C. 1961. **Man and Dolphin.** New York. Doubleday.

Lilly, J.C. and Miller, A.M. 1961. Vocal exchanges between dolphins. **Science,** 134: 1870-1873.

Lindauer, M. 1971. **Social Communication Among Bees.** 2nd ed. Cambridge,Mass. Harvard Univ. Press.

Macphail, E.M. 1982. **Brain and Intelligence in Vertebrates.** Oxford, Clarendon Press.

Marler, P. 1970. A comparative approach to vocal learning: song development in the White-crowned sparrows. **J. Comp. Physiol.** 71: 1-25.

Owings, D.H. and Leger, D.W. 1980. Chatter vocalizations of Calif. ground squirrels: predator and social-role specificity. **Z. Tierpsychol.** 54:163-84.

Patterson, F.G. and Linden, E. 1981. The Education of Koko. New York.
Holt, Rhinehart and Winston.
Pepperberg, I.M. 1981. Functional vocalizations by an African Grey
parrot (Psittacus erithacus). **Z. Tierpsychol.** 55:139-160
_____. 1983. Cognition in the African Grey parrot: preliminary
evidence for auditory/vocal comprehension of the class concept.
Anim. Learn. Behav. 11: 179-185.
_____. 1985. Social modelling theory: a possible framework for
understanding avian vocal learning. **Auk,** vol. 102, 854-864.
Premack, D. 1976. **Intelligence in Ape and Man.** Hillsdale, N.J., Erlbaum
Regis, E. 1985. **Extraterrestrials, Science, and Alien Intelligence.**
Cambridge, Cambridge Univ. Press.
Reiss, D. 1983. Pragmatics of Human-Dolphin Communication . Ph.D.
Dissertation. Temple Univ. Phila. Penna.
Richards, R.G., Wolz, J.P. and Herman, L.M. 1984. Vocal mimicry of
computer-generated sounds and vocal labeling of objects by a
bottlenosed dolphin (Tursiops truncatus)
J. Comp. Psychol. Vol. 98 No. 1. 10-28.
Rumbaugh, D.M. 1977. **Language Learning in a Chimpanzee:The Lana Project.**
New York, Academic Press.
Sagan, C. ed. 1973. Communication with Extraterrestrial Intelligence
(CETI), Cambridge, Mass. MIT Press.
Seyfarth, R.M., Cheney, D.L. and Marler, P. 1980. Monkey responses to
three different alarm calls: evidence for predator classification
and semantic communication. **Science,** 210: 801-803.
Simpson, G.G. 1964. **This View of Life.** New York. Harcourt.
Tavolga, W.N. 1983. Theoretical principles for the study of
communication in cetaceans. **Mammalia,** 47, No. 1.
Tayler, C.K. and Saayman, G.S. 1973. Imitative behavior of the Indian
Ocean bottlenose dolphin (Tursiops aduncus) in captivity.
Behaviour, 44: 286-297.
Tipler, F.S. 1980. Extraterrestrial intelligent beings do not exist.
Q. Jour. R. Astr. Soc. 21, 267-281.

SETI AND THE PHYSICAL SCIENTISTS MISCONSTRUAL OF EVOLUTIONARY BIOLOGY

E.J. Coffey
481B Nornsey Road
London N19-3Q1 England

It was the total alienness of the creature, during the strange encounter, which was most startling for me. Why it should be so I cannot quite say for I had long known of its existence: the long spindly legs, the pink colouration of its delicate body, that extended neck, the strange apparatus with which it strained the water for food from that curiously inverted posture.

Knowing of such a creature from books, or TV programmes, however excellently portrayed, could not prepare me for an actual encounter. For what did that encounter show me? Here was a creature supremely indifferent to my concerns, assuming that such an anthropomorphism has any value at all in such a context. It moved away, towards the nearby pool of water, to join its fellow kind, or to preen further. Yet what remains with me, above all else, is this: the mystery, the utter mystery of that creature's existence; that, whatever its effect upon me, the nature of its experience, or even whether it had any, would forever remain beyond my grasp.

A fact too easily forgotten by the physical scientist supports of SETI is that it is our own human constitution which biases us against appreciation of how truly alien extraterrestrial creatures will be: physically, behaviourally, and cognitively. We cannot suppose they will have our concerns, or that what is important to us could be of any real significance to them. We will not find ourselves mirrored in them for, as with the flamingo, we will long ponder their strangeness and, gazing forth upon the universe, eternally wonder why they will never by like us.

G. Marx (ed.), Bioastronomy – The Next Steps, 265.
© 1988 by Kluwer Academic Publishers.

CAN WE COMMUNICATE WITH ALIENS?

V. Csányi and Gy. Kampis
Department of Behaviour Genetics, Eötvös University
Jávorka Sándor 14, Göd 2131 Hungary

ABSTRACT. Theoretical problems of communication between intelligent systems are discussed in a framework provided by ethological studies of human and animal intelligence. Intelligence is understood as an ability of a system to perform goal-directed behaviour on the basis of internal models of its environment. Communication is defined here as an exchange of information related to these internal models. This is shown to be possible only between systems which share a common environment and construct similar environmental models. Therefore, a direct meaningful communication with extraterrestial intelligence is highly improbable. Some possibilities to achieve this are, nevertheless, imagineable.

1. INTRODUCTION

The question in the title of this paper can be put forward also to cultural anthropologists and ethologists and both have the necessary knowledge to provide answers. The subject of this paper is: What answer can be given if the aliens are neither animals nor humans of an unknown culture, but specimens of an extraterrestrial intelligence? If we generalize this question, we may ask, how and to what extent is communication possible among intelligent beings?

2. INTELLIGENCE: AN ABILITY OF MODELLING THE EXTERNAL WORLD

In the realm of sciences of the animal and the human, the otherwise gloomily defined term <u>intelligence</u> can be described in relation of the nervous system to the external world.

For the nervous system of the simplest animals, the external world consist entirely of a few stimuli. The motor apparatus of the animals is not activated directly and exclusively by external stimuli reaching the receptor, but also by an interposed interneuron--type network. The two together shape and elicit the response reaction.

267

It has been shown elegantly by Donald MacKay (1) that this interneuron-network forms an internal representation of the external world of the animal. This internal representation of the environment serves as the basis of purposive behavior of any animal being. Such goal-directed behavior is an essential component of intelligence; usually we consider an intelligence the higher the more complex ways it can achieve its goal.

The internal representation of the environment can be considered as an abstract construction, which is essentially a model, more precisely, a dynamic model of the environment. Under the term 'model' we mean a system-theoretical concept. That is, a model is always a simpler system in which the components and the interactions of components reflect the components and some interactions of a more complex system. Model building, therefore, is always a kind of simplification and a special identification between two different systems, of which one is the model and the other is the system being modelled. Now the way a model is used is to let it operate in its domain; based upon its operation, predictions can be made on the behavior of the system being modelled.

The environment of an animal is an enormously complex system, and its simplified model constitutes the internal representation constructed by the animal brain. This model captures environmental factors and interactions, which are most important for the survival and reproduction of the animal.

The most important biological function of the animal brain is to construct this dynamic model of the environment, to maintain and operate the model continuously, and to use the data obtained by this operation for predictions, in the interest of the survival and reproduction of the animal (for a detailed discussion of this conception, see (2) and (3).

More concretely, a neural model is not only a simple projection, but a kind of complex reconstruction containing also instructions of the possible behavior of the organism in response to the stimuli of the external world. The animals' activity is influenced by expectations and internal analyses of situations based on the model formed in the nervous system. It is used by the brain as an internal reference in the control of various actions, for example, in eliciting fear, orientation, attack and defence, and avoidance of predators.

Examining the internal structure of models in animal brains, one finds that such models are functionally closed structures built from neural interconnections and neural excitation patterns as components. These components may be determined either genetically or through experience. From a functional viewpoint, components can be characterized by triplets of key - reference structure - action. A key is related to perception and to key stimuli in the ethological sense. An action is understood as a possible behaviour pattern and its activity feed-back; whereas a reference structure is related to the internal state, to motivational systems and, primarily, to mechanisms of memory.

In higher animals the environmental model is primarily built on the individual experiences of the organism. Also in higher animals, the formation of an environmental model, to a great extent, involves the internal representation of the animal itself. This process has culminated in the emergence of consciousness in apes and man.

3. COMMUNICATION AND CORRESPONDANCE

Animal communication conveys information about internal states and intentions of the animal, and less often about the environment. This communication and the information provided by it becomes   an organic part of the structure of models in animal brain, but   its influence is rather limited. Animals do not, i.e. cannot communicate either the internal structure of the model, or their individual past experiences and future expectations.

A new organizational level of mental models was reached with the emergence of man. Human communication, with the aid of human language, is able to transfer messages not only on intention and internal state, but also on the own structure of models. In human language, unlike in all form of animal communication, an openended, abstract, digital code is used.

With the appearance of this language it became possible to form supermodels in human groups, as it can be found in all   cultures. Individuals of the group have identical (or almost identical) components but share only partly the set of components which belong to other individuals, and no individual has a complete set. Experiences and expectations of any individual become used collectively by the group. The predictive value of these models can be checked by any other group members; and with the death of individuals their experience does not vanish. Instead, independently of the individual, it has an own history on the level of the collective memory of the group. This forms a new organizational level of model-building activity. For further details on this, the reader is referred to (3) and (4).

Let us now turn to a more precise characterization of these phenomena. Note that the definition of communication was left open so far. We understand communication as an exchange of information between two systems which have dynamic representational models of their environment. In this definition the term "information" is what is crucial.

This concept will be used here in two different senses, based on ideas outlined elsewhere (5). In the first interpretation, it is   a knowledge of something. We call this nonreferential information, because it does not work in the system which provides it. It can be only   a component of the model built by an observer. In the other interpretation, information can be defined as a structural arrangement of building block which is able to act in the given system. This is called referential information, and it belongs to the given system as a reference frame and has some function within this frame.

A mental model is, according to the definition given previously, a dynamic representation of the environment. We suggest that it   is, therefore, a dynamic storage of nonreferential information, i.e. knowledge.

What we would like to emphasize now is that, in a communicational act, a component-transformation takes place between models which belong to different systems. That is, knowledge as it appears in one model, gives rise, through many transformations, to some actions (referential information) performed by the system. These actions are received and sensed by the other system. This builds up new components in its internal model, and through the functioning of this model, this is transformed to new nonreferential information. That is, generally it is not possible to exchange nonreferential information immediately.

These definitions allow the formulation of what we call the principle of correspondance. We speak of correspondance if two systems can exchange referential information so that it can be handled as nonreferential information in both systems. In other words, if a model--building system, as a source, is able to produce components during communication, that stand in a proper relation to the structure of the model within the receiver so that they have a nonreferential value in it; and vice versa. The degree of correspondance depends on the overlap between the referential information content of the given component in the two systems, that is, on the similarity of chains of internal effects evoked by the components in the models within the two systems.

Now our principle states that a correspondance is necessary to any communication. The higher the correspondance, the more effective is the communication.

If the degree of correspondance is low, the communicational act can still provide some information, because the receiver will "interpret" the components in its own model. This process leads to different nonreferential information (different knowledge), and the effectivity of such communication is low. For instance, a dog can learn to sit down on the command "sit!", but the correspondance is low in this case, although the command is effective from the viewpoint of the human. To us, the word "sit" means many different things, to the dog it is a sign followed by punishment if not some particular behaviour is performed after it. Although a dog can be conditioned in this way, the overall effectivity of this communication is fairly low and this comes along with inherent limitations.

4. COMMUNICATION WITH ANIMALS, HUMANS, AND ALIENS

It follows from the discussed characteristics of models and communication, that there are various inequivalent classes of communication between intelligent systems.

Communication between animals was already mentioned. This is a case in which the participants do have similar models for the representation of their environment. Therefore, transferred components have a high chance to work in the internal model of the receiver. Due to the differences of the systems, however, this does not always lead to correspondance. Also, because this communication is not able to convey information on the models themselves, there is no way to improve the degree of correspondence and it is rigidly determined by the differences of participants. We shall call this 1. type communication.

Human to human communication is a very special case. Here we
have the exchange of components which are parts of the same supermodel,
which are functional in both systems and the correspondence of which
is extremely high, due to an enormous biological and cultural similarity
of speakers. This enables an effective immediate exchange of
nonreferential information which is encoded into linguistic terms, and
understood by the receiver. This is, in turn, possible because of the
possibility of communication on the models themselves and because of
socialization that dimininshes their variety and bootstraps the
linguistic communication system. In other words, the right rules for
the processing of linguistic messages are acquisited by the participants
during a common biological and cultural ontogenesis. This type of
communication can be called II. type communication.

What about human to animal communication? This is essentially
again a I. type communication with low correspondance, in which the
exchanged components usually bear different meanings for the participants,
as in the example of the training of a dog.

And now we have reached our ultimate subject, the problem of
communication between humans and extraterrestial intelligence.

We have to realize that the possibility of II. type communication,
which is the most meaningful and desirable form of communication, is
very low. It can happen only if the body, organization, and brain
mechanisms of aliens are essentially identical to those of humans.
Even translation is possible only when there are no barriers of
linguistic communication, that is, when participants have subsets of
sufficiently common cupermodels - such as in the case of human societies.
The possibility of I. type communication, of course, exists in both
directions. Either the humans or the aliens might emit components
which can be used for some function by the other party, no matter what
these components actually are in the original system. Correspondance is
low in this case and the value of communication of this type is not
very high.

But why wouldn't there be common components with an alien, which
can serve as a starting point of communication? To a natural scientist,
it might be plausible to assume that e.g. mathematics provides such
a universal basis. Unfortunately, however, at a closer look we find
that the usual way of reasoning applied in mathematics is a specifical
trait of humans, having its roots in human culture. In fact the whole
of mathematics is subordinated to the language and social medium, in
which it is formulated. To put it more clearly: to have even 'integer
numbers', a fairly simple mathematical concept, possibly we have to
have fingers to count with, and - more importantly - we have to live in
a society where numerosity has a fundamental importance in the typical
practical activity. From a logical point of view, this does not
necessarily fulfil in any case. Moreover, mathematical concepts form a
universe, the structure of which shows laws that are products of
culture, and not of the external universe. And, finally, even if we
have a common mathematics (nonreferential information) with aliens,
how should we communicate it (i.e. transform it into referential
information), and why to think that the aliens re-construct it by

using the <u>same</u> transformation?

Ultimately, however, there is one possibility. Suppose that both parties are really <u>willing</u> to communicate (what this means in terms of system structure is another question). In this case they can look for situations in which <u>the level of correspondance can be rised</u>. This is what we would call <u>III. type communication</u>. By performing joint experiments and observations, and by performing actions in a common environment, there emerges the possibility of formation of a <u>common language</u> which has high correspondence. This type of communication is, of course, possible only among beings which form very abstract models of their environment and which are willing to build models (and supermodels) together with others. These are, unfortunately, very special requirements.

If these conditions are not met, only I. type communication is possible, which the humans can be either a source or a receiver of, depending on whether, in intuitive terms, the other intelligence is higher or lower than ours.

REFERENCES

1. MacKay, D.M. 1952: 'Mindlike behaviour of Artefacts',
    <u>Brit.J.Phil.Sci</u>. <u>2</u>, 105-121.

2. Csányi, V. 1982: <u>General Theory of Evolution</u>, Publ. House of
    the Hung. Acad. Sci., Budapest.

3. Csányi, V. 1987: <u>Evolutionary Systems</u>, Duke University
    Press, Durham, to be published.

4. Csányi, V. and Kampis, G. 1987: 'Modelling Society: Dynamical
    Replicative Networks', <u>Cybernetics and Systems</u> <u>18</u>, to
    appear.

5. Kampis, G. 1987: 'Some Problems of Systems Descriptions II.:
    Information', <u>Int.J.General Systems</u> <u>13</u>, 157-171.

CETI: HISTORICAL RETROSPECTIVE OF INTERACTION OF CULTURES

I.S. Licevitch

Institute for Oriental Studies, Academy of Sciences
Moscow, USSR

Discussing the problem of the extraterrestrial civilizations we must begin with a statement that there is not a single real extraterrestrial civilization that we know of so far. The only civilizations that we know something about are the terrestrial civilizations, and many of them have existed throughout history. Therefore, it would be quite legitimate to extrapolate the available data on contacts between the terrestrial civilizations on a possible future contact of mankind with the ETI.

It should be noted that human societies at rather early stages of development felt an anxiety to search for and explore faraway, strange lands, inhabited by strange human communities – which we now call civilizations. As early as in the 2nd century B.C., for example, the first Chinese Emperor Cin Shih-huang sent a many thousand-strong sea expedition to explore the Pacific Ocean in the hope to find the islands Penglai and Yingchang, allegendly inhabited by creatures of a higher order. The inhabitants of those islands were believed to have an ability to fly, never died, and could foresee the future and were generally superior to the humans. The Emperor very much hoped to learn from them a recipe of the elixir of immortality. Of course, his men never found it.

In the Middle Ages the Europeans were very much interested in a certain "Kingdom of Presbiter Ioann", which was believed to be situated somewhere in the heart of Asia, or possibly, in India. Most exciting legends were told about that state, allegedly ruled by a king-priest. The European rulers believed he could help them fight the moslims, and two Popes – Alexander III in 1177, and Innocent IV in 1245 sent their envoys to find the kingdom. As early as in the 15th century the Portugese sea travellers dreamed to reach the mysterious country.

However, in the epoch of the great geographic discoveries another legend gained popularity – the one about Eldorado – a country, where people were making everything,

273

G. Marx (ed.), Bioastronomy – The Next Steps, 273–277.
© *1988 by Kluwer Academic Publishers.*

including kitchen utensils, of gold. It was believed there was an abundance of gold there, while people living there were gentle and generous. Possessed by a great desire to find the rich land, the Conquistadors explored the most remote areas in South America, many times thought they found the land, and yet, it all remained just a legend.

There were many real contacts, though: the discovery of America, the conquest of India and South-East Asia, many human communities were discovered on the islands in the Pacific. Everything was going there according to the same pattern: a more developed civilization always conquered a less developed one, the population was enslaved, the local labor force was exploited, as well as the local natural resources. That pattern was not too fine for the local population, and yet that is how it all happened in human history. However, thinking about the search for ETI, the three above-mentioned legends come to mind: the Penglai legend, the Eldorado legend and the one about the Kingdom of Ioann. Everyone was talking about these lands, many travelled to find them, but it was all in vain. As we see it, human aspirations have not always been rewarded.

All the three legends, unfortunately, prove the idea that the search for contact with faraway civilizations in the past centuries was never unselfish in character — all those who were seeking for contact hoped to get something — be it gold, immortality or military aid. As for the initial ideas about a possible object of contact, the image has always been over-idealized. Otherwise, could the search be worth the effort?

Now that our planet has been totally explored, space became a natural new area for the application of the exploration instinct. Hence, the idea of searching for contact with the extraterrestrial civilizations is sort of a logical continuation of the great geographical discoveries. But what's most important is this: the hypothesis about the ETI contains the same two basic features. First it is a real interest, connected with the hope to gain something from the discovery. Second it is the same idealization of the object of contact, stimulating the searching effort.

Reports, delivered at the Bjurakan conference (1971) spoke about possible contact being "really beneficial", also about "help in solving terrestrial problems and increasing the life of our civilization", etc. Let no one speak about material gains, which were so attractive for the discoverers of the past. We are definitely hoping to gain something far more important — a giant volume of information, which will make it possible to perform an impressive leap forward, without having to accumulate our own experience for many future centuries. It seems, however, that it is here that we start idealizing our future partner in contact.

It is believed that the object of contact will be a

highly developed civilization in terms of technology: outlining the strategy of the search N.S.Kardashev says, with good reason, that an encounter with our brothers in reason, undergoing the same period of development as we do, is "highly improbable" [1]. So what can the Earth offer them, if we exclude, from the very beginning, vivisection, exploitation and experiment? ETI, according to our pattern, is given a role of a patient philanthropist, and that means that acknowledging a giant gap between us and ETI in the field of science and technology, we believe that the behavioral norms can not be radically different. In our view, morally, an ETI is same as we are – only better.

UnfortunatelHy, in human relations there seems to be only one stable psychological pattern: "one of us – alien". Which does not seem too encouraging. As for the so-called "eternal" and "common" moral norms, it should be noted that directly related to the survivability of mankind as an entity, many of these norms, as the conditions for the survival improve, tend to weaken. We have witnessed a so-called "sexual-revolution", which proclaimed the abandon-ment of all sexual taboos; respect for the elderly, which has existed for many centuries as a universal norm, is being replaced by state-sponsored social security pattern, which is a lot different from how it used to be in the past; as the accessibility to knowledge increases, the authority of teacher and knowledge as it is, is decreasing. It would seem that on a person from the 18th century our life, featured on TV (even without scenes of violence or sex), would seem as a true nightmare: thundering rock groups, floating in clouds of liquid nitrogen, long-haired men singing in female voices, flooded by multi-color flashing lights; children's TV programs full of witches, water sprites and other evil spirits. We do not believe that the moral norms of our times are lower than they were in the past – not at all – however, for a man from the past our life would seem blasphemous and shameless. And who can guarantee that the moral norms of the technically-developed ETI, perhaps, progressive and reasonable – would not seem shameless to the Homo Sapiens of today?! Even more so, that this Homo Sapiens will surely be inferior to his ETI partner. And, unfortunately, throughout history, the inferiors have never enjoyed all the moral benefits to the full ...

There is another aspect to contacts worth to recall. When Captain Cook discovered the Hawaii, he forbade his sail to land, as he was aware that half of them had syphilis. However, the hospitable native women, expecting a lot from their God-like visitors, made the ban inefficient. A little

[1]   'CETI   Problem   (Contact   with   extraterrestrial civilizations) Moscow, 1975 "MIR", pp. 315, 311, 317.

later sailors brought leprosy to the island — the consequences were briliantly described in Jack London's novel. Millions of Indians that inhabited the American continent died of measles and other diseases, against which their organism had no protection. When we picture ourselves an extraterrestrial civilization we somehow refuse to admit it may have any negative features, including diseases. Which is not too dialectic — you have to pay for progress, and usually, the cost matches the achievement. A relatively low efficiency of antibiotics, which first were regarded as a panacea for all diseases; the increased number of cancer diseases, the problem of AIDS — this all showed that the victory over the diseases on the planet is nothing else but a grave illusion. So, I am afraid that if we chance to find a different reason in the Universe, there is a strong chance that we will also find a biological danger, no less than the one the Spanish brought to the New World.

Naturally, the moral, or, let's say "infection" aspect of contact is questionable. However, there also exists a third aspect, which to me seems as absolutely real. Any contact suggest interactions, and as far as we are talking about civilizations, this will be a cultural interaction. Information which will come from a community at a higher stage of development, will inevitably destroy the existing vision of the world, common stereotypes of thinking and behavior, will result in changes in human psychology — all in all, it will affect everything connected with the spiritual nucleus of the civilization. The collapse of the stereotypes will be a profound one, and the negative consequences will depend first of all on the length of the information perception period and the degree of the lag. In that sense the analogies, known in human history, are most discouraging. A high degree of cultural development (with no genocide, of course) guarantees the stability of a civilization. The Tatar yoke in Russia and the four conquests of China failed to wipe out both Russia and China because of the low cultural level of the invadors, who to a considerable degree, assimilated with the conquered nations. However, when the European civilizations came into contact with the less developed civilizations in pre-Columbian America, island civilizations in the Pacific, and natives in Australia, the result was quite different. The destruction of the old, as a rule, lasts not so long as the creation of the new. Dominated by the outside information, the native world was destroyed, common values ceased to exist, and with no roots existing any more, people found themselves in sort of a psychological vacuum, with nothing to support them. A practically spontaneous loss of roots inevitably decreases viability of a human society, deprives a person of a sense of integrity and makes it easier for a community of dissolve in the others. It is fair to suggest that a possible contact with the extraterrestrial civilizations will have a most far-reaching consequence — in any case we will cease to be what we are now. And possibly, most acceptable in that respect is the theory of a so-called "space reserve", which promises a gradual and less painful movement forward. Incidentally in the history of our planet the discoverers have always been the superior civilizations, and not the opposite. So our common desire to find highly developed ETI looks like a paradox against the background of the entire historical experience.

The aforesaid will, probably, make you feel the speaker is too conservative, but, on the other hand, it all may serve as a good material for further discussion. To conclude I would like to make one more reference to the past, and remind you of an interesting custom, which, as the historian Svetoni said, existed in Rome in the period of the republic. Every time a new triumpher entered Rome, greeted by enthusiastic crowds, there was always a man at the end of the triumphal procession who would shout and tell people about the not-s-attractive negative sides of the triumpher. This man had to be endured, for the sake of truth. So I would like to see that in the modern triumphal procession of science, more people would speak about the fact that science yields not just sweet fruit. Every discovery has a certain price, which mankind inevitably has to pay, whether he likes it or not. It was the New World that paid for the discovery of the New World, but will it always be like that? It seems that already now we should realize well that a hypothetical cosmic interaction will have both pluses and minuses. This does not at all means that there should be no contact — however, we should be expecting it fully aware of the possible consequences.

SOME BIOLOGICAL IMPLICATIONS ON DRAKE'S FORMULA

Mircea Pfleiderer

Institute for Astronomy
Technikerstrasse 15, Innsbruck 6020, Austria

ABSTRACT. Biological considerations contribute heavily to decreasing the
estimates of Drake's number and the possibility of galactic colonization.

The life of ETIs is governed by the same laws of nature as ours is.
Without claim or even intention for completeness I want to mention some
relevant fundamentals of terrestrial life which seem to be of sufficiently general nature to be applicable to any expectable form of life
and, thus, to ETIs.

a) *Symbiosis* has evolved early in life's history, and there is not
a living thing – except some bacteria, algae, protozoa, and mesozoa –
which does not depend on some kind of symbiosis, i.e., which could live
without it. Most of these necessary symbioses are with bacteria. A special form is "cultural symbiosis" within members of a civilization.

b) *Specialization* occurs very early in evolution as a means of fil-
ling effectively given ecological niches. Once established, it has con-
sistently prevented adaptation to ecological changes and thus turned out
to be a dead end of evolution.

c) *Secondary regression* attacks organs which are no longer needed
for living. They become less effective and smaller and eventually they
often vanish completely, thus diminishing the ability for adaptation to
ecological changes. No animal or plant is without some form of it. A
special form is "cultural degeneration" which means loss or deterioration of old knowledge or manual ability in a changing civilization.

d) *Instinct and different forms of learning*. All (higher) animals
must have a stock of instinctive complex behaviour for survival. Additional learning increases the adaptation capability: Lower forms of learning (facilitation, sensitation, habituation, association, imprinting)
are very common and successful, while the survival value – in the long
run – of higher forms (operand conditioning, insight, reasoning) is by
far less certain. A special form of intelligent information processing
is "complex comprehension" in which a realm of complex information is
extremely condensed into a one-bit or few-bits conclusion. Such kind of
"knowledge" is ambiguous and cannot be stored in a simple chain of bits.

Galactic colonization. Why it has not yet, at least not visibly,
taken place, has, e.g., been discussed by Hart, by Papagiannis, by Drake
(1980, *Strategies for the Search for Life in the Universe*, Reidel). Detailed biological argumentation was essentially missing from the discus-

G. Marx (ed.), Bioastronomy – The Next Steps, 279–280.
© *1988 by Kluwer Academic Publishers.*

sion. Biology may well inhibit any extended space travel, the main reason being that space colonies are necessarily a very small closed ecological system, as mentioned by Goldsmith & Owen (1980, *The Search for Life in the Universe*, Benjamin/Cummings).

There are at least three kinds of dangers to such a small system: (i) Breakdown of some biological symbioses, (ii) technical breakdown from loss of knowledge learned by complex comprehension, and (iii) loss of abilities by secondary regression within the small gene pool. All will – by amplification via chain reactions – eventually be fatal, unless the system size were in time increased again by the colonization of a new planet. The recovery time cannot be as short as 10^3 years (as frequently suggested) but must be comparable to typical evolutionary times of at least several 10^5 years.

It follows that the colonization of our Galaxy would, if possible at all, need several billion years and cannot be expected to have already taken place, even if ETIs were abundant. The assumption in Drake's formula that all ETIs evolve independently is most probably true.

The intelligence factor f_i: For evolution as a kind of "learning" how to live in ecological niches, adaptation is needed rather than intelligence. All plants and most animals do not show a hint of any form of intelligence. Lowest forms of it (or, more strictly speaking, of learning ability) have, however, been invented many times, even in different phyla. Higher forms are essentially restricted to Vertebrates. In most cases, evolution seems to have stopped at moderate levels, indicating the existence of optimum levels (the optimum height depending, however, on many other circumstances). Even high levels have evolved polyphyletically – which is indicative of survival value –: some Cetacea, higher Primates. Nevertheless, the actual survival value is not at all obvious or even probable in the long run: many species of *Homo* and of Simians became extinct, and all present Simians were endangered species even before the impact of Man. The present *Homo* is excessively specialized on intelligence, with corresponding loss of other abilities.

Analogous to Rood & Trefil (1981, *Are We Alone*, C.Scribner's Sons) I am splitting up f_i into several subfactors: (1) Does some intelligence evolve? Yes, very probably – this subfactor is about 1. (2) Does it survive? Yes. (3) Does higher intelligence evolve? Polyphyletic evidence: Yes. (4) Does it survive? The evidence is uncertain. (5) Does high-level intelligence evolve (Cetacea, Primates)? The evidence is quite uncertain. (6) Does it survive? As soon as manipulation ability is also achieved, adaptation to the natural environment is no longer needed. If it is lost too early, the survival chance is low. Combination with other specialization: Still lower survival chance. The answer is therefore more on the "no" side: The subfactor is $\ll 1$. (7) Evolution of very high intelligence (Man)? Most uncertain. (8) Does it survive? Any technical society is endangered by overmanipulation of the environment, resulting in its fast change, and the loss of ability to adapt to that change. Probably "no". Note that this argument has nothing to do with the possibility of a destruction of the environment which we at present face.

Conclusion: Only a very large lifetime of advanced technical civilizations could compensate those subfactors which are probably $\ll 1$. It seems that the pessimistic estimates $N \simeq 1$ are also the realistic ones.

REGIONAL JURISDICTION IN OUR GALAXY
(A POSSIBLE EXPLANATION FOR THE ABSENCE
OF EXTRATERRESTRIAL SIGNALS)

M. D. Papagiannis
Astronomy Department, Boston University
Boston MA 02215 USA

ABSTRACT. If there is a relatively large number of 10^5 - 10^6 advanced
civilizations in our galaxy, most of them with histories of millions of
years, then they all must belong to an intercommunicating network. Also
for purposes of contacting new emerging civilizations, they must have
divided the whole galaxy into regional jurisdictions centered around
each active member of their galactic society of stellar civilizations.
Therefore, instead of hoping to receive messages from a large number of
signaling civilizations, we ought to anticipate only one strong signal
from our nearest civilization in whose jurisdiction we happen to belong.
Hence the absence of extraterrestrial signals, rather than implying the
absence of any advanced civilizations in our galaxy, may simply mean
that our nearest civilization did not yet have the time or the
willingness to communicate with us.

1. SETI AND THE DRAKE EQUATION

The probable number N_a of advanced civilizations in our galaxy, with
which we could establish radio contacts, is given by the so called Drake
equation,

$$N_a = R_* \cdot P_a \cdot L_a \qquad (1)$$

where R_* is the rate at which new stars are being born in our galaxy, P_a
is the probability (actually the product of many independent
probabilities) that an advanced civilization will emerge around a star,
and L_a is the average longevity of advanced civilizations. Though there
is a wide diversity of opinions on the value of N_a (see, e.g., Part I in
Papagiannis, 1980), the values most frequently used are: $R_* \sim 20$ new
stars/year, $P_a \sim 10^{-2}$ - 10^{-3}, and $L_a \sim 10^6$ - 10^7 years, which when used
in equation (1) yield that the number of advanced civilizations in our
galaxy that must be of the order of $N_a \sim 2 \times 10^5$.

Given the fact that our galaxy has about 2×10^{11} stars, the value
$N_a \sim 2 \times 10^5$ means that on the average there will be one advanced
civilization per million stars. Furthermore, since the average distance
between stars in our stellar neighborhood is about 5 light years, the
dimensions of a volume with 10^6 stars must be approximately 500 light

281

G. Marx (ed.), Bioastronomy – The Next Steps, 281–285.
© 1988 by Kluwer Academic Publishers.

years and therefore the most probable distance to our nearest advanced civilizations must be of the order of 100-200 light years.

Based on these computations, Frank Drake started in 1960 (Project Ozma) the Search for Extra-Terrestrial Intelligence (SETI). In the next 28 years we managed to accumulate more than 150,000 search hours, in more than 50 search projects, that have used radio telescopes in practically all of the technologically advanced countries of the world (USA, USSR, Australia, Canada, France, Germany, Holland, England, Argentina, and Japan). We also now have two SETI dedicated facilities (the Ohio SETI Program and the META Project of Harvard and Smithsonian) and an active parasitic program (SERENDIP II of the University of California-Berkeley), which collectively accumulate close to 20,000 search hours per year (Papagiannis, 1985a,1985b).

The fact that no signals have been detected so far is beginning to be of some concern, though practically all of these searches have been conducted at selected "magic" frequencies and especially around the hydrogen line at 21 cm. NASA is now getting ready to address this problem through the NASA SETI Program which will have two components, the Targeted Search and the Sky Survey. The targeted search will study about 800 - 1,000 specific targets (primarily stars like our sun), with a high spectral resolution (~1 Hz) in the whole frequency range of the "water hole" (1.2 - 2.0 GHz). The sky search, on the other hand, will survey the entire sky with a frequency resolution of about 32 Hz, but will cover the entire frequency range of the earth's microwave window (1 - 10 GHz). The NASA SETI Program is expected to start in the early to mid 1990's and will take about 10 years to be completed.

Thus in 10 - 15 years we will reach the stage where we will have devoted to SETI close to a million search hours and we will have carried out a fairly comprehensive search for radio signals, both in terms of sky positions and in terms of frequency coverage. A success, of course, would be a truly spectacular achievement that will open broad new vistas for our civilization. But what if our searches were to produce no positive results? Are we to conclude that we must be one of very few if not the only technological civilization in our galaxy, or might there be a different explanation for the absence of extraterrestrial signals? In the sections that follow we will try to present an explanation that would justify the absence of signals, even if our galaxy were to have 10^5 - 10^6 intercommunicating advanced civilizations.

2. SEARCHING AND SIGNALING

When new stellar civilizations develop the technology to communicate with other stellar civilizations, they are more likely to begin with searches for signals, which are much easier to do, rather than with the transmission of signals to unknown targets in the galaxy, which is a much more difficult undertaking. However, if a comprehensive search over a wide range of radio, optical,infrared, and ultra-violet frequencies were to prove unsuccessful, it seems very unlikely that such a civilization would be willing to enter the transmitting phase, which we think is probably done by the more advanced civilizations. The obvious reason would be that having failed to detect any signals, they would probably conclude that there are no other advanced civilizations in the galaxy active in interstellar communications, and therefore it would be a great waste of resources to begin to transmit for thousands, or even for millions of years to probably non existing listeners. Hence a galaxy with only a few advanced civilizations is likely to be a quiet galaxy in terms of interstellar communications.

On the other hand, if our galaxy does indeed have 10^5 - 10^6 advanced civilizations with life-spans of 10^6 - 10^7 years, they must all be in contact with each other, having established a network of intercommunicating stellar civilizations. Therefore, when a new civilization will establish contact with an older member of the galactic society, most probably it will also be given something like a telephone book of our galaxy that will contain the locations of all of the advanced civilizations, as well as the particular frequencies and modes of communication needed to establish contacts with them. Thus the newcomer will be invited to join the whole galactic community of advanced civilizations.

For such an interconnected network of stellar civilizations, it would make no sense to have all of its members transmit signals in all directions for potentially emerging new civilizations. A possible solution could have been to set up a single beacon, probably near the center of the galaxy, to signal on behalf of all of them to all emerging new civilizations throughout the entire galaxy. This, however, is not a fair distribution of responsibilities, because only one civilization would have to shoulder this task, since the transfer of resources over interstellar distances is economically prohibitive.

Consequently the simplest, fairest, and hence the most probable solution would be to divide the whole galaxy into regional jurisdictions centered around each one of the members of this galactic community. In this manner all the members would share equally in this task, and each one would be responsible for only the new civilizations that would emerge inside its own jurisdiction. It would also make contact easier, less expensive and much faster since the distances involved could only be in the range of 10 - 1,000 light years. Furthermore, advanced technological civilizations with space borne astronomical observatories will be able to study all of the solar systems in their own jurisdiction and will know what stars have planets with liquid water on them. They probably will also know where life has emerged and has converted the reducing atmosphere of its planet to an oxidizing one with free oxygen. Finally since the life-spans of these advanced civilizations are expected to be in the range of 10^6 - 10^7 years, they may also be able to dispatch space probes to the solar systems in their jurisdiction where life has originated, and thus to keep a close eye on its evolution, and later on to follow from close range the possible appearance of higher intelligence and advanced technology.

If such a system of regional jurisdictions is actually in effect in our galaxy, which indeed seems inevitable if our galaxy has been having 10^5 - 10^6 advanced civilizations for several billion years, then all we need to do would be to patiently wait for a clear message from the advanced civilization that is in charge of the galactic jurisdiction in which we happen to belong. Signals sent by other advanced civilizations to new emerging ones in their own jurisdictions would be too directive and too weak (due to the much larger distances that would separate us) to be intercepted by us. Furthermore, the initial message that would be much easier to recognize would probably last for only a relatively short period, especially if it is transmitted by their probe in the solar system of the new member. Subsequent communications between the two civilizations would probably be carried out at the more sophisticated system of the galactic network, which would be far more difficult to recognize. Consequently, in either case it would be very difficult to intercept signals intended for other civilizations.

3.   THE LACK OF SIGNALS

If indeed it is only our nearest advanced civilization that will try to contact us, then one can think of at least two reasons why we might have not yet received any messages from them.

3.1 Not Enough Time

If the distance of our nearest advanced civilization is of the order of 100 - 200 light years, as it seems to follow from the Drake equation, it would take them 100 - 200 years to find out that we have finally reached the capability to receive radio messages. This they can learn either through a highly sophisticated eaves-dropping process, or through a message from a probe they might have placed in our vicinity. Based on the information received they may decide to send us a message, either directly from their solar system, or by asking their probe to begin to interact with us. In either case, their response will also take 100 - 200 years, and hence the whole process will take 200 - 400 years.

Given the fact that we reached the stage of transmitting and of receiving high frequency radio signals with reasonably large antennas only in the last 30 - 40 years, it becomes obvious that a message could have already reached us only if our nearest advanced civilization were less than 20 light years away, i.e., one of the 1,000 or so nearest stars. But as we said, most probably it will be 5 to 10 times farther away, and hence we are not likely to hear from them for a few more centuries.

3.2 Not Yet Ready to Join

It is also possible that the galactic community has a rule, or an ethic, which new advanced civilizations must satisfy before they will be invited to join the community.

We are all aware of the many problems that our civilization is now facing (overpopulation, pollution, depletion of natural resources, destruction of the environment, the danger of a major nuclear war, etc.). It seems quite probable that all new technological civilizations are bound to go through such a major crisis, because technology accelerates immensely all processes and in many cases may also lead to self-destruction (Papagiannis, 1984). This then will be the critical test that will separate the selfish and materialistic civilizations that are ultimately bound to self-destruct, from the more intellectual and spiritual civilizations that are more likely to overcome these explosive problems of the newly acquired technology.

It is possible, therefore, that the galactic ethic requires that all new advanced civilizations must first prove that they have managed to overcome all these materialistic problems, and then they will be invited to join the community of stellar civilizations. The obvious reason is that the senior members of this community, with their long life-spans of millions of years, must have reached high levels of maturity and essentially zero levels of growth in all of their materialistic activities, including population growth, energy usage, etc. (Papagiannis, 1988). Therefore, they would not like to have as partners civilizations that would disrupt this serene equilibrium.

4. CONCLUSIONS

The final line of this analysis is that if after a long and comprehensive search we were not to find any extraterrestrial signals, we should not necessarily conclude that we must be all alone in the galaxy, because the division of our galaxy into regional jurisdictions seems to explain, as discussed above, the absence of any signals from other advanced civilizations. But it also says that in due time, especially if we would manage to overcome our current technological crisis, we are bound to receive a message from the headquarters of our galactic jurisdiction to join the community of stellar civilizations of our galaxy.

REFERENCES

Papagiannis, M.D., Editor, Strategies for the Search for Life in the Universe, Part I, D. Reidel Publ. Co., Dordrecht, The Netherlands, 1980.

Papagiannis, M.D., 'Natural selections of stellar civilizations by the limits of growth' Q. Jl R. Ast. Soc., 25, 309-318, 1984.

Papagiannis, M.D., Editor, The Search for Extraterrestrial Life: Recent Developments; (IAU Symposium 112), D. Reidel Publ. Co, Dordrecht, The Netherlands, 1985.

Papagiannis, M.D., 'Recent progress and future plans on the search for extraterrestrial intelligence', Nature, 318, No. 6042, 135-140, 1985.

Papagiannis, M.D., 'The evolution of technological civilizations within the limits of their solar systems', Bioastronomy: The Next Steps, ed. G. Marx, (IAU Colloquium 99), D. Reidel Publ., Dordrecht, Holland, 1988.

ONE MORE SOLUTION TO THE FERMI PARADOX

M. Subotowicz
Institute of Physics, University Marie Curie-Sklodowska
Lublin, Poland

Terrestrial scientific-technical civilization (STC) produces the power of about 10^{13} watt , Dyson's-type STC energy output — about 4×10^{27} watt . The probability of the existence of the Dyson-type STC is negligible. There are well known the difficulties of the realization of the manned interstellar flight. No terrestrial-like STC could fulfill the energy, power, financial and man-power demands to realize the interstellar flight. We have not observed astroengineering phenomena and any other signals that could be realized by Dyson-type STC. This could constitute the indirect proof that Dyson-type STC controlling the energy of amount 4×10^{26} watt does not exist, at least in the vicinity of Solar System up to 200 light years apart.

In CETI following principles should be fulfilled: anticryptography, mediocrity and partnership. We contradict to the third one when being convicted that "they" will send message to us but we will not. They may have similar troubles in sending beacons to those of us.

As the first approximation we should accept that ET STC are similar to the terrestrial one in their energy output (about 10^{13} watt). We cannot mutually communicate because of the large interstellar distances and very large energy demand to send beacons isotropically. In that case the only possibility to discover terrestrial-like STC is to eavesdrop on it looking for the leakage radiation that represents a result of the complex communication and transportation network spread over the planet like our own UHF – TV signals. The terrestrial-like STC should develop detectors of extreme sensitivity (say 10^{-28} Wm^{-2} to 10^{-30} Wm^{-2}) t eavesdrop in this leakage from other STCs. It may be that no one ET STC will intend to be engaged in the very costly and energy consuming transmission of beacons, say 10^{12} to 10^{14} W.

The existing radio telescope are capable of discovering the microwave signals form a distance of abut 15

287

G. Marx (ed.), Bioastronomy – The Next Steps, 287–288.

l.y. of a power comparable to the terrestrial leakage radiation (television).

The most intensive terrestrial leakage signals originate from several US Ballistic Missile Early Warning Systems (BMEWS). These signals make a large proportion of an terrestrial leakage radar radiation. ETI equipped with the Arecibotype telescope could detect BMEWStype radar as far away as 15 l.y. The sphere of this radius contains only 40 stars.

The continuously working isotropic transmitter of the range 1000 l.y. should be supplied with the power 2×10^{16} W. The sphere of a diameter of 1000 l.y. contains about 10 solartype stars.

The possible future of the terrestrial-type STC is: self-destruction in nuclear war, demographic explosion and shortage of food, raw materials, energy producing sources and land for agriculture, genetic degradation, over-stabilization of socialpolitical systems, survival via stagnation. The growth curve of the terrestrial STC is growing rapidly to the saturation. On can predict a slow growth rate for energy production. And we are 14 ranges of magnitude apart from the energy production rate of the Dyson-STC,

This is the reason why we accept that the overhelming majority of STC in our Galaxy or all the existing STC (if any) are the terrestrial-like STC with the energy production rate like that of the present terrestrial STC.

This conviction may present also the other solution to the Fermi paradox. Instead of "We are alone!" [1] we could also accept that "All the existing STC (if at all) are like us!" [2].

The consequences of the above point of view is that the search strategy is the following: to look for the leakage radiation eavesdropping STCs with the ultimate sensitivity of the radiotelescopes: 10^{-28} to 10^{-30} W.m^{-2}!

References:

[1] A.R. Martin and A.Bond (1984) "Is mankind unique in the Galaxy"
    paper IAA-84-239, (IAF Congress in Lausanne, 1984).
[2] M. Subotowicz (1985) "Possible existence in the Galaxy of terrestrial-like civilizations only"
    paper IAA-85-477, (IAF Congress in Stockholm, 1985) and JBIS, 39 (1986) 499-502.

THE EVOLUTION OF TECHNOLOGICAL CIVILIZATIONS WITHIN THE LIMITS OF THEIR OWN SOLAR SYSTEMS

M. D. Papagiannis

Astronomy Department, Boston University

Boston MA 02215 USA

ABSTRACT. Advanced stellar civilizations are expected by the Drake equation to live for 10^6-10^7 years. Since the interstellar transport of resources is unrealistic, they would manage to have such long lives only if they would adhere to a practically zero growth in all of their matter and energy related activities. The social implications are discussed.

1. BOUNDED INSIDE THEIR OWN SOLAR SYSTEMS

Though occasional interstellar missions cannot be excluded, interstellar commerce would be impossible due to the immensely high transportation costs and the long periods (centuries) involved. Consider a case, e.g., with an interstellar velocity $V = 0.1c$, and a mass ratio $M = 100m$ of the mass M of the spaceship and the fuel, to the mass m of the cargo. Since the spaceship must first accelerate to $V = 0.1c$ and at the end slow down again to $V = 0$, it follows that the minimum energy E needed would be:

$$E = 2(^1/_2 MV^2) = MV^2 = (100m)(0.1c)^2 = (100m)(0.01c^2) = mc^2 \qquad (1)$$

which is actually the theoretical conversion of energy into matter. As a result, the basic energy cost per gram of transported cargo would be:

$$c^2 = (3 \times 10^{10})^2 = 9 \times 10^{20} \text{ ergs} = 2.5 \times 10^7 \text{ kWh} = \$1,000,000/\text{gm} \qquad (2)$$

Given the fact that gold costs only about \$14/gm, a cost of $\$10^6$/gm would make interstellar commerce essentially impossible.

The obvious implication is, that stellar civilizations would have to live within the matter and energy limits of their own solar systems.

2. THE NEED FOR A PRACTICALLY ZERO GROWTH RATE

The advent of technology seems to initiate a rapid growth in all matter and energy related activities, because it accelerates immensely all processes. The population of the earth, e.g., is now 5×10^9 and it is growing at about 2% per year with a doubling period of 35 years, while 2,000 years ago it was only 10^8 and was growing at about 0.05% per year. If the present rate were to continue, in 30 doubling periods, i.e., in $30 \times 35 = 1,050$ years, the population of the earth would become:

289

G. Marx (ed.), Bioastronomy – The Next Steps, 289–290.

$$(5x10^9) \cdot (2^{30}) = (5x10^9) \cdot (10^9) = 5x10^{18} \qquad (3)$$

and have a total mass ($\sim 4.5x10^4$ gm/person) of about $2.3x10^{23}$ gm. Given that about 15% of our bodies is carbon, this result implies that we will need to incorporate into human bodies more carbon than is available in the entire crust of the earth, or in all of the carbonaceous asteroids.

The same is also true for energy. Our present energy consumption is 10^{14} kWh/year = $1.2x10^{20}$ erg/sec, and its growing rate is 3% per year that implies a doubling period of only 23 years. Since the total output of our sun is $4x10^{33}$ erg/sec, it follows that at the present rate in 45 doubling periods, i.e., in 45x23 = 1,035 years, our energy consumption will reach the level of the whole energy output of the sun,

$$(1.2x10^{20}) \cdot (2^{45}) = (1.2x10^{20}) \cdot (3.3x10^{13}) = 4x10^{33} \text{ erg/sec} \qquad (4)$$

which obviously is the ultimate limit for our solar system.

From these two examples it becomes clear that upon the advent of technology, stellar civilization must curtail rapidly their rates of growth, or they will soon (within centuries) self-destruct by reaching the limits of growth for their entire solar system (Papagiannis, 1984).

But even a modest growth is unacceptable. If these stellar civilizations are to survive for 10^6-10^7 years, as it is required by the Drake equation (Papagiannis, 1988) in order to have a reasonable number (10^5-10^6) of them around, they must avoid reaching these limits even in 10 million years. Since our present rates will bring us to the ultimate limits in just about 1,000 years, they must maintain growth rates that are at least 10,000 smaller. This would imply growth rates of the order of 0.0002% per year, i.e., doubling periods of 350,000 years, which for all practical purposes would be equivalent to zero growth.

3. SOCIAL IMPLICATIONS

An active civilization that will manage to maintain a practically zero growth rate, must be a very stable one and must consist of extremely conscientious people who have stopped being interested in materialistic possessions and pleasures. If they are continueing to thrive, they must have discovered other interests that are not intensive in energy and raw materials. Therefore they must have risen to much higher intellectual and spiritual levels, and they must have eliminated completely even the idea of a war, which is the most energy and matter consuming activity of a technological civilization

It would be a great uplift, therefore, if we were to make contacts with such civilizations and learn from them how to handle the problems that are now threatening the future of our young technological society.

REFERENCES

Papagiannis, M.D., 'Natural selections of stellar civilizations by the limits of growth' Q. Jl R. Ast. Soc., 25, 309-318, 1984.

Papagiannis, M.D., 'Regional jurisdictions in our galaxy, a possible explanation for the absence of extraterrestrial signals,' Bioastronomy: The Next Steps, ed. G. Marx, D. Reidel Publ., Dordrecht, Holland, 1988.

THE GALACTIC CENTER: A NICE PLACE TO VISIT — BUT YOU WOULDN'T WANT TO LIVE THERE

Antony A. Stark
Bell Laboratories, Holmdel, NJ 07733 USA

ABSTRACT. Astronomical observations of the galactic center region show a deep gravitational potential well that is partially filled with dense gas. Gas at the very center is less dense and in a peculiar ionization state. This unstable situation probably results in bursts of lethal radiation every few million years.

Many galaxies have "dragons" in their centers. A very small region at the very center of the galaxy produces a large fraction of the total energy output, and exhibits a variety of strange energetic phenomena. These "active galactic nuclei" occur in galaxies classified as N type, Seyfert, radio galaxies — and in extreme cases, quasars. The beast at the center is probably a black hole, possibly containing millions of solar masses, although some of the less energetic dragons could result from a deep but non-singular gravitational potential. Time-variability is common, and it seems likely that they turn on and off in times that are short compared to the age of the universe. This paper will briefly review the astronomical evidence indicating that there is a "sleeping dragon" at the center of our own Galaxy (Figure 1), and that our "dragon" has been awake only recently. This is by no means a review of galactic center observations; it is a brief argument on the uninhabitability of the galactic center region. The galactic center is complex and beautiful, and the observations have become very detailed (see the recent review by Genzel and Townes 1987).

The galactic center is full of stars. The stars can be seen directly in near-infrared light (Becklin and Neugebauer, 1968), and their presence can be inferred by the kinematics of gas orbiting around the center. Density increases inwards, $\rho \propto R^{-0.75}$, down to scales at least as small as the limits of observation, about 0.03 pc. The observed orbital velocities are approximately constant at about 150 km s^{-1} down to this smallest observed scale (Lacy *et al.* 1980), which implies that the mass interior to this point is a few million solar masses, all of which may be in a black hole. As far as we can tell, the density increases inwards, and may become infinite. On the other hand, it is important to realize that the escape velocity to infinity from the innermost observed point is only about 2000 km s^{-1} (cf. Oort 1977). It is possible that the escape velocity from the very center is not much larger than this and is therefore much less than the speed of light. The observations neither forbid nor require the existence of a massive black hole in the galactic center.

291

G. Marx (ed.), Bioastronomy – The Next Steps, 291–293.

Fig. 1. There is a sleeping dragon in the galactic center.

The galactic center is also full of interstellar gas in all phases: molecular, atomic, and ionized. The molecular gas is observable in millimeter-wave transitions of molecular lines, and in far-infrared dust emission. The atomic gas is observable in the 21-cm line of HI (e.g. Liszt and Burton 1980). The ionized gas is observable in infrared emission lines and radio recombination lines. The gas orbits the center but the orbits are not quiescent. Radial motions are comparable to the orbital azimuthal motions, and much of the gas is well out of the galactic plane (e.g. Bally *et al.* 1987). Figure 2 shows dense molecular gas surrounding the nucleus.

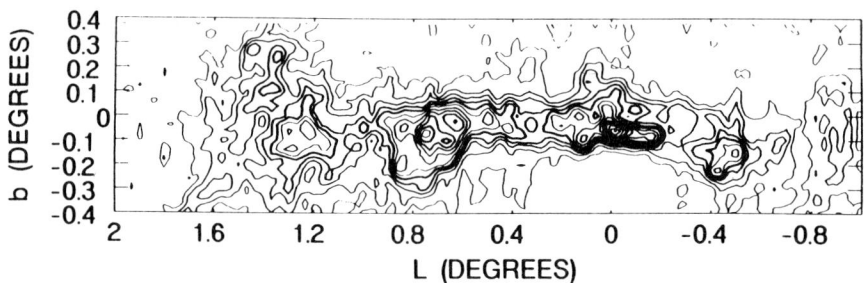

Fig. 2. Emission in the J=2→1 line of CS, integrated over all velocities (Bally *et al.* 1987). This line is a tracer of dense gas. The galactic center is near (0, 0).

The dense gas pervading the galactic center region extends very nearly all the way into the center. About 2 pc from the center is a broken ring of molecular gas, seen in tracers of dense ($n \approx 10^5 \text{cm}^{-3}$) gas like HCN (Güsten *et al.* 1987). Inside this is ionized gas, with an unusual temperature and ionization state. The gas is ionized by several million solar luminosities of ultraviolet radiation (Lacy *et al.* 1982, Becklin *et al.* 1982). However, the effective temperature is low, because high-ionization state lines of common atoms are relatively weak (Lacy *et al.* 1980). This seems to allow only two possibilities:

(1) The ultraviolet radiation comes from a black hole at the center. This would imply much greater luminosities in the past and in the future, because the black hole is currently immersed in low-density gas, and is surrounded by high-density gas, an unstable situation. A million years ago, the dense ring of molecular matter may have been touching the black hole, with spectacular results (Cowie *et al.* 1978).

(2) The ultraviolet radiation comes from a cluster of massive stars. This star cluster cannot currently contain any stars more massive than late-O type, and yet there must be over 100 OB stars to provide the observed luminosity. Lacy *et al.* (1982) suggest that there was a burst of massive star formation about ten million years ago, so that the most massive stars have already died, leaving only late-O and B stars. A burst of star formation is a situation in which a great many stars form in a short time. This would presumably be fueled by the accumulation of gas at the bottom of the potential well, gas which would be blown out or ionized by the newly-formed massive stars.

This is not good news for the hypothetical extraterrestrials living at the galactic center. It seems that the best current astronomical evidence indicates a sleeping dragon of some sort at the galactic center. The deep potential well filled with dense gas is unstable on timescales of a million years or so. Every few million years, the dragon awakes, and sterilizes the galactic center with a burst of fire.

REFERENCES

Bally, J., Stark, A. A., Wilson, R. W. and Henkel, C. 1987 *Ap. J. Suppl.*, **65**, 13.

Becklin, E. E., and Neugebauer, G. 1968 *Ap. J.*, **151**, 145.

Becklin, E. E., Gatley, I., Werner, M. W. 1982 *Ap. J.*, **258**, 134.

Cowie, L. L., Ostriker, J. P., and Stark, A. A. 1978 *Ap. J.*, **226**, 1041.

Genzel, R., and Townes, C. H. 1987 *Ann. Rev. Astron. Astrophys.* **25**, 377.

Güsten, R., Genzel, R., Wright, M. C. H., Jaffe, D. T., Stutzki, J., Harris, A. I. 1987 *Ap. J.*, in press.

Lacy, J. H., Townes, C. H., Geballe, T. R., and Hollenbach, D. J. 1980 *Ap. J.*, **241**, 132.

Lacy, J. H., Townes, C. H., and Hollenbach, D. J. 1982 *Ap. J.*, **262**, 120.

Liszt, H. S., and Burton, W. B. 1980 *Ap. J.*, **236**, 779.

Oort, J. H. 1977 *Ann. Rev. Astron. Astrophys.* **15**, 295.

SUPERCIVILIZATIONS AS POSSIBLE PRODUCTS OF THE PROGRESSIVE EVOLUTION OF MATTER

N.S. Kardashev

Space Research Institute
Profsoyuznaja 84/32, Moscow 117810, USSR

V.S. Strelnitskij

Astronomical Council of the Academy of USSR
48 Pjatnitskaja Str., Moscow 109017, USSR

ABSTRACT. The hypothesis of possible existence of supercivilizations (SC's) is considered in the light of modern ideas about the progressive evolution of matter in the Universe. The existence of SC's presupposes the fulfilment of at least 2 conditions: 1) Reason is a sufficiently probable "solution" in the course of progressive evolution of matter in the Universe; 2) the main property of Reason - aspiration for maximum information about the Universe - isn't lost at the subsequent, "post-reason", stages of progressive evolution. If the seconde condition is fulfilled, civilizations tend to become SC's and can, in principle, last an indefinitly long (cosmological) time. Some cosmological limitations for the space location of SC's are analysed, in particular, the possibility of local gravitational collapse of our part of the Universe, impelling civilizations to escape from this region. This can serve as a possible explanation of the Fermi paradox.

1. INTRODUCTION

One hope to detect the extraterrestrial intelligence (ETI) is connected with the possibility of existence of supercivilizations (SC's), postulated about one quarter of a century ago [1,2]. The SC hypothesis is based on the assumption that the main immanent property of any civilization is the aspiration for collecting and utilizing maximum information about the surrounding Universe and about itself [3]. To obtain new information civilizations have to spend energy and this impels them to look for more and more powerfull energy sources. So can arise the SC's of types II and III, having mastered the energy of their star or galaxy respectively.

295

G. Marx (ed.), Bioastronomy – The Next Steps, 295–302.
© 1988 by Kluwer Academic Publishers.

In this communication we would like to discuss at
first the scientific validity of the SC hypothesis, proceeding from the concept of civilization as a product of a
special sequence of development processes - the "line of
progressive evolution" (Sections 2,3). Then, in Section 4,
some cosmological limitations for the possible residence
of SC's will be considered.

2.   UNIVERSE-PROGRESS-REASON

The SC hypothesis is a result of extrapolation to the future of some properties and tendencies of our civilization-
the only one known to us. This extrapolation is peculiar
in that it is done for time intervals exceeding by orders
of magnitude the time of existence of the "extrapolation
basis" - our civilization. The validity of such extrapolation will probably somewhat increase if we follow the prehistory of our civilization (which lasted much longer than
its history), because we can hope to feel already in this
prehistory the embryos of the appropriate tendencies.
    It is widely accepted now that the terrestrial civi-
lization arose as a result of a particular sequence of evolutionary processes which is named "progressive evolution".
The very idea of progress, is old: "No single idea has be-
en more important than, perhaps as important as, the idea
of progress in Western civilization for nearly three thousand years" [4]. However, almost all this time the idea of
progress could not have any scientific ground and was in-
separably linked with the religious mythology. It was "se-
cularized" only in the middle of the 18 century, and the
first, who contributed to this, was probably A.Turgot, who
identified the progress with the gradual selfimprovement
of the human knowledge, ethics and social organization.
Thereupon the concept of progress, as oriented selfdeve-
lopment from simple to complex, from less to more perfect,
has itself experienced an astonishingly regular evolution
[5]. In 18-19 centuries this concept was spread down
to the biological level of organization (Buffon, Lamarck,
Darwin). In 19-20 centuries Haeckel, Oparin, Haldane, Miller, Urey, Eigen and others have contributed to spread the
idea of progressive evolution down else - to the prebiological, chemical evolution. In the last half of our
century, due to the achievements of the evolutionary astrophysics, the idea of progress went out of the biosphere -
the theory of progressive complication of elements in the
early hot Universe and in the interiors of stars appeared.
    Thus, the concept of progress has been gradually
spread to more and more low levels of organization of matter. But going thereby farther and farther into the past,
Man has reconstructed the chane of events, which had led to
his own appearance! Now this chain is clearly seen as a

quasi-continuous line of gradual complication of a small
part of cosmic matter, finding itself under proper conditions, by natural selection of stable and capable for subsequent development hierarchical structures - from elementary particles in the hot young Universe up to the thinking brain and civilization.

The reality and genetic connectedness of this line of
progressive evolution is confirmed by the regular change
of some important parameters along it. For example the degree of structural complexity (defined by the number of hierarchical levels in the system), as well as the number of
stable varieties representing a structural level, monotonically increases along the sequence: elementary particles - atoms - molecules - biological macromolecules, whereas the binding energy, securing the integrity of the systems, monotonically decreases in the same direction.

It was pointed out in [6],that speaking about "progress" or "regress" of an evolving system we always ascribe
to the system some aim (or quasi-aim if the system is inanimate). What is then the quasi-aim of the line of progressive evolution? It seems impossible to formulate for it
a more "pithy" aim,than: by trial and selection among the
objects of a given structural level create the next, higher level, capable for subsequent evolution. Accordingly,
the criterion of progress will be: the degree of proximity
to the higher organization level, and it is clear, that
such criterion can be applied only retrospectively, after
the higher level has been created.

One can present arguments [6] against the existence of
a common criterion of progress for all the evolutionary
processes proceeding in the Universe, as well as against
the idea, that progress is the "main direction" of evolution in the Universe.The progressive evolution forces its
way amidst the cosmic processes of quite different kind,
for which the consept of progress-regress is deprived of
any sense, but with which this line can, however, be inse-
parably linked (by the exchange of mater, energy, entropy).
The course of events along this line of development does
not contradict the laws of Nature, but probably can not be
deduced from them, because it is determined to a great degree (at least, on some stages) by a freak of chance. Somewhat enlarging the neat remark of J.Monood [7] about the
origin of the biosphere, one can say, that the line of progressive evolution "is not obliged to exist, but has the
right to exist".

The role of chance is probably especially high at the
stage of biological evolution. The randomness of mutations
and of sexual recombinations make the bioevolution a stochastic process. It is true, that the presence of a stocha-
sity does not by itself presuppose a total undeterminism,
unpredictability. However, most contem porary biologists

consider the biological evolution to be a creative, highly unpredictable process [8]. Such a complicated, interde-
pendent system as an aggregate of simultaneously evolving
species must have in its mathematical description many
"bifurcation points" at which the choice of the "solution" is determined by a pure chance. Nevertheless, some
"canalization" of the biological evolution (in the sense
of interdiction of certain evolutionary ways by its prehistory) is admitted by many specialists [9] . Unfortunately, no adequate mathematical description of biological
evolution has been created yet. It is not excluded that
such questions as the degree of evolution canalization
and the probability with which its partial canalization
leads at a certain stage to the "invention" of the intelligence, could not be solved at all with the "material"
of only one example - the terrestrial biosphere. This uncertainty just makes the problem of search for extraterrestrial life and intelligence an urgent and rather well-
grounded scientific problem.

In any case, the today's level of knowledge let us
to use the fact of partial "canalization" of bioevolution as a sufficient ground for the hypothesis, that Reason
is a natural and sufficiently often realized "solution"
in the course of progressive evolution of matter in the
Universe.

3. THE POST-REASON EVOLUTION

If it is true that the quasi-aim of the line of progressive evolution is the creation of higher and higher levels
of organization, then the contemporary stormy social evolution can be regarded as a final step in formation of a
new evolutionary level ("intellectual"?) on the top of
biological evolution. This transition is realized by the
selective search for stable and flexible forms of social
organization, capable to secure a maximum release of the
creative potential of each human personality.

How will the subsequent levels of evolution look out?
Will they conserve and amplify the known properties of the
intelligence (in particular, its aspiration for information), or will they be something fully different? Nobody
can answer this question with any certitude. However, taking into account the unprecedented adaptive, stabilizing
potential of the intelligence we have right to suppose
(as one of the possibilities), that in its main properties the intelligence will conserve itself for an indefinitely long time. Such "perpetuation" of the intelligence
would not mean the cessation of progressive evolution, because the Reason itself becomes an arena for creative progressive evolution, which it only transfers from the "noxious real-time environments" to the "innocuous memorized

environments" [10] .

It is obvious, however, that the initial information
for the "unnocuous" intellectual creation Reason can
draw only from the "noxious" surrounding Universe, and
this will require more and more energy, impeling civilizations to evolve into SC's.

Utilization of the information about surroundings
(for "working-out the preserving reactions") is a general
property of the living matter, which can be traced back
to the most primitive forms. At the level of Reason the
evolution finds an especially effective mode of working-
out the preserving reactions: the abstract analysis of information - creation of models of the surroundings permit-
ting a forsight of possible issues of the natural processes and of the human actions. The role of informative processes has continously grown along the "progressive trunk"
of the biological evolution, coming up to Man (improvement of the nervous system), and it is continuing to grow
in the course of social evolution. According to O. Toffler
[11] our civilization is experiencing just now the "information wave" of its evolution, the third one, after the
two previous - agrarian and industrial. The informative
processes become dominating, the information (but not the
material resources) gradually becomes the main "capital"
of the civilization. It is natural to suppose that in the
future the role of information in the life of our civilization will only grow.

4.   SC's AND COSMOLOGY

Considering the infinite diversity and non-stationarity
of the civilizations themselves, it seems rather clear
that civilizations have no internal reason for cessation
or limitation of their activity in acquiring and storing
information. On the contrary, intelligence, as a means of
forstelling, predicting reflection of reality, always tries
to enlarge the spatial and temporal horizons of its possessions. It is to be stressed that the study of the Universe on the largest, cosmological scales is needed, contrary to the currant prejudice, not only for satisfaction of
the scientists' insatiable thirst for knowledge. It will
be shown below on a concrete example that a real necessity can arise later to utilize the obtained cosmological
knowledge just for "working-out the preserving reactions".

Having no fatal internal limitations for their activity, the civilizations can, however, meet in the course of
their development with some limitations of external nature. Such possible cataclysms, as the explosion of the parent star or galaxy nucleus, are probably predictable and
avoidable somehow by a well-developed civilization. Let
us discuss here some possible limitations, connected with

the cosmological evolution of the Universe.

The problem of the evolution of intelligence on very long, cosmological time scales was analysed by F.Dyson [12] within the limits of the model of uniformly expanding Friedman Universe filled with the ordinary matter. In the closed model expansion is followed by contraction, which eventually leads to superhigh temperatures and densities, and thus to the obvious limitations for the duration of the life and the reason. For open models the time span from the beginning of the expansion is finite, but the Universe will expand unlimitedly long in the future. Dyson has shown that in this case the appearing civilizations can exist infinitely and in infinite time they can acquire and create an infinite quantity of information. Any two civilizations at any distance between them eventually find themselves inside a common cosmological horizon and, consequently, can establish a contact with each other.

However, the modern cosmology assumes that the Universe consists not only of ordinary matter. At early stages of the Universe the superrelativistic matter dominated and, according to the inflation theory, at the earliest stages there was only relativistic vacuum.

The most interesting topologically seems to be the model of chaotically inflating Universe, proposed by Linde [13]. According to this model, the Universe is infinite in time in the past, as well as in the future, homogeneous and isotropic on the average, is described by a flat metrics and its density doesn't change with time. The most of the Universe volume is filled with the highly excited vacuum, providing an exponential expansion. The density fluctuations of the vacuum are found to be unstable and this gives birth to expanding bubbles - the mini-universes. Within the bubbles the process of expansion leads to the breaking of the symmetry of vacuum, its transformation into high-temperature matter of elementary particles and radiation, which eventually becomes the ordinary matter surrounding us in the form of stars and galaxies. After decay of the vacuum within the bubble, its dimensions do not continue to grow exponentially they begin to grow according to a power-low (proportionally to

$t^{1/2}$ for the relativistic stage and to $t^{2/3}$ - for the nonrelativistic stage). At the same time the distances between the bubbles, continue to grow exponentially. This just secures the constancy of the average density of the Universe, which is probably close to that of the excited vacuum (the Planck density: $5 \cdot 10^{92}$ g/cm^3).✶

What is then the fate of an individual bubble on cosmological time scales and what is the fate of the life and the reason in the bubble? At the early stages it seems

to be quite analogous to the cases analysed by Dyson. Yet
the bubble has a "wall". Apparently, near the wall the
process of the vacuum decay will take place, which is the
most powerfull resevoir of matter and energy. We believe
that such favorable possibilities to utilize matter and
energy would impel civilizations to localize themselves
near the bubbles' walls.

Furthermore, new bubbles can be born near or on the
wall of an already existing bubble. This makes it possible, in principle, to travel from one bubble to another
and to meet very old supercivilizations, which may have
had indefinitely long time for development in the older
bubble.

Unfortunately the probability to be born just near
the bubble's wall is quite small, so most civilizations
have to wate for a long time for the moment, when their
horizon will intersect the nearest wall, and the wall will
become visible as a growing "spot" of the black-body radiation on the sky.

Let us now consider one more question: the possibility of the local collapse, which may take place in any cosmological model. The expansion of the space can give place
to a local contraction due to a local excess of density,
produced by the fluctuation of the barion / antibarion as-
symetry. It is quite possible that our Galaxy is placed
near the center of such a region [14-16]. Its boundary
corresponds to a red shift of 4-5. Beyond these red
shifts we don't observe quasars, galaxies or protogalaxi-
es. This gave the ground to suppose [14-16], that the bari-
on/antibarion asymmetry is much smaller there than here.
If it is true, the expansion of our part of the Universe
can readily give place to the contraction. The same result
can be produced otherwise: if the vacuum with negative
density was conserved in our Methagalaxy (our bubble).
This is possible if the density of the lowest state of the
vacuum has a negative value, which corresponds to a negative Λ-term in the cosmological equations [17]. Fortunately, there is a principal possibility to leave the dan-
gerous region before the collapse beginns, on the stage
of expansion [18].

Doesn't the mass escape of civilizations from our
part of the Universe explain the Fermi paradox - the ab-
sense of visible manifestations of SC's in the local part
of the Universe?

And don't we also have to seriously study the possibility of the danger of local collapse and the possibility to migrate to an ever-expanding part of the Universe,
if such a danger is confirmed?

In any case, it is evident, that the cosmological
aspect has to be taken into account, when we consider such
a potentially long-living phenomenon as SC.

302

REFERENCES

1. Kardashev N.S., 1964, Astronom. Zh., 46, 282.
2. Dyson F.J., 1966, Prosp. Mod. Phys., N 4, Ed. Marshak R.E., N.Y., Willey, p. 641.
3. Kardashev N.S., 1969, in: " Vnezemnye Tsivilisatsii", Ed. Kaplan S.A., M., Nauka, p, 25.
4. Nisbet R., 1980, History of the idea of progress, L., Heinemann.
5. Strelnitskij V.S., 1986, in: "Problema Poiska Zhisni vo Vselennoi", Eds. V.A. Ambartsumian, N. S. Kardashev, V. S. Troitskij, M., Nauka, p.50.
6. Strelnitskij V.S., 1987, in: "Vselennaja, Astronomija, Philosophija", Ed. Martynov D.Ya., Kasutinskij V.V.,M., Isdatelstvo MGU, in press.
7. Monod J., 1970, Le hasard et la nécessité, Ed. du Seul. Paris.
8. Mayr E., 1978, Sci. Amer., 239, N 3, 46.
9. Vorontsov N.N., 1980, Zhurnal Vsesojusnogo Himicheskogo Obschestva, 25, 295.
10. Calvin W.H. (see this volume).
11. Toffler O., 1982, The third wave, A.Bantam Book.
12. Dyson F.J., 1979, Rev. Mod. Phys., 51, 447.
13. Linde A.D., 1986, Phys. Lett., 175 B, 395.
14. Kardashev N.S., Blome H., Priester W., Comments on Aph. (in press).
15. Dolgov A.N., Kardashev N.S., Nature (in press).
16. Dolgov A.N., Illarionov A., Kardashev N.S., Novikov I,D. Comments on Aph. (in press).
17. Kardashev N.S., 1986, Astron. Zh., 63, 839.
18. Novikov I.D., Frolov V.P., 1986, Physika Chernych Dyr, M., Nauka.

*To some extent this model is a modern variant of the steady-state universe model by Hoyle (mnras, 108, 372, 1948). In their model the creation of matter by the "c-field" secured the constant density of the expanding universe. In the Linde's model the excited vacuum is the cause of the expansion of the universe, it conserves its density during expansion (like a cosmological model with a constant positive Λ-term) and gives birth to the normal matter, whose average density also remains constant.

Session 6

PROSPECTS FOR DETECTING TECHNOLOGICAL CIVILIZATIONS

The exciting search for direct signals attracted the most speakers. Several talks discussed various aspects of the recent ambitiosus NASA SETI program. Restricting only to these: Kent Cullers *described three-pulse and multiple-stage continuous-wave detection algorithms to achieve increased computational efficiencies in the targeted SETI. Algorithms operating on pulse triplets, the use of Hanning windows and intra-spectra will lead to improved sensitivity.* Jill Tarter *presented field-test results with the targeted search multi-channel spectrum analyser, describing tests with a 74000 channel prototype such as an (unsuccessful) attempt to relocate Pioneer 9 spacecaraft, pulsar studies, etc.* Ivan Linscott *discussed artificial signal detecation in terms of the strategy to be adopted for signals at a S/N level around unity and the use of filtering to rejecat radio-frequency interference (RFI) etc. A part of the signal-processing hardware has been realised on a single VLSI chip, substantially reducing the scale and cost of the full 120 M-channel spectrometer.* Edward Olsen *described the NASA Sky Survey program, both its goals and the problems faced. The latters include driving the antenna faster than usual, interference problems, both RFI and intermediate- frequency interference (IFI) within the detection system. The ten-million channel system has to discern 10-30 "events" out of 10^{11}. Techniques developed in this program for RFI rejection may prove useful in radioastronomy generally.* Charles Seeger *discussed counter-measures for coping with radio frequency interference (RFI) in broadband SETI. One has to plan on delving to the level of $10^{-26} - 10^{-27}$ W m^{-2}. A major concern is that the deep-space microwave window is rapidly being filled with RFI. Some sites are better tan others with regard to RFI. A pragram of RFI monitoring in the 1-10 GHz is almost complete and appears to be the first general survey of this kind available.* Michael Klein *emphasised the importance of pursuing SETI now because of increasing RFI. For thorough SETI purposes the RFI survey needs to be extrapolated to determine the fraction of the frequency band occupied by RFI to 130 db down. We are unsure of this extrapolation as we are concerning how the number of microwave communication channels will grow. The possibility of sites subject to less RFI, perhaps Arecibo, Australian sites, etc. needs to be considered.*

Ronald D. Brown

BEING OPTIMISTIC ABOUT SETI

D. Schwartzman

Geology & Geography Department, Howard University
Washington DC 20059, USA

L.J. Rickard

E.O. Hulburt Center for Space Research, Naval Research Lab
Washington DC 20375, USA

ABSTRACT. We briefly review the history of alternating optimistic and
pessimistic expectations of searches for extraterrestrial radio
communications. We point out that a proper consideration of such
searches should distinguish between those anticipating beacons from
advanced civilizations and those anticipating leakage radiation from
emergent civilizations. The latter provide more reliable constraints
on the rate of emergence of civilizations, but require a much greater
observational effort than usually contemplated.

1. HISTORY: OPTIMISM

Early discussions of SETI were fairly sanguine. The first
comprehensive analysis of the SETI enterprise, "Intelligent Life in the
Universe" (1966) by I. S. Shklovskii and C. Sagan, laid out the
standard structure for subsequent arguments. (Indeed, this book may be
considered the "bible" of SETI, being still in print and commonly used
for courses on the subject.) The number of technological civilizations
available for communication is factored into terms representing
specific unknowns:
$$N = R^* \times f_p \times n_e \times f_l \times f_i \times f_c \times L,$$
where R^* is the rate of star formation, f_p is the fraction of stars
which have planetary systems, n_e is the average number of planets per
planetary system which fall in a "habitable" zone, f_l is the fraction
of habitable planets on which life arises, f_i is the fraction of
planets with life on which intelligence arises, f_c is the fraction of
intelligent species that evolve to a technological stage that enables
interplanetary communication, and L is the lifetime of such
technological civilizations. This equation is variously known as the
Drake or Green Bank equation. Of all its terms, only R^* is supported
by observations extending beyond our solar system even now. Certainly,
Shklovskii and Sagan were unabashedly optimistic in guessing the other

305

G. Marx (ed.), Bioastronomy – The Next Steps, 305–312.
© 1988 by Kluwer Academic Publishers.

terms, to estimate the likely number of advanced civilizations in our
Galaxy at "perhaps" 10(6).

Shklovskii and Sagan estimate fp=1, ne=1, fl=1, and fc=0.1, giving
N=0.1L. The critical factor is L, taken by Shklovskii and Sagan to be
10(7) as "an average of all technical civilizations, both short-lived
and long-lived". Drake (1974) adopted numbers yielding N/L=1 and
N=10(4), which would put the nearest detectable civilization at a
distance of 1000 l.y..

Other optimists, like Morrison (1974) following von Hoerner,
argued that "there is probably a great feedback system in this
communications circuitry [between civilizations]. Namely, if we build
any link, it is extremely likely that this action increases the time
that the link endures."

Even when thoroughly unsatisfied with adopted values for the
factors in the Green Bank equation, researchers were inclined to err on
the side of optimism. Thus, MacGowan and Ordway (1966):

> We cannot estimate the distance to the nearest societies
> with any degree of accuracy, but we do not have any firm
> reason for pessimism. This implies that we should greatly
> increase our efforts to detect extrasolar intelligent
> signals that may emanate from our neighbors in space.

By 1975, some 14 observing programs had been attempted, involving
radio astronomers from the U.S., the U.S.S.R., Australia, and France.
Roughly 25 additional short-term programs were conducted in the next
decade. In addition, several long-term monitoring programs have been
begun. These include university programs such as that operated by
Robert Dixon at Ohio State; independent programs such as Project
Sentinel, which is funded through private grants; and even individual
amateur efforts (see, e.g., Adler 1983).

2. HISTORY: PESSIMISM

In 1975, Michael Hart provoked a major controversy among SETI
speculators by reflecting on the absence of signals in the radio
astronomical searches in combination with the absence of any other
accepted evidence indicating the presence of extraterrestrials on earth
either now or in the archeological past. Shklovskii and Sagan had
already pointed out that, even under fairly pessimistic assumptions
about the speed and capacity of interstellar travel, a single
spacefaring civilization could have spread throughout the Galaxy in a
time much shorter than the age of the Galaxy. If expansion into space
is a consequence of the growth of any of the technological
civilizations that emerged before us, then the probability is near
unity that they have either visited the solar system already or left a
beacon nearby. The only way to account for the absence of
extraterrestrials is to adopt a value for N of order unity (i.e., we
are alone). The argument can be extended by replacing the colonizing
aliens with self-replicating machine probes, which could fill the
Galaxy very rapidly (Tipler 1980, 1981).

Speculation on Hart's paper led to the formation of the "N=1 school", researchers who focussed on arguments suggesting the uniqueness of our own technological civilization.

Several authors argued that the origin of life is highly unlikely because of the exceedingly small probabilities for the random assembly of polypeptide sequences (e.g., Monod 1971 and even Shklovskii 1978). However, recent research on problems of chemical evolution relating to the origin of life tend to argue in the opposite direction. For example, Eigen and coworkers have demonstrated that the coupling of autocatalytic cycles ("hypercycles") could well produce self-replicating macromolecules in relatively short times, rather than the billions of years that would be required by random assembly (cf. Field 1985). Similarly, the phenomenon of RNA catalyzing its own synthesis (cf. Lewin 1986) supports the idea that the appearance of life may not be particularly unlikely. Shapiro (1986) has explored this complex debate in greater detail.

Another N=1 argument contends that the Earth's biosphere has been rather lucky in its climatic history. Hart, using models of atmospheric evolution, argues that the continuously habitable zone in the solar system is very narrow, ranging only from 0.95 to 1.01 AU. The zone narrows for stars of different temperatures, suggesting the extreme rarity of habitable planets (Hart 1978, 1979). However, several climatologists have cautioned against drawing definitive cosmic conclusions from these models, given the uncertainties in the input parameters (Schneider and Thompson 1980, Sawyer 1984). Furthermore, Hart's models ignore possible feedback controls on climate involving carbon dioxide.

One of the most expansive of these feedback mechanisms is known as the Gaia hypothesis (Lovelock 1979). Briefly, this is the idea that the habitability of the Earth is maintained by the development of feedback loops mediated by organic evolution in open system interaction with the atmosphere, the hydrosphere, and the crust. It is known, for example, that the coupling of terrestrial carbon and oxygen biogeochemical cycles leads to an increase in the stability of the system (e.g., concentration of constituents) even though it is far from chemical equilibrium (Lasagna 1980). It would be encouraging to determine that specific feedback mechanisms have been strengthened by natural selection - e.g., in the evolution of microbes and marine organisms that are involved in the geochemical thermostat linked to weathering and atmospheric carbon dioxide balance (Walker and Hays 1981). It would then be likely that alien biospheres are similarly regulated, enhancing the prospects for the emergence of life elsewhere (Schwartzman 1981; Tang 1982).

Another argument is that the origin of intelligence has very low probability because of the highly contingent nature of evolution. The classic presentation of this point of view was made by Simpson (1964). Within the N=1 school, Simpson's argument has been espoused by Tipler (1980, 1981) and Ornstein (1982). The basic counterargument, supplied by Bieri (1964) is that evolution is often convergent on commonly desirable structures. For example, Salvini-Plawen and Mayr (1977) have shown that photoreceptors have originated independently in at least 40 and possibly as many as 65 different phyletic lines. However, Ornstein argued that the apparent similarity of the cephalopod and vertebrate eyes is not due to convergent evolution. Salvini-Plawen and Mayr (1977) and Gould (1985) have largely refuted Ornstein's arguments. Russell (1981) has argued that progressive encephalization appears to have been an evolutionary process with a fairly regular pattern for the last 225 million years (i.e., since the emergence of mammal-like reptiles), suggesting that the emergence of intelligence may be a repeatable phenomenon in the Galaxy.

So far, the arguments of pessimism have been biochemical, climatological, and evolutionary. Perhaps the most depressing argument in the N=1 school is in some sense sociological. Suppose we factor the Green Bank equation into only two terms, so that N, the number of communicating civilizations, is equal to AL. A is now the product of all factors that can, in principle, be resolved by astrophysical or biological study (the number of habitable planetary systems in the Galaxy, the probability of initiating microbial life, the probability of evolving intelligence, etc.). L, the average lifetime of communicating civilizations, is the one factor that cannot be inferred indirectly (beyond the lower limit given by our own survival, which is too low to be significant). If we assume that present equipment is sufficient to detect beacons of reasonable power originating at great distances (e.g., Murray et al. 1978), and that the characteristics of such signals are perfectly anticipated, then the SETI experiment is principally a test to determine L.

If we assume that the hypothetical communicating civilizations are planet-bound (as did the early SETI analysts), and make reasonable estimates of the factors entering into A, then it usually turns out that large values of L (say 100,000 years [von Hoerner 1973]) are still possible, even given the absence of beacon detections to date. However, if the N=1 school argument is correct that expansion into the Galaxy is as much a natural activity of technological civilizations as the building of communications beacons, then the result changes. Individual civilizations with fairly short intrinsic lifetimes spawn colonization waves, sequences of colonies in which the newest are formed before the demise of the previous and then survive long enough to create more colonies. Thus while each civilization may have a small value of L, the process of colonization may have arbitrarily large apparent values of L, and it is these values that are the relevant factors for the Green Bank equation. In short, in order that the number of communicating civilizations be low, the lifetimes of all civilizations must be too short to sustain colonization. This argument does not suggest a long future for the human race; and the more optimistic the choice of factors in A, the more discouraging our prospects appear.

It may be that all extraterrestrial civilizations have adopted a policy of concealment, for reasons that become the subject of the debate. There are actually quite a few interpretations of the present negative results that do not require N=1, which may be generally characterized as "zoo" hypotheses, following Ball (1973; cf. also Bernal 1967; Sagan 1973; Kuiper and Morris 1977; Schwartzman 1977, 1981; Sagan and Newman 1983; Brin 1983; Wilson 1984).

But it may still be said that the N=1 school controls the focus of

the debate. Arguments tend to be about whether long-lived
civilizations would be interested in colonization, or in setting up
beacons, or in talking with less advanced civilizations like ourselves
- i.e., speculations about the motives of hypothetical beings. At
least one SETI researcher (P. Palmer, private communication) has
pointed out the unhappy similarity of SETI arguments of this form to
alchemical theorizing, as analyzed by Jung: When facts are few,
speculations are most likely to represent projections of individual
psychologies.

3. PROSPECTS

In reaction to the arguments of the pessimists, Sagan (1982) organized
an international petition in support of a systematic radio-telescopic
search for one or two decades, at a cost of a few million dollars a
year:

> It has been suggested that the apparent absence of a
> major reworking of the Galaxy by very advanced beings,
> or the apparent absence of extraterrestrial colonists
> in the solar system, demonstrates that there are no
> extraterrestrial intelligent beings anywhere. At the
> very least, this argument depends on a major
> extrapolation from the circumstances on Earth, here
> and now. The radio search, on the other hand, assumes
> nothing about other civilizations that has not
> transpired in ours.

Yet, it is important to note that both discussions of SETI,
optimistic and pessimistic, have generally framed the issue as a search
for advanced civilizations, defined as those capable of providing
beacons for detection with the intent of at least announcing their
presence. We wish to point out an important consequence of
distinguishing searches for advanced civilizations from those for
emerging civilizations, the latter being defined by the absence of
beacons and thus the necessity of searching for leakage radiation (as
discussed extensively by Sullivan et al. 1978). A negative result for
the former tells us essentially nothing, because it returns us to the
standard polemic between optimists and pessimists. A negative result
for the latter, on the other hand, may be quite profound, for it
requires no presumptions more extreme than the cosmological principle:
If we are not anomalous, then intelligent civilizations in the Galaxy
should evolve through similar technological stages. It is not
difficult to define how we should appear from a distance, and thus how
to search for our analogs (cf. Sullivan and Knowles 1985). Negative
results for such searches would set definite limits on the product of
all terms in the Green Bank Equation preceding L, i.e. on the rate of
appearance of civilizations that use radio wavelength communications
(Schwarzman and Rickard 1986).
We emphasize that the assumption of beacons, even those of modest
power which we could generate ourselves (as proposed by Subotowicz

1986), vitiates the significance of an unsuccessful search. Negative
results can only be useful if we avoid all questions about the
intentions of ETIs.

The problem with the immediate application of searches in this
direction is the difficulty of setting useful limits with current
instrumentation. If we write the rate of emergence of civilizations
analogous to our own as (N/L), assuming a homogeneous distribution
through a cylindrical galaxy, and adopt Sullivan's (1978) estimate of
roughly 200 pc as the distance at which a Cyclops-style array could
detect our own military radars, then it is hard to exclude even the
most optimistic limits on N/L (see table below). Larger arrays,
capable of extending the detection limits to include the whole galaxy,
would be needed to set truly meaningful limits on N/L.

| | Number of Detectable Emergent Civilizations in | |
| | Cyclops Volume | Whole Galaxy |
N/L (# per yr)	(assuming L = 100 yr)	
10.	0.08	1000.
1.	0.008	100.
0.1	0.0008	10.
0.01	0.00008	1.
0.001	0.000008	0.1

Note that our own civilization has been in the detectable category less
than 50 years.

The detection of emergent civilizations anywhere within the Galaxy
would require an effective telescope diameter some 25 times that of the
Cyclops array. Such a system would presumably have to be constructed
in near-Earth space, with a design for gradual expansion as resources
become available, and would have an enormous cost. For comparison,
note that the proposal by the U. S. Naval Research Laboratory for a
space-based array roughly one-tenth the size of the Cyclops array
estimates a cost on the order of 50 billion dollars. Rough scaling
suggests that the desired SETI array would cost the equivalent of a
decade's worth of the world's current military budget (presently 900
billion dollars per year). Needless to say, an Earth willing to make
such a transfer of resources would be well on its way to becoming a
unified planetary civilization.

An argument commonly heard in speculations on why advanced
civilizations choose to conceal themselves is that they require a
certain level of maturity of our planetary civilization before we may
be allowed to "join the club" (e.g., Bernal 1967). One could easily
imagine, then, that the beacons of advanced civilizations would be
designed to be weak, comparable to the level of leakage radiation, in
order to demand such a concerted level of effort for detection. Thus,
even if one prefers to think in terms of advanced civilizations, it may
still be wiser to plan SETI for emergent civilizations. Of course, the
implicit assumption that planet-wide cooperation and social maturity
are well correlated may be optimistic; the galactic club may have to
supplement its precautions with long-term surveillance of candidate
members, e.g. with Bracewell probes.

SETI should be seen as a scientific research program of long
duration. Current and projected efforts to detect beacons from
advanced civilizations are a noble attempt worthy of support in view of
the value of any positive result as well as the scientific and
technological side benefits (cf. Tarter 1984). Within the next few
decades, we may reasonably expect meaningful determinations of fp, ne,
and perhaps fl on the basis of data from the next generation of space
observatories, beginning with the Hubble Space Telescope (cf. Burke
1986). As we improve the astrophysical constraints on these terms in
the Green Bank Equation, it may well become possible to set useful
limits on the terms involving intelligent life regardless of whether
signals from extraterrestrial civilizations, emergent or advanced, are
ever detected!

4. REFERENCES

Adler, J. 1983, Newsweek (January 31), 64.
Ball, J. A. 1973, Icarus, 19, 347.
Ball, J. A. 1980, Amer. Scientist, 68, 656.
Bernal, J. D. 1967, The Origin of Life (London: Weidenfeld & Nicolson).
Bieri, R. 1964, Amer. Scientist, 52, 452.
Brin, G. D. 1983, Q. J. Roy. Astr. Soc., 24, 283.
Burke, B. F. 1986, Nature, 332, 340.
Drake, F. D. 1974, in Interstellar Communication, eds. C. Ponnamperuma
        and A. G. W. Cameron (Boston: Houghton Mifflin), 118.
Field, R. J. 1985, Amer. Scientist, 73, 142.
Gould, S. J. 1985, The Flamingo's Smile (New York: W. W. Norton).
Hart, M. H. 1975, Q. J. Roy. Astr. Soc., 16, 128.
Hart, M. H. 1978, Icarus, 33, 23.
Hart, M. H. 1979, Icarus, 37, 351.
Kuiper, T. B. H., and Morris, M. 1977, Science, 196, 616.
Lasagna, A. C. 1980, Geochim. Cosmochim. Acta, 44, 815.
Lewin, R. 1986, Science, 231, 545.
Lovelock, J. E. 1979, Gaia (Oxford: Oxford University Press).
MacGowan, R. A., and Ordway, F. I. III 1966, Intelligence in the
        Universe (New Jersey: Prentice-Hall).
Monod, J. 1971, Chance and Necessity (New York: Knopf).
Morrison, P. 1974, in Interstellar Communication, eds. C. Ponnamperuma
        and A. G. W. Cameron (Boston: Houghton Mifflin), 168.
Ornstein, L. 1982, Physics Today, 35, 27.
Russell, D. A. 1981, in Life in the Universe, ed. J. Billingham
        (Cambridge: MIT Press), 259.
Sagan, C. 1973, Icarus, 19, 350.
Sagan, C. 1982, Science, 218, 426.
Sagan, C., and Newman, W. I. 1983, Q. J. Roy. Astr. Soc., 24, 113.
Salvini-Plawen, L. v., and Mayr, E. 1977, Evolutionary Biol., 10, 207.
Sawyer, C. 1984, Icarus, 57, 135.
Schneider, S., and Thompson, S. 1980, Icarus, 41, 456.
Schwartzman, D. W. 1977, Icarus, 32, 473.
Schwartzman, D. W. 1981, SETI-81 Conference, Tallinn, preprint.
Schwartzman, D. W., and Rickard, L. J 1986, in Proceedings of the

37th Congress of the International Astronautical Federation, Innsbruck.

Shapiro, R. 1986, Origins, (New York: Summit Books).

Shklovskii, I. S. 1978, Social Sciences, 9, 199.

Shklovskii, I. S., and Sagan, C. 1966, Intelligent Life in the Universe, (San Francisco: Holden-Day).

Simpson, G. G. 1964, Science, 143, 769.

Subotowicz, M. 1986, J. Brit. Interplan. Soc., 39, 499.

Sullivan, W. T., III, Brown, S., and Wetherill, C. 1978, Science, 199, 377.

Sullivan, W. T., III, and Knowles, S. H. 1985, in The Search for Extraterrestrial Life: Recent Developments, ed. M. D. Papagiannis, (Dordrecht: D. Reidel), 327.

Tang, T. B. 1982, J. Brit. Interplan. Soc., 35, 236.

Tarter, J. 1984, Acta Astronautica, 11, 387.

Tipler, F. J. 1980, Q. J. Roy. Astron. Soc., 21, 267.

Tipler, F. J. 1981, Q. J. Roy. Astron. Soc., 22, 279.

Walker, J. C. G., and Hays, P. B. 1981, J. Geophys. Res., 86, 9776.

Wilson, T. L. 1984, Q. J. Roy. Astron. Soc., 25, 435.

FERMI PARADOX AND ANTERNATIVE STRATEGIES FOR SETI: ANTHROPIC PRINCIPLE AND SEARCH FOR SOLAR ANALOG

M. Fracassini, L.E. Pasinetti Fracassini, S. Rosazza, A.L. Pasinetti

Dipartimento di Fisica, Universita' di Milano
Via G. Celoria 16, Milano 20133, Italy

ABSTRACT. The Anthropic Principle, a new trend of the modern cosmology, claims that the origin of life and the development of intelligent beings on the Earth is the result of highly selective biological processes, strictly tuned in the fundamental physical characteristics of the Universe.

This principle could account for the failure of some programs of search for extraterrestrial intelligences (SETI) and suggests the search for strict solar analogs as a primary target for SETI strategies. In this connection, we have selected 22 strict solar analogs and discussed their choice.

REFERENCES

Barrow, J.D., Tipler, F.J. 1980, The Anthropic Cosmological Principle, Oxford University Press.

G. Marx (ed.), Bioastronomy – The Next Steps, 313.
© *1988 by Kluwer Academic Publishers.*

THE SETI RELATED USE
OF LARGE ASTRONOMICAL ARCHIVES

R. Albrecht
Space Telescope European Coordinating Facility*
European Southern Observatory
B. Balázs
Department of Astronomy, Eötvös University
Kun Béla tér 2, Budapest 1083 Hungary

ABSTRACT. Large archives of scientific data in computer processable form are coming into existence right now. This paper examines how such archives can be used for SETI-related projects. Particular emphasis is given to software techniques which will make the large scale, automatized cross-search of archives possible.

1. INTRODUCTION

Data gathered during space missions are special in that they are not easily reproducible, they are taken from a special vantage point, and they are enormously expensive expressed in money per data point. Along with the fact that such data are usually available in digital representation, this has led to large collections of such data. Collections, rather than archives, because they mostly consist of reels of magnetic tapes on shelves. It has not been easy to gain access to the data because of the missing data base management.

Computer industry has recently produced commercially viable bulk storage media which make the retrieval of data and the scientific re-use possible and attractive. These devices typically use optical disks with a capacity in the Gigabyte range. Although the retrieval speed is somewhat lower than the speed of magnetic disks, this disadvantage is more than offset by the fact that an enormous amount of data can be kept on line, eliminating the need for time consuming tape operations. This technology will, for instance, be used to keep all data produced by the Hubble Space Telescope (HST) on line.

Archives under development include IUE, IRAS, EXOSAT, HST, ROSAT. Large ground base observatories are beginning to collect data, place them in an archive, and, after a suitable period, make them publically available. Thus, we can expect to soon have available an enormous wealth of astronomical data of potential use for SETI.

*) Affiliated to the Astrophysics Division, Space Science Department, European Space Agency

G. Marx (ed.), Bioastronomy – The Next Steps, 315–317.

2. ACCESSING DISTRIBUTED DATA BASES

The following considerations are not unique to SETI, they apply to any
scientific program which needs to access archives located in different
places, on different computers, different software systems, and accessible through different networks. These problems are being tackled right
now by organizations like the US National Space Science Data Center
(NSSDC) and the Space Telescope Science Institute (STScI). The European
Space Agency (ESA) is in the progress of establishing the European Space
Information System (ESIS). Experimental implementations of systems which
allow user friendly access to data bases residing in different locations
are under way. STARCAT (Space Telescope Archive and Catalogue), developed at ESO/ST-European Coordinating Facility (ST-ECF), already provides
access to the major astronomical catalogues, the SIMBAD data base in
France, and will, in the future, provide access to the IUE archive.
STARCAT provides a uniform and consistent user interface, while the
problems of network connections and different query languages are being
handled internally (Russo et al., 1986).

Another problem is the transfer of the actual data. As long as the
amount does not exceed the range of 10^4 bytes they can be transferred
through a network. The situation is different for image data: these
frames are very large, so even though it might be possible to keep them
available on-line on suitable media, transmitting them across a network
is usually not possible. The solution is to allow for de-archiving requests, so the data of interest get dumped on magnetic tape (or on a CD-
ROM) and shipped to the scientist who requested them. Another interesting possibility is to also allow a limited amount of remote processing
in order to check whether or not data sets which are of potential interest are indeed interesting once the pixels are being examined.

3. SEARCH STRATEGIES AND COMPUTERIZED ARCHIVE SEARCH

We do not intend to deal with possible phenomena in the data, from which
the existence of ETI's could be inferred. Some strategies exist already
and are being employed in the search for artifacts in radio data. Considerable work will have to go into the development of search strategies
suitable for data which were not taken explicitly to examine them for
ETI related phenomena. We assume, however, that any SETI related program
will require a large amount of data to be examined for a set of criteria. The common approach is to do this interactively, with the aid of
computers. The advantage of this is that the researcher can examine the
data very thoroughly, and, in fact, change the search strategy dynamically, based on operational experience gained during the search. There is
also a significant disadvantage: once the body of data to be examined
exceeds a certain size, processing with human intervention becomes pro-
hibitively cumbersome.

Another complication is the need to cross-query different archives
or data bases, to use one archive to identify candidates for the occurrence of certain phenomena, which then can be looked for in a different
archive, or archives. Intermediate processing is implied through the use

in these data bases of different units, scales, coordinate systems, identifications, etc.

Once the on-line availability of large bodies of data is ensured, procedures have to be developed which allow the efficient examination of the data to check for the occurrence of phenomena which were predicted by a search hypothesis. These procedures have to minimize human intervention, and at the same time retain the possibility of dynamic adaptation.

A relatively recent product of the software market are expert systems. These software systems have the potential of solving the problems of the computerized examination of large bodies of data. Expert systems attempt to simulate the human reasoning and decision making process; they are quite successful, provided the problem domain is narrowly scoped and properly defined. Already expert systems are being used to optimally calibrate data by generating software procedures for image processing systems (Johnston, 1987). Experiments to use expert systems for spectral calibration have been started (Heck et al., 1987). Other applications in astronomy are being investigated (Albrecht, 1987).

For automatic classification we can also iterate the classification criteria. It is precisely this area in which expert systems could be used in a SETI context. Not only would it be possible to examine large data sets for ETI phenomena in a consistent manner, it would also be possible to apply "what-if" search strategies and automated reasoning, something which is impossible to do for humans for such large data sets. For the non computer scientist these software techniques might appear farfetched at this point in time. However, the technology is available and is awaiting application. Certainly by the time the HST science data archive will have reached a representative size, the use of expert systems will be accepted software methodology.

What needs to be done now is to identify the data bases which will become available within the next five years, to investigate whether their contents will be usable in a SETI context, and in which way a SETI related investigation could proceed. Since the approach outlined above will require considerable efforts, it should be coordinated with those organizations which will maintain archives, in order to make sure that they are available for computerized SETI. We also have to develop search hypotheses which are optimized for the expert system approach, which includes the possible automatized generation of new search hypotheses.

References

Albrecht, R., Adorf, H.M., Fosbury, R., Johnston, M., Rampazzo, R., 'Applications of AI Technology in Astronomical Research II: What can be done, What should be done', BAAS Vol.19, No.2, p. 743, 1987.

Heck, A., Rampazzo, R., Murtagh, F., Albrecht, R., 'Rule-based Classification of IUE Spectra', Preprint ST-ECF, 1987.

Johnston, M., 'An Expert System Approach to Astronomical Data Analysis', Preprint ST-ECF, 1987.

Russo, G., Albrecht, R., Ochsenbein, F., 'Optical Mass Storage for a 10 Terabyte Archive', SPIE Vol. 696, p. 366. 1986.

ARTIFICIAL SIGNAL DETECTORS

I. Linscott, J. Duluk, J. Burr and A. Peterson

Stanford University
Stanford CA 94305, USA

1. Introduction

The detection of an artificial, extraterrestrial, continuous wave (CW), or pulsed signal is a primary objective of the NASA Search for Extraterrestrial Intelligence (SETI). In support of that goal, research in signal detection at Stanford University has focussed on the development of high performance computer architectures that achieve optimum sensitivity to pulsed and CW signals. Our architectures integrate parallel signal processing with matched filter computation in arithmetic processors and associative memories suitable for implementation as custom VLSI circuits. Using these devices as the computation engines in signal detectors makes matched filter pulse and CW detection both feasible and cost effective for a wideband ETI search of the microwave spectrum.

This talk will cover the strategy for detecting pulsed and CW signals, and the implementations which capture efficient signal processing methods in custom VLSI devices.

Starting with Drake's Project Ozma in 1959, there have been some 47 separate radio searches for ETI signals. Many of them are still under way and all were designed to detect CW signals. None however, have searched for signals from pulsed beacons. The most sensitive searches have placed extreme demands on the frequency stability of the signal, because the most detectable signals are those with the most narrow bandwidths.

In principle, the minimum possible bandwidth for a signal is the reciprocal of the signal's duration. By its nature a CW signal has a very long duration and an ultra-narrow bandwidth. Thus, a CW signal is best detected with a receiver of very narrow bandwidth tuned to the right frequency. For greatest sensitivity, the bandwidth of the receiver should match the signal's, but in the past, a practical limit of the order of 1 Hz, has been chosen many physical and economic considerations other than signal duration.

The physical conditions that limit bandwidth are predominantly frequency drift from accelerations along the line of sight, and multipath scattering in clouds of interstellar hydrogen, while economic considerations constrain the ability to try all the combinations of choices among possible frequencies and drift rates.

The lack of a cost effective means of dealing with the huge number of choices among

319

G. Marx (ed.), Bioastronomy – The Next Steps, 319–335.
© *1988 by Kluwer Academic Publishers.*

frequency and drift has impaired the development of detectors with minimum bandwidth and therefore optimal sensitivity. The special signal processing engines we have developed solve this computational bottleneck by capturing the matched filter computation in VLSI devices. A modest number of these devices can then be employed to keep up with the input data rate.

The input data rate is one of the most important factors in designing a SETI signal detector, where emphasis is placed on supporting high rates. In most cases the input data rate determines the bandwidth of the detector.

2. Signal Detection Strategy

A strategy that achieves near optimal sensitivity manages signal detection as a two stage process. A staged approach is appropriate because the conditions that limit the sensitivity of a pulse or CW detector to narrow band signals are partly physical and partly economic. To see how each stage strikes a cost effective balance, the limitations will be considered separately.

The constraints on minimum bandwidth are predominantly frequency drift from accelerations along the line of sight, and multipath scattering in clouds of interstellar hydrogen. The accelerations, caused by planetary rotation and orbital motion, will continuously shift the received frequency of a signal resulting in a sweep, or chirp, at a rate of up to approximately 1 Hz/sec. The rate depends both on acceleration, a, and frequency, f, as

$$\frac{df}{dt} = \frac{af}{c}$$

where t is time and c is the speed of light. The total drift rate will have a range of 0.1 Hz/sec to 10 Hz/sec over the microwave window for terrestrial like orbits and rotations.

During its passage through the interstellar medium, a signal's bandwidth is increased by multipath scattering in the interstellar clouds of ionized hydrogen, (Drake and Helou, 1978), and arrives with a minimum bandwidth of the order 0.001 Hz to 0.1 Hz. The low end of this limit is appropriate for the approximately 1000 stars in the targeted search portion of the SETI Microwave Observation Program. This is mainly due to the close proximity of these stars, and the scarcity of interstellar clouds in our galactic neighborhood.

The fundamental problem of the microwave search is to sift through the thousands of megahertz of the microwave spectrum looking for spectral features on the order of one

Hertz or less in width. In our two stage method, the first stage is optimized to match the frequency and bandwidth of a hypothetical signal. The second stage recognizes signals by searching for combinations of frequencies throughout the observation.

2.1. First Stage

In the first stage, a multichannel spectrum analyzer (MCSA), is used to match a hypothetical signal to ideally one of 120 million narrowband filters. These filters span the 10 MHz input bandwidth of the MCSA encompass 2 polarizations, 5 octave resolutions, 3 overlapped spectra, and closely match a broad range of pulse widths and drift rates.

As many channels as possible are produced in parallel to shorten the search. These channels are matched to the physical characteristics of an essentially signal, but may not be optimally matched to the signal's drift rate and bandwidth. However, the match is expected to be good enough to permit a second stage of signal recognition to succeed.

2.2. Second Stage

The greatest sensitivity to a CW signal is obtained by a filter that follows the signal's drift in frequency with a bandwidth comparable to the intrinsic bandwidth of the CW signal. So many choices are possible for the filter's drift rate and minimum bandwidth that a novel approach and special implementation techniques are needed to make CW and pulse detectors practical and affordable.

Building a matched filter for each choice would entail having a filter for each of 10 million frequency bins in the MCSA, and of the order of 2000 drift paths for each bin, for a total of over 20 billion filters. Such a solution would prohibitive if a separate device were required to compute each filter.

Efficient algorithms that searching for CW and pulsed signals have been developed at the Ames Research Center. (Cullers, Linscott, and Oliver, 1985) Experimental versions of these algorithms have been written to run in real time on a VAX 750.

The criteria for successful detection is that a signal processed by these techniques will survive a postprocessing threshold test with a probability of 50%, at a threshold level such that the rate of false alarm is of the order of 10^{-12}. The objective is to meet these criteria for: (1) CW signals whose signal to noise ratio (SNR), is -6 db, and (2) pulses whose SNR is +4 db or stronger in the high resolution channels of the MCSA.

3. Pattern Detectors for the MCSA

3.1. Integration Methods for CW

The integration of successive samples from the MCSA can be done easily by adding together the power in each sample. This is referred to as incoherent integration. Although easily performed, the method improves the SNR by only the square root of the number of samples, assuming the number of samples is large. Strategies for incoherent integration have been developed that provide for economies in both computation and memory use, (Cullers, et. al. 1985). An alternative method is coherent integration where the SNR gain is linear in the number of samples integrated.

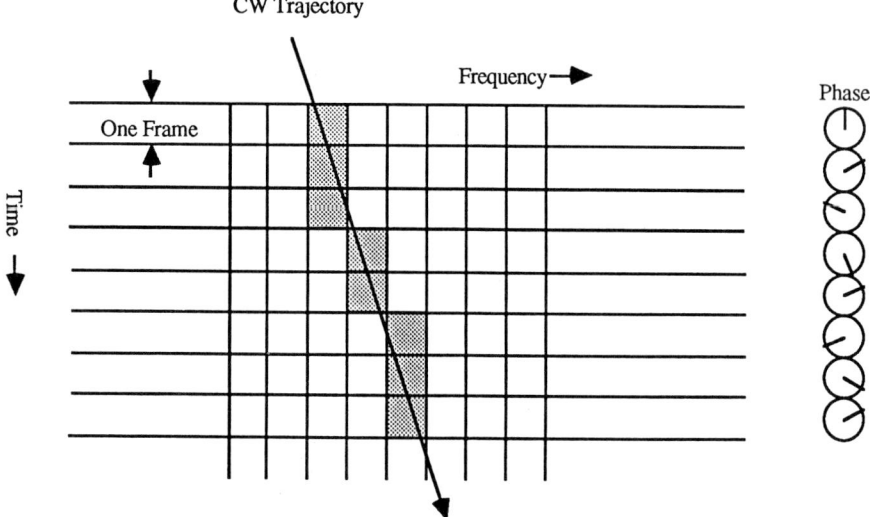

Figure 1 : CW Signal Drifting Along in the Time–Frequency Plane

A coherent integration technique, suitable for VLSI implementation is an attractive means for the detection of narrowband, drifting CW and pulsed signals. This technique is applicable to data sampled from the high resolution outputs of the SETI MCSA. In these bins a weak signal may be present but because these bins are only approximately matched to the signal's bandwidth the signal is not as yet detectable.

The term coherent integration refers to the addition of waveform samples that are compensated for phase rotation according to a model of a signal's temporal behavior. Coherent integration is equivalent to processing a signal through a matched filter. The terms coherent integration and matched filter will be used interchangeably.

Coherent integration is performed by a filter matched to a CW signal's initial frequency and drift rate. The matched filter is implemented as a digital signal process on

N samples from a single resolution sequence chosen from the high resolution bins of the MCSA. This sequence has a staircase appearance and is illustrated in Figure 1 .

As the signal drifts, a bin will be chosen at each time frame along the path. The bin's center frequency is closest to the hypothetical trajectory of the CW signal. An N–sample sequence chosen in this manner improves the frequency resolution under appropriate assumptions of stability. The resolution of a CW signal is limited by the minimum bandwidth due to interstellar scattering, B_S. Thus, as long as the sequence duration is less than $1/B_S$, the frequency resolution improves.

Within each path there will be N possible initial frequencies, and N possible final frequencies. Consequently, N^2 choices are possible, each corresponding to a linear trajectory within the original path. A CW detector that follows a signal coherently for N time frames must be able to evaluate a matched filter not only for each path but for each of the N^2 trajectories along that path. By limiting the drift rate to 1 Hz/sec, or 1 bin/frame, there are $2N-1$ paths to choose from, and the total number of matched filters required becomes $N^2(2N-1)$, or of the order of N^3 matched filters for each of the 1 Hz bins in the MCSA.

To manage a computation of this magnitude N matched filters must be evaluated for each path, or $(2N-1)N$ matched filters evaluated in parallel. Each matched filter performs the following computation:

$$ y_{pq} = \sum_{n=0}^{N-1} x_n W_N^{r_n(p,q)}, \qquad (n,p,q) = 0,1,\ldots,N-1. $$

All values are complex, the x's are the input samples, p and q label the initial and final CW frequencies, and the W's are the complex phase exponentials of the matched filter. Each matched filter is characterized by a unique phase rotation sequence that is represented by a phase function $r_n(p,q)$.

A high precision implementation of these matched filters is conceptually straight-forward but expensive. A total of N^2 complex multiply-accumulators (MACS), are needed to perform the computation, each with its own stored version of the phase rotation sequence. A much less expensive process is possible for limited precision MACS, especially when the precision is 4-bits or less.

However, both the gain of the matched filter and the variance of processed noise are effected when the precision of the MAC process is reduced. These effects reduce the signal to noise ratio (SNR), of the matched filter process. This effect implies that longer sequences are needed to recover the lost SNR. Longer sequences require more complex processors. Thus an optimum precision exists for which the largest SNR is produced with the least processor complexity. A minimum complexity for a matched filter CW processor appears possible for data, represented in polar notation, with approximately 4 bits of magnitude precision and 2 bits of precision in phase.

As an example of the CW detection process, fix the matched filter length $N = 32$ samples. Pick a bin from the 1 Hz channels of the MCSA. Select a drift path 32 samples long. Send those samples sequentially into all 32^2 matched filters, all operating in parallel on a single device. Each filter forms a 32–pt coherent integration using the proposed quadrant phase rotation and increment/decrement accumulation method. At the completion of the accumulation, the power P, in the accumulator is formed from the accumulator's real component reA, and the imaginary component imA,

$$ P = (reA)^2 + (imA)^2 $$

The power P, possess an exponential statistic, the mean value of which is proportional to the sum of the power in the signal and the noise. If a CW signal is present along one of the paths covered by the set of matched filters, then the power output from the filter that follows the drift path will exceed on the average the mean power level from all other paths.

The design goal is to detect a CW signal with a -6 db SNR. In particular, if the strength of the CW signal is -6 db relative to the noise level in each of the 32 samples, and the noise possesses a zero-mean, Gaussian distribution of unit variance, then the excess power level will be high enough that a 50% threshold test will exclude all but 1 in 7000 of the outputs from filters processing noise alone.

The false alarm rate from a 32–point matched filter is at this point 1 in 7000. To decrease the false alarm rate to 1 in 10^{12}, repeat the accumulation on successive groups of 32-points. Each 32-point group is now considered a single stage in a multi-stage detector. The strategy will be to threshold the power level from the matched filters, saving only 1 in 1000 results on the average, and combining the thresholded results from successive stages.

The intent is to save the direction along with the complex accumulator for the filter values over threshold. Then the detector will search for repeated occurrences of over threshold filters along the same path. The search shares many of the same properties of the SETI pulse detector, and could be carried out for example in the CAM Processor that is currently being developed for detecting pulses.

A combined false alarm rate of 1 in 10^{12} is possible using the joint probabilities of 4 or 5 of the 32–point accumulation stages. However, thresholds that are lower than for a single stage will be needed to achieve both the desired false alarm rate and a 50%

probability of detection. The precise threshold depends on the search strategy that combines the results from multiple stages. For example, assuming that each stage's result must be over threshold. the threshold on each stage must be low enough to permit $(0.5)^{1/D}$ of the filters to pass, where D is the number of stages needed to reach the combined false alarm rate.

The requirement that all stages must be over threshold is unnecessarily restrictive. In analogy with pulse detection allowing missing stages in the search still achieves the desired false alarm rate while greatly reducing the search population because, in this case, the threshold can be increased and the population of survivors is exponential in the threshold level. Samples which survive the threshold process are referred to as "hits".

The management of hits is best done in a separate processor closely coupled with the first. The first process performs: 1) computation of N^2 matched filters; 2) power computation; 3) threshold test; and 4) reporting the values and trajectory information of the survivors to the second processor. It is the second processor that searches for continuity along promising paths.

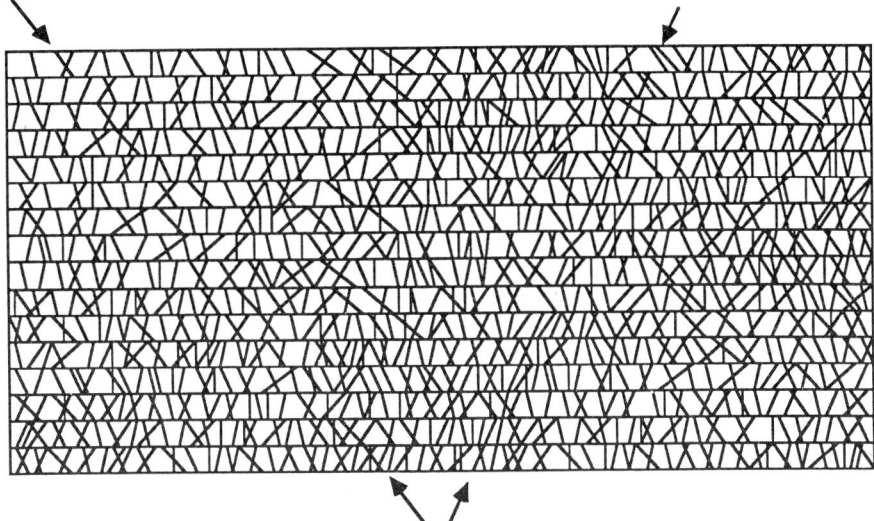

Figure 2 : Above–Threshold Matched Filter Outputs in the Time–Frequency Domain

The second processor needs to find successive, or nearly successive, hits along the same path. In Figure 2 , two such paths are pointed out by pairs of arrows. Notice these paths are not necessarily precisely straight and also have some segments missing. The CAM Processor pulse detector is ideally suited for searching along the same path especially when some of the data segments are missing or not precisely aligned.

In particular, if all the hits are transformed from the time–frequency domain to the slope–intercept domain, colinear hits will lie in a cluster in this latter domain. If the CAM is loaded with this slope–intercept data, then CAM queries can be performed to locate clusters of hits and therefore find the desired signals. The CAM queries are efficient and can typically be performed in a few processor cycles. When a cluster is found, these values are further accumulated in either a coherent or incoherent manner. Figure 3 shows a sample slope–intercept domain data set including an obvious cluster which could be an ETI signal.

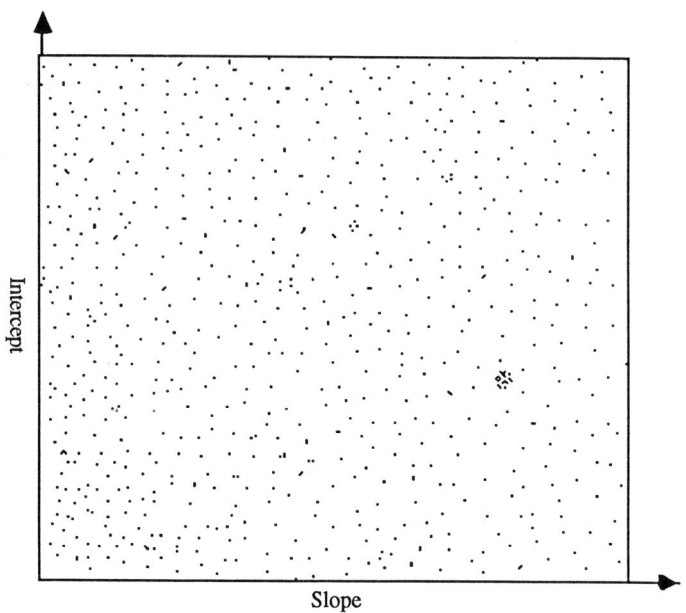

Figure 3 : Above–Threshold HIts in the Slope–Intercept Domain

An interesting prospect emerges for long observations where up to 1000 time frames are sampled from the MCSA. In this case up to 32 stages of matched filters can be computed from these samples. This is 5 to 6 times longer than is required to reach a sensitivity of -6 db, but should permit a proportionately higher sensitivity for drifting CW signals.

The upper bound for detectable signal strength could reach -14 db by sending those survivors of the threshold test lying along statistically promising paths back through the 32–point matched filter accumulators.

The strength of the proposed approach for a CW detector lies in the use of VLSI to capture a simple signal process in dense logic circuitry and through replication manage an otherwise formidable computation. The solution to CW detection will be able to detect signals earlier and with higher sensitivity than non matched filter detectors. The two tiered detection strategy which combines the Matched Filter Processor with a CAM Processor results in a compact, efficient design that is cost effective.

3.2. Pulse Detector

The MCSA periodically produces both an amplitude spectrum and a power spectrum of the input signal. A series of frequency spectra, or frames, are processed together as one observation period. Since frames are successive in time, the data from one observation period can be viewed as power (or amplitude) as a function of both frequency and time. A three-dimensional representation of an example set of data is shown in Figure 4 . This data is further processed before being passed on to the pattern detector. This added processing is a simple threshold operation for the pulse detector. The pulse detector examines the entire observation period and looks for a regular pattern amidst the noise.

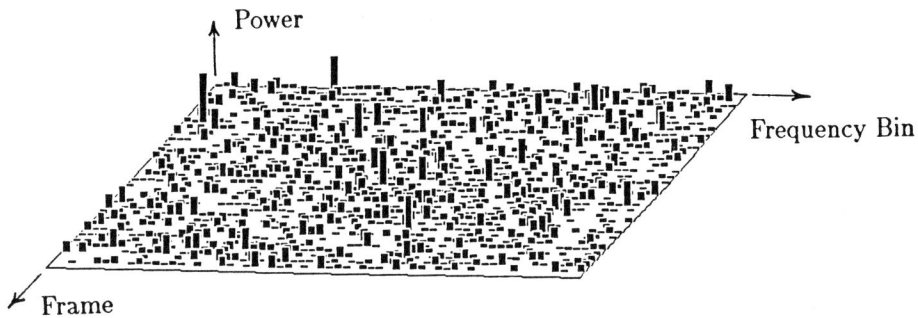

Figure 4 : Three–Dimensional Representation of output from the spectrum analyzer

3.2.1. Pulse Train Detection Algorithm

In addition to CW, the MCSA's pattern detectors are designed to search for regular pulses of narrowband signals.

The input to the pattern detector is the output from the spectrum analyzer. When searching for pulse–modulated signals, the last operation performed on the data before it is passed to the pattern detector is a form of data reduction and is accomplished by a thresholding algorithm. Data points having power above the threshold, referred to as "hits", have their frame, frequency bin, and power saved as an ordered triple for further processing by the signal detector. Figure 5 shows a realistic set of data points above threshold as a function of frame and frequency bin. In this diagram, a pulse–modulated signal appears as a set of evenly spaced colinear points. A pulse train of this type is included in Figure 5 .

Locating pulse trains is a formidable task since, for every pair of hits that could be the first two points in a pulse train, the location of a possible third hit must be extrapolated. The set of all hits must then be searched to see if the projected location in the frame–frequency domain is above threshold. If the third point is above threshold, a search is done for a fourth point. If found, the process of searching for the next pulse in the train is continued until either a hit is not found, or enough pulses are included to declare it a valid pulse train.

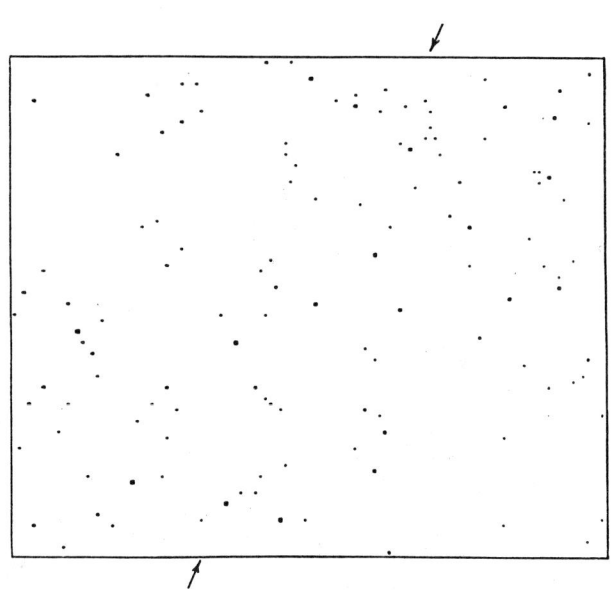

Figure 5 : Above–Threshold Pulses in the Frame–Frequency
Domain. The arrows show the line of a pulse train.

A tremendous amount of searching for hits within projected domains is done. Rather than sequentially search through the data or use some sort of complicated set of linked lists, the memory containing all of the above–threshold data could be directly interrogated for a data point at a specific frame and frequency. This is precisely the function of a conventional content addressable memory (CAM). Thus, a CAM could determine the existence of the next pulse in a pulse train in a single memory cycle.

An attractive approach to pulse detection is based upon a new kind of Content–Addressable Memory (CAM) developed at Stanford (Duluk, et. al., 1986). It is expected to yield well over an order of magnitude improvement in both space and cost when compared to conventional architectures. Approximately four CAM-based processors would be needed to perform real-time pattern detection on the data from

a 1000–second observation period. These processors will share an array of approximately 100 custom VLSI CAM chips. This new architecture is made possible only
through the use of state–of–the–art VLSI technology.

3.2.2. CAM–Based Pattern Detector

Conceptually, the new CAM architecture performs operations on an N-dimensional
space which is sparsely populated with data points. Arbitrary shapes can be constructed in the N-space and used to bound subsets of data points in the space.
Operations such as intersection and union can then be performed on these subsets,
the results of which can also be tested to see if they contain any data points. Data
points that belong to a subset can be read out of the memory for external processing.
By building these functions into hardware, manipulation of spatially oriented data
can be processed faster than ever before.

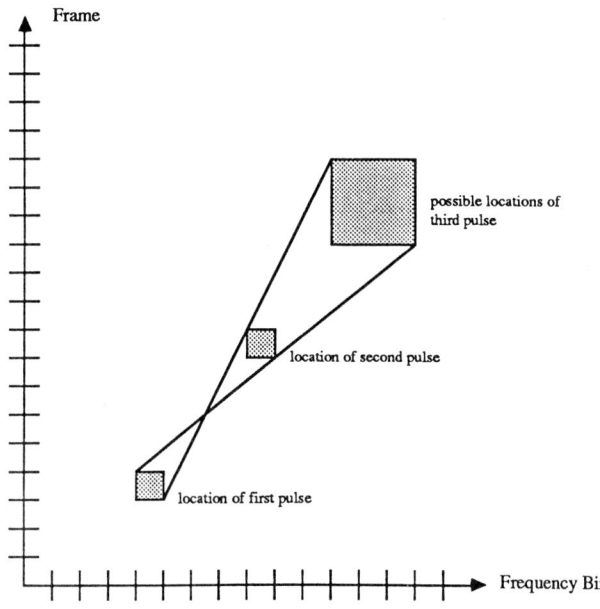

Figure 6 : Roundoff Effect Requires 3x3 Search Area

The problem of signal detection becomes more complicated once quantizing effects in
frame and frequency are included. That is, each hit's position in the time–frequency
domain has been rounded off. Therefore, when a pulse train is extrapolated to possibly include another hit, a 3x3 region must be searched in order to insure that the
extrapolated point is not missed due to roundoff error. This roundoff effect is illustrated in Figure 6 . Because a 3x3 window must be tested, nine searches into the
set of hits are required. With a conventional CAM, nine memory cycles would be
required for these nine searches. It is preferable for a CAM to test all points in the
window in one or two memory cycles. In fact, to help with the other parts of the

algorithm, it would be ideal to have the capability for testing any region of arbitrary shape in a small number of clock cycles. This "generalized spatial processor" is the machine under development for SETI pattern detection.

3.2.3. The CAM-Based Search for Pulses

A comparison operation is referred to as a "query" because it can ask, for example, "Which hits are located at a point with frame greater than 198 and frequency greater than 279 Hz?" The query is one of a set of CAM functions that allow an arbitrary shape to be constructed in the three–dimensional space of frame, frequency, and power by a sequence of queries.

A simple form of a SETI algorithm is illustrated in Figure 7 . In step one, a rectangular region is constructed using two queries.

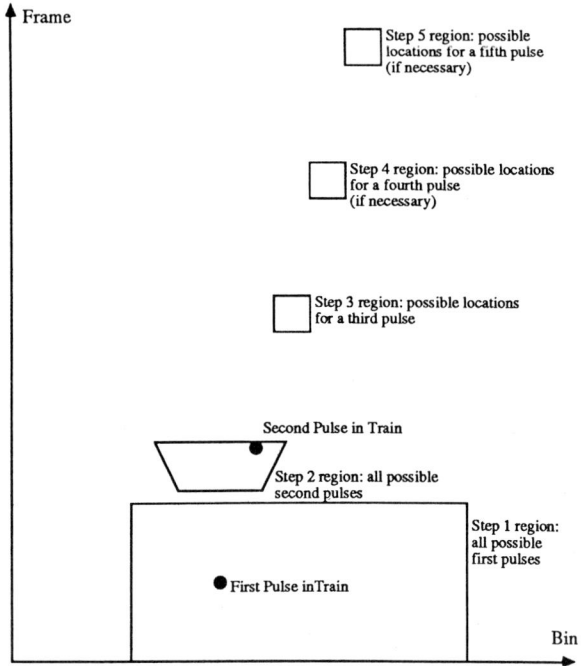

Figure 7 : Search For a Pulse Train.

This region contains all the possible first data points for all pulse trains. One data point is read from the CAM and will be assumed to be the first pulse of a train. The second step queries a region containing all possible second pulses in the train by constructing a trapezoid. The sides of the trapezoid are determined by the maximum drift rate. The top and bottom of the trapezoid are determined, respectively, by the maximum and minimum allowable period of the pulse train. A data point is read from this region and, if one is found, it is assumed to be the second pulse in the train.

Now that two pulses are known, the third step is to construct a 3x3 window which contains all possible locations of the third pulse in the train. If this window does not contain any data points, the pulse train is discarded, and another second data point is selected from the region constructed in step 2. Step 3 is then repeated. If the window does contain a data point, it is read out of the memory and used in step 4, which extrapolates the pulse train to a fourth point.

This process continues until either the pulse train is discarded or until it is declared a valid pulse train. When all the data points have been read out of the region constructed in step 2, a new first pulse of a train is selected by reading a data point out of the region constructed in step 1. When all points have been read out of this region, processing is complete.

The above discussion of the search algorithm was simplified for the sake of brevity. The full version includes, for example, provisions for missing pulses. It also takes advantage of the fact that pulse trains with fewer pulses in the observation period are required to have a larger average power per pulse.

4. Field Tests

A prototype version of the MCSA has been built to test the signal processing concepts and signal recognition sensitivities. This version consists of a single instance of each of the major MCSA functional components. The full 10 Mchannel MCSA could be constructed from replicas of the prototype modules if so desired, (Linscott, Chen, and Peterson, 1982).

The 73,728-channel, single polarization prototype is presently being used in field tests at NASA's Goldstone Tracking Station-13 in California's Mojave desert. Together with Station-13's 26-meter antenna, the MCSA prototype has detected the 1 watt signal from Pioneer-10, a spacecraft now outside the boundaries of the solar system and whose signal possesses many of the attributes expected in an artificial signal of extraterestrial, intelligent origin.

The organization and implementation of the bandpass filters in the MCSA will briefly reviewed. Figure 8 is a block diagram of the functional components of the MCSA which consists of the two bandpass filters followed by the final DFT which brings the resolution down to 1 Hz.

Each bandpass filter produces a bank of filter channels. The bandpass filter signal processing is Allen Peterson's technique which first takes a "coarse" convolution of the sampled data sequence with the filter's impulse response, and then follows the coarse convolution with a Fourier transform. The convolution is performed in a weighting

332

network, as is shown in Figure 8 . One weighting network apiece for the two filter banks are used.

Next the convolved sequences are Fourier transformed, which has the result that the channels of this DFT represent successive filters in a bank. The DFT's are computed using an efficient decomposition algorithm based on prime number factorization developed in the thesis of Professor Peterson's student, Shankar Narayan. The DFT algorithms run on a high performance, microcoded signal processor, designed at Stanford by another of Professor Peterson's students, Patrick Barkhordarian.

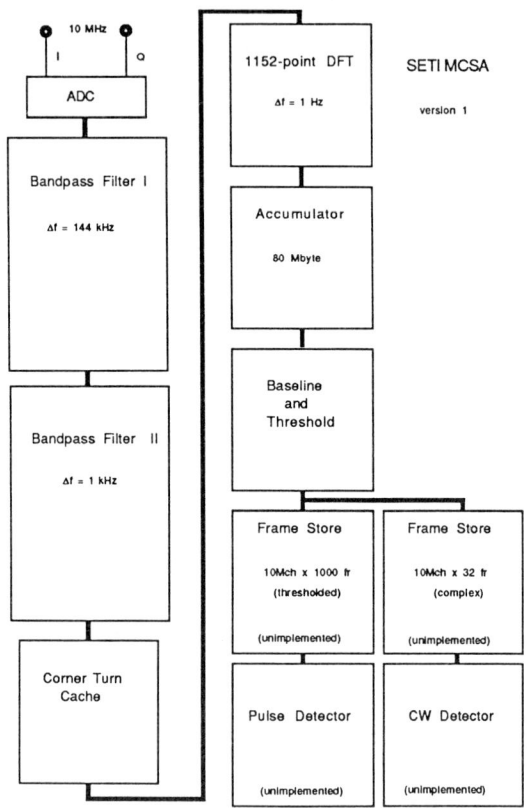

Figure 8 : Block Diagram of the MCSA.

The key to a successful design for the multiple resolution MCSA was to use microcoded signal processors. The same hardware was used at each bandpass filter level, as well as the final DFT, each running it's own version of DFT code appropriate to that level. The architecture of the DFT Processor optimizes performance for the short length DFT's associated with the prime factor DFT algorithms.

The DFT Processor is highly concurrent, where many arithmetic and register transfer operation must be simultaneously specified. A great deal of difficulty was experienced hand coding even the simplest of the short DFT's on this processor and a microcode compiler was written to manage the concurrency and optimize the code. The compiler has since been used to generate code for all three levels of the MCSA.

The prototype MCSA consists of a single instance of the hardware components needed to compute the bandpass filters and the final DFT, as well as perform simple baseline modeling and thresholding of the high resolution bins. The MCSA currently consists of 24 wirewrap boards in a single 6-foot high equipment rack, and is controlled through a DMA channel by a VAX 11/750.

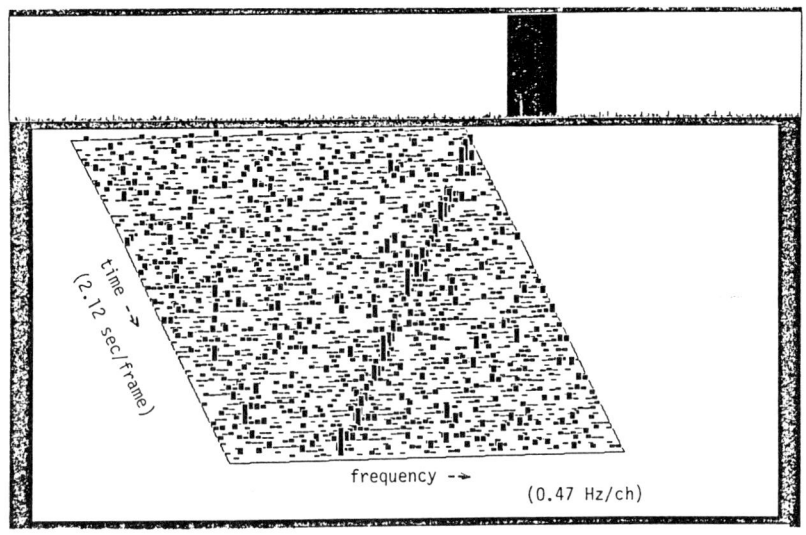

Figure 9 : MCSA output showing the signal from Pioneer–10

The MCSA is now at JPL's Goldstone Tracking Facility in the Mojave. It is attached to the Venus Site's 26-m antenna and is undergoing field test. The signal from Pioneer–10 was acquired using the MCSA at the Venus Site. Pioneer–10 was approximately 40 AU from earth, and the spacecraft's signal seen in the output of the MCSA and shown in Figure 9 , appears slightly brighter than the background noise, which is at -184 dbm.

The MCSA spectra in Figure 9 was obtained from the display on our SUN graphics 68000 workstation. The workstation is linked via an ethernet to the VAX, and is dedicated to displaying MCSA spectra. The display in Figure 9 , is managed by an icon driven window based graphics program, developed at Stanford by Ernst Kimler, and consists of three windows viewing spectra from different MCSA resolutions. The top window is used for the 144 bands from bandpass filter 2. The middle window is for 1152 channels from one of the 144 bands, selected by the user. The bottom window displays a subset of the 1152 frequency channels, again selected by the user. This display uses a a pseudo grey scale to represent signal power, and maintains a scrolling time history for 100 frames in the vertical.

5. VLSI Developments

The research at Stanford University has produced a design for a real-time spectrum analyzer that takes an input signal of 10 MHz bandwidth and produces a spectrum with 1 Hz resolution (ten million channels or frequency bins). The MCSA was designed using TTL compatible chips that were commercially available in 1982.

The current MCSA design, if constructed to produce the full ten million channels and handle two polarizations, would require 976 printed circuit boards full of chips and fill over 20 racks. Fortunately, the design contains only 13 different board types. However, the main processor would be replicated 680 times, and so is an ideal candidate to be implemented as a Very Large Scale Integrated (VLSI) circuit. This VLSI circuit, called the SETI DSP Engine, was designed at Stanford and was fabricated through the DARPA-sponsored MOSIS fabrication clearing house. (Waller, 1986)

This chip is able to outperform the original processor board by a factor of 5 because of a faster clock rate and special architectural features which can only be realized by custom VLSI. Since eight processors can be put on a single board, a total performance–space improvement of a factor of 40 is achieved.

5.1. SETI DSP ENGINE

5.1.1. Winograd and Prime Factor Transform

In the MCSA, a key part of spectrum analysis is done by performing long-length Discrete Fourier Transforms (DFTs) of blocks of data. There are several techniques for efficiently performing this computation, one of which involves using the Prime Factor Transform (PFT). The underlying idea of the PFT is to map a one dimensional DFT of length N into a two dimensional DFT of size pxq, where $N = pq$ and p and q are relatively prime. The two dimensional DFT of the array is then computed by taking p DFTs of length q, and then taking q DFTs of length p. The one dimensional transform is found by an appropriate index permutation mapping. The PFT as described is easily generalized to the case where there are more than two factors. (Nussbaumer, 1982)

If the dimensions of the array (p and q) are not relatively prime, then it becomes necessary to use a generalization of the PFT called the Winograd Transform (WT). In the WT, it is necessary to introduce some additional "twiddle" factors between successive transforms of non-relatively prime length. Asymptotically, the number of operations required in the WT is on the order of $N \log N$. (Winograd, 1977)

5.1.2. Short Length Transforms

The use of the Winograd Transform reduces the problem of evaluating a long transform to one of taking many short DFTs. Work by Rader (1968), Narayan (1981) and others has yielded special purpose algorithms which can be used to calculate transforms of length p, if p is small. These algorithms have approximately twice as many addition operations as multiplication operations. The most important property of this family of algorithms is their three–step nature: 1) adding and subtracting the inputs producing the first intermediate results; 2) performing multiplications on the first intermediate results, producing the second intermediate results; and 3) adding and subtracting the second intermediate results to produce the final answers. This inherent pipelining of the algorithm into three stages suggests that it is an architecture which can be implemented efficiently in VLSI.

5.1.3. Chip Architecture

The chip is designed to directly implement the three–stage algorithm for short transforms. The chip does not contain an on-chip controller due to the lack of silicon area. However, this is not a disadvantage since many SETI DSP Engines will be used in parallel to form the data path of a Single Instruction Multiple Data (SIMD) machine. Hence, one controller can generate instructions for all chips running in parallel.

Generating the set of instructions to implement a set of short DFT algorithms has been done in a nearly optimal way such that all DSP Engine resources are busy almost every clock cycle.

The DSP Engine, was designed as a graduate student research project and succeeded both in capturing the most replicated portion of the MCSA and increasing the speed of the matched filter signal processing. A new version of the MCSA, incorporating the DSP Engine, is being developed that will substantially reduce the scale and cost of the full 120 Mchannel spectrometer.

The second stage of signal detection uses the matched filter outputs from the MCSA. The signal detection strategy here involves sending samples from all the filter channels into separate pattern detectors to search for either pulses or drifting CW signals. The pattern detectors are organized around signal recognition algorithms developed in collaboration with members of the SETI Institute and the Ames research center.

The signal recognition algorithms have been demonstrated to achieve near optimum sensitivity for pulsed and drifting CW signals.

SEARCH FOR ARTIFICAL COSMIC RADIO EMISSION

A.V. Arkhipov
Institute for Radio Astronomy, Ukranian Academy
Kharkov, USSR

The search for extraterrestrial civilizations through the
analysis of continuous isotropic radioemission at frequencies
from 10^2 to 10^3 MHz is discussed. The probable appearance of
an extraterrestrial civilization accessible at these frequencies for modern radio-astronomical facilities from a distance
of ∼20 pc is derived. A search for similar radio sources has
been performed. Fortunately, such objects admittedly lie in
the vicinity of four solar-type stars favourable for intelligent life. The probability of an accidental projections of a
radio sources to the vicinities   *of   such*   stars is
only one chance per several hundred. Therefore, these stars
are promising objects for SETI.

An extraterrestrial civilization can be detected by leakage
of the radio emission accompanying industrial activities and
communication processes, as well as space transportation. It
seems quite probable that a highly developed extraterrestrial
civilization may posses a lot of radio-transmitters. Conse -
quently, the integrated radio emission could be almost non -
directive and continuous, both in time and frequency coverage.
The most promising range for eavesdropping extends from 100
to 1000 MHz. Lower frequencies are unfavourable because of the
galactic radio noise, while higher frequencies may prove incon-
venient, since the intensity of artificial radio emission drops
abruptly there (one can see that from the example of the ter -
restrial human civilization [19]).

What may be a possible appearance of an extraterrestrial
civilization that could be accessible for modern technical
means from the distance of ∼20 pc at this frequency range?
It is necessary to note that 20 pc is the average distance of

G. Marx (ed.), Bioastronomy – The Next Steps, 337–342.

the faintest sun-like stars in the catalogue employed [3].
The most extensive modern radio surveys have flux density
limits of $\sim$1Jy at these frequencies. Then the estimated
power of radio emission from an extraterrestrial civilization can be assumed $\gtrsim 2.2 \times 10^7$W. Assuming that civilization to allocate to electromagnetic emission about
$3 \times 10^{-4}\%$ of its total power consumption [1,16,19] , we
can estimate the total power of this extraterrestrial civilization about $\gtrsim 10^{25}$W. It is nearly $\gtrsim 100$ million times as
mach as the influx of solar energy to the Earth. Certainly,
the civilization cannot consume that enormous energy on an
Earth-like planet. Therefore, it would be necessary for that
civilization to remove its power industries far from the inhabited planets. V.S. Troitsky notes that a civilization
cannot consume more than 10^{23}W within its planetary system
because of ecological problems [5] . One of these is the
danger of an accident like an explosion of fuel reserves.
It is also necessary to decrease the pollution of the interplanetary environment. Besides, powerful processes could
possibly cause essential climatic changes on the planets.

The typical dimension of a planetary system is about 100
astronomical units [14,15] . Accordingly the characteristic
displacement of the industrial and power producing zone from
the star should be $\sim$1000 astronomical units. This corres -
ponds to $\sim$1' from the distance of 20 pc. According to M.H.
Hart intelligent life can arise around stars belonging to
spectral classes F7V to KOV [11] . This is a rather conservative estimate. Therefore, we can observe some civilizations
as discrete radio sources in close vicinities ($\sim$1') of the
stars like these. Of course, this is only some model representing a real situation. But since it seems more or less
reasonable the question arises: whether there are real as -
sociations of stars with radio sources?

The author has been searching for radio sources located
closer than D=130" to stars of different types. For this
purpose, "The Molonglo reference catalogue of radio sources" has been used [13] . The survey was performed at the
frequency of 408 MHz. It contains K=12141 discrete sources

of listed flux densities above than 0.7 Jy. The catalogue
indicates celestial coordinates with the standard error
typically lying between 3 and 10 arcsec. In addition, the
photometrical and spectral catalogue of stars brighter
than magnitude 6.25 has been also used [3]. Approximately
4050 stars are in the zone of the radio survey. Nine of
these are closer than 130" to their respective radio
sources (Table 1). What is an unexpected result that 4
sources are located in the vicinities of four yellow
dwarfs having spectra from F8V to KOV in accordance with
the prediction of M.H. Hart [11]. The number of such stars
is only 4% among objects from the stellar catalogue employed. The probability of more than 3 accidental projections
of such stars to the vicinities of radio sources is

$$\omega = 1 - \sum_{i=0}^{3} \left[n! \, p^i (1-p)^{n-i} / i! \, (n-i)! \right] = 3 \times 10^{-4}$$

where n=156 is the number of F8V-KOV stars covered by the
radio survey; p=πD^2 K/S = 1.93 x 10^{-3} is the probability
of one accidental projection (S=7.85 sr is the area of the survey). If we take the vicinity radius of 2', then there are
3 such stars. Now the probability of more than 2 accidental
projections is 2 x 10^{-3}. There are only 2 yellow dwarfs nearer than 1.5'to the radio sources. Then the probability of
more than 1 accidental projection is 9 x 10^{-3}. It was sug -
gested that the disposition of yellow dwarfs was quite accidental because of their relatively small distances from the
sun. They are located nearer than 42 pc. But their scale
height is 340 pc above the galactic plane [6].

Since there is only one chance per several hundred for
these accidental projections, the associations of the star
HD 21899, HD 100623, HD 187691, HD 187923 and the radio
sources 0328-415, 1132-325, 1948+102, 1949+115 are of interest for SETI. It is necessary to note that their stellar
spectra, average distance from the Sun (21 pc) and the radio
emissions flux density (0.85-2.02 Jy at 408 MHz) are all in
accordance with the model predictions.

Table 1.

Radio sources from 130" - vicinities of the stars.

Star	Spectral type	Distance from the Sun	Radio source	Flux at 408 MHz	Distance from the star
HD21899	F8V	≈12pc	0328-415	1.64Jy	1.'7
HD37022	O6I	≈2000pc	0532-054	40.30Jy	0.'2
HD41841	A4V	125pc	0604-230	0.78Jy	2.'0
HD100623	KOV	9.5pc	1132-325	0.86Jy	1.'2
HD101154	KOII	166.7pc	1135-021	0.81Jy	1.'5
HD106625	B8III	76.9pc	1213-172	2.76Jy	1.'0
HD129956	B9V	≈67pc	1442+009	1.28Jy	0.'8
HD187691	F8V	21.7pc	1948+102	0.85Jy	1.'3
HD187923	G2V	41.7pc	1949+115	2.02Jy	2.'2 (129."6)

Spectral types and trigonometric distances from the Sun are quoted after the stellar catalogue [3] . The approximate distances have been calculated from the visible [3] and absolute [6] V-magnitudes of the stars.

Table 2.

Flux density from the radio sources.

Star	Radio source	Frequency (MHz)	Flux density (Jy)	Reference
HD21899	0328-415	408	$1.64^{\pm}0.17$	13
	PKS0328-325	2700	$0.19^{\pm}0.03$	7
HD100623	1132-325	408	$0.86^{\pm}0.05$	13
	Anonymous	1415	$0.13^{\pm}0.03$	8
	PKS1132-325	2700	$0.21^{.\pm}0.02$	17
HD187691	4C10.60	178	$2.4^{\pm}0.5$	10
	1948+102	408	$0.85^{\pm}0.05$	13
HD187923	4C11.59	178	$2.4^{+}_{-}0.5$	10
	1949+115	408	$2.02^{-}_{+}0.07$	13
	DA493	1420	$2.8^{+}_{-}1.0$	9
	DA493	3200	< 0.5	9

The attempts to identify these radio sources with extra-galactic objects have failed [1,7,17]. Can they belong to some known type galactic radio sources? The radio emission from flare star is highly sporadic. But the standard errors in flux densities of these radio sources are typical [13]. Any sign of variability is unknown. Moreover, none of the known flare stars has been identified with radio sources from the Molonglo catalogue [2].

The attempts to identify the four objects disscussed with pulsars have been unsuccessful as well. It is important to note that practically all active pulsars lying at near dis - tances, up to 200pc, are known (there are only 8 objects [4]). Besides, pulsars associated with stars make up only ∿1% of the total number. Therefore, the association of four unknown nearest (distance of ⩽ 42pc) ones with stars is almost in - credible.

It is also doubtful that the radio sources could be compact HII regions. Their radio spectra are obviously nonthermal (Table 2), with spectral indices close to 1. Hence the sources associated with the yellow dwarfs either are a rare chance, or an unknown type of galactic objects.

Recently and independently J. Heidmann discovered another similar radio source near a class G star [12]. He admits that it can be either a new type of a powerful radio star or a manifestation of artificial radio emission.

The stars HD21899, HD100623, HD187691 and HD187923 are favourable for intelligent life, considering their variability, multiplicity and age [1,18]. Hence the neighbour radio sources are promising objects for SETI.

ACKNOWLEDGMENTS: The author would like to acknowledge helpful discussions with L.M. Gindilis, V.V. Rubtsov, V.F. Shvartsman and V.G. Surdin. It is a pleasure to express a special gratitude to L.N. Litvinenko for his encouragement and advice.

342

REFERENCES

1. Архипов А.В. О вероятных местах расположения внеземных цивилизаций.-ИРЭ АН УССР.Препринт №303. Харьков, 1986.

2. Гершберг Р.Е. Вспыхивающие звезды малых масс.-М.:Наука, 1978.

3. Комаров Н.С., Драгунова А.В., Карамыш В.Ф., Орлова Л.Ф., Позитун В.А. Фотометрический и спектральный каталог ярких звезд.- Киев: Наукова думка, 1979.

4. Манчестер Р.,Тейлор Дж. Пульсары.- М.: Мир, 1980.

5. Троицкий В.С. Почему не обнаружены сигналы внеземных цивилизаций?-Земля и Вселенная,1981, №1, с.63-65.

6. Allen C.W. Astrophysical Quantities. The Athlone Press, London, 1973.

7. Bolton J.G., Shimmins A.J. The Parkes 2700 MHz survey (Fifth part).-Austr. J. Phys. Astrophys. Suppl., 1973, No.30, p.21.

8. Ehman J.R., Dixon R.S., Kraus J.D. Ohio survey between declinations of 0° and 36° south.- Astron.J., 1970, vol.75, No.4, p.351-506.

9. Galt J.A., Kennedy J.E.D. Survey of radio sources observed in the continuum near 1420 MHz, declination -5° to $+70^o$.- Astron. J., 1968, vol.73, No.3, p.135-151.

10. Gower J.F.R., Scott P.F., Wills D. A survey of radio sources in the declination ranges $0.7^o$to 20° and $40^o$to 80°.- Mem. Roy. Astron. Soc., 1967, vol.71, part 2, p.49-144.

11. Hart M.H. Habitable zones about main sequence stars.- Icarus, 1979, vol.37, No.1, p.351-357.

12. Heidmann J. Emission radio anormale provenant de la direction d'une etoile.- Comptes rendus hebdomadaires des seances de l'Academie des sciences, 1986, t.303, Serie II, No.1, p.47-49.

13. Large M.J., Mills B.Y., Little A.G., Crawford D.F., Sutton J.M. The Molonglo reference catalogue of radio sources.- Mon. Not. R.Astr. Soc., 1981, vol.194, No.2, p.693-704 and Microfishes MN 194/1.

14. McLoughlin W. IRAS and Vega.- Spaceflight, 1983, vol. 25, No. 12, p. 438 .

15. McLoughlin W. Possible new solar system.- Spaceflight, 1985, vol.27, No.2, p.72.

16. Sassin W. Profiles of change.- The Unesco Courier, 1981, July, p. 9-12.

17. Shimmins A.J., Bolton J.G. The Parkes 2700 MHz survey (Sixth Part).- Austr. J. Phys. Astrophys. Suppl., 1974, No.32, p. 1-55.

18. Soderblom D.R. A "Short List" of SETI Candidates.- Icarus, 1986, vol,67, No.1, p. 184-186.

19. Sullivan W.T., Brown S., Wetherill C. Eavesdropping: The radio signature of the Earth.- Science, 1978, vol. 199, No.4327, p. 377-388.

THE QUASAT SATELLITE AND ITS SETI APPLICATIONS

C. Maccone
Via Martorelli 43
Torino 10155 Italy

ABSTRACT. The QUASAT radioastronomy satellite of the European Space Agency (ESA) is presently undergoing its Phase–A Study. It will be an antenna of about 15 meters diameter, orbiting the earth on an elliptical orbit optimized for high quality imaging with the VLBI networks in Europe and the US. The satellite's four operating wavelenghts are : 1.35 cm (water maser), 6 cm, 18 cm (OH maser) and 92 cm. The possible applications of Quasat for Seti investigations are considered in the present paper.

1. DEVELOPMENT OF THE QUASAT PROJECT.

The Quasat satellite for radioastronomical studies (Ref.[1]) is one of the qualifying projects presently being planned by the European Space Agency (ESA). The story of Quasat goes back to 1982, when a group of European, American and Australian radioastronomers, led by Dr. R. T. Schilizzi, submitted the Quasat proposal to ESA in response to a call for a new mission. The project was jointly studied by ESA and NASA from its inception until 1986, and the coordination of the study effort was ensured by three joint meetings of the ESA and NASA study teams. Also, on June 18–22, 1984, a Workshop was held in Gross–Enzersdorf (Austria) to allow the scientific community to discuss about Quasat : over 80 scientists from 12 countries attended this meeting, whose proceedings were later published by ESA (Ref.[2]).

However, on January 28, 1986, the Shuttle's tragedy occurred. This influenced the further development of the Quasat project in that NASA pulled out of it due to funding shortages. So ESA had to face the decision of whether to go on on its own or to give up the project. At its meeting held in October 1986, the ESA Astronomy Working Group (AWG) decided to set up a so–called "Tiger Team" to study new ways for proceeding with the mission. The Team was given the task of re–analysing the scientific goals with a view to defining a feasible, yet scientifically rewarding, mission that would correspond to a medium–sized project of ESA's Horizon 2000 long–term program.

G. Marx (ed.), Bioastronomy – The Next Steps, 343–349.

The proposals of the Tiger Team appeared in January 1987 (Ref.[3])
and may be summarized as follows :
1) An Ariane launcher is going to put into orbit Quasat in addition
   to another yet undefined satellite (dual launch) ;
2) In this event, the Quasat orbit is essentially forced to be the
   Ariane's Geostationary Transfer Orbit (GTO) (apogee altitude =
   36000 km, perigee altitude = 5000 km), modified by means of a
   solid booster to change the orbit inclination to 30 degrees ;
3) Only one receiver (left circularly polarized) per frequency is
   to be installed ;
4) Given a total link bandwidth of 64 MHz, the remaining single
   polarization receivers can operate at twice the original
   bandwidth of 32 MHz, thus increasing their sensitivity by a
   factor of SQRT(2) ;
5) As a consequence of the increased sensitivity, the antenna
   diameter can be decreased from 15 to 10 meters, leaving the other
   scientific goals of Quasat unchanged ;
6) Finally the Quasat antenna was to be changed from the center-fed
   type to the offset type, thus revolutionizing all the previous
   studies about the satellite .

While the Tiger Team was elaborating the above new scheme, Quasat
was selected by ESA for Phase-A Study early in 1986 (Ref.[4]).
Proposals made by several European industrial trusts for this study
were then reviewed by ESA, and, in the autumn of 1986, the Phase-A
study was assigned to Aeritalia Space Systems Group of Turin, Italy
(Ref.[5]). After the kick-off meeting, on June 30 - July 1, 1987,
some 50 experts from Europe, the USA, and Australia, gathered at
Aeritalia to pave the way for the Quasat final configuration.
During this meeting the number of possible candidate antennas was
lowered from seven to two : the wrap-rib antenna by the US company
Lockheed, and the inflatable antenna by the Swiss company Contraves.
Which one will actually be selected will depend on the conclusions
of the Phase-A Study, that are bound to be completed by the spring
of 1988. Another important point that was settled during the meeting
was the return to the center-fed configuration for the antenna, so
that the offset configuration proposed by the Tiger Team was now
definitely discarded. This choice enables the engineers to plan a
satellite having a cylindrical symmetry around the main paraboloid
axis. The spacecraft will be located just behind the paraboloid, and
will contain all the circuitry and the apogee kick motor. Two solar
panels and two 50 cm antennas for the VLBI link will be attached to
the spacecraft by means of expandable booms. The feed of the main
paraboloid will actually consist of three coaxial feeds, tuned to
the three observing frenquecies prescribed by ESA to be at 1.35 cm
(water maser), 6 cm, and 18 cm (OH maser) in wavelenght. The prototype
of this technologically advanced three-frequencies feed has already
been built and tested for ESA by the CSELT laboratories of Turin,
Italy. Moreover, a fourth operating frequency of 92 cm in wavelenght
will be tuned to a further outer coaxial feed around those above.
In conclusion, the Quasat basic radioastronomical parameters and VLBI
SNR are shown on the following table.

Band	Tuning bandwidth[a] (GHz)	Instant. BW per pol.(max.) (MHz)	Syst. Temp. (K)	SNR on 1 Jy source for integration time of 300 sec and BW of 32 MHz 100 kHz (spectral line)	
1.35cm(K)	22.0 -22.5 (19.8 - 24.2)	32	125	22	1.2
6 cm (C)	4.72- 5.02 (4.25- 5.02)	32	100	54	-
18 cm (L)	1.60- 1.73 (1.54- 1.78)	32	75	65	3.8
92cm	0.324-0.327	4	150	13 (bandwidth 4 MHz)	-

a: Wideband extremes for bandwidth synthesis mapping are given in parentheses assuming slightly degraded feed/receiver performance is allowed.

2. ADVANTAGES OF ARRAYS OVER SINGLE DISHES FOR SETI.

SETI, the Search for Extra-Terrestrial Intelligence, has been pursued over the past 25 years, mainly in the U.S., by some radioastronomers in cooperation with a few other experts (Refs.[6]and[7]). However, no proposals appear to have been made so far in order to use the Quasat satellite for SETI investigations. This circumstance amounts to missing a unique opportunity, inasmuch as two, at least, of the Quasat operating wavelenghts may be of interest to SETI : the 1.35 cm (water maser) and the 18 cm (OH maser) lines. Moreover, the following additional advantages for SETI would be provided by Quasat with respect to the present "single dish" SETI searches :
1) An array is a better match to the source since the signals are expected from a single direction ;
2) By combining the simultaneous output of several antennas oriented in the direction, the array pattern has narrow principal maxima ;
3) There is a strong radio-frequency interference rejection if the signal is processed by multiplication rather than by addition;
4) Signals from any direction within the diffraction beam of the largest antenna will be detected, and the array provides also directive information on the scale of λ/D, where D is the largest telescope separation distance ;
5) Multiplying systems, especially, provide good baseline stability, improving weak signal detection ;
6) Presence of interference fringe phase offers an unbiased estimator for weak signals ;
7) Strong radio-frequency interference rejection implies minimum time wasted on false alarms ;
8) Adding more collective areas increases sensitivity .

The foregoing advantages of telescope arrays over single dish techniques for SETI have been pointed out by William J. Welch, of the University of California at Berkeley, in Ref.[7], Appendix L. This author considered only arrays on the ground, not in space. Yet his way to process the ETI signal in order to remove the effect of the delay $\tau(t)$ may be applied to Quasat as well, and we shall briefly describe it to find the required hardware for Quasat.

Let the ETI signal be $f(t)$. By aid of an oscillator at ν_0 and of a low pass filter, a translation of the signal to baseband is performed, and let $f_\varrho(t)$ denote the baseband time function. The relationship between these two time functions can be shown to be

$$f(t) = f_\varrho(t)\cos(2\pi\nu_0 t) - \hat{f}_\varrho(t)\sin(2\pi\nu_0 t)$$

where $\hat{f}_\varrho(t)$ is the Hilbert transform of $f_\varrho(t)$ (Ref.[8], pp. 118-123):

$$\hat{f}_\varrho(t) \equiv (1/\pi) \int_{-a}^{a} [f_\varrho(u)/(t-u)]\,du$$

Suppose now that $f(t)$ is received at either of a pair of antennas, while $f(t - \tau)$ is received at the other. Let the signal be processed by multiplication as shown below :

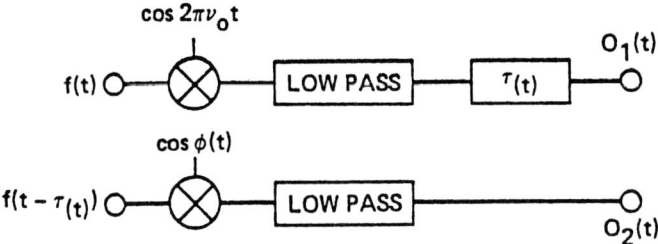

And finally suppose that the multiplying function $\phi(t)$ has the form

$$\phi(t) = \phi(0) + \int_{0}^{t} 2\pi\nu_1(t')dt' \; ; \quad \nu_1(t) \text{ is close to } \nu_0 .$$

Then the standard baseband translation tecnniques yield the outputs:

$$O_1(t) = (1/2)f_\varrho(t - \tau).$$

$$O_2(t) = \text{Low Pass}\,[(1/2)f_\varrho(t - \tau)\cos 2\pi\nu_0(t - \tau)\cos \phi(t) - (1/2)\hat{f}_\varrho(t - \tau)\sin 2\pi\nu_0(t - \tau)\cos \phi(t)]$$

$$= (1/2)f_\varrho(t - \tau)\cos[2\pi\nu_0(t - \tau) - \phi(t)] - (1/2)\hat{f}_\varrho(t - \tau)\sin[2\pi\nu_0(t - \tau) - \phi(t)].$$

We see that $O_2(t)$ can be made equal to $O_1(t)$ by setting

$$2\pi\nu_0(t - \tau) - \phi(t) = 0 .$$

In turn, this and $\phi(t)$, yield the unknown $\nu_1(t)$ by differentiation

$$\nu_1(t) = \nu_0(1 - \dot{\tau}) \quad \text{and} \quad \phi(0) = -2\pi\nu_0\tau(0) .$$

The equation $\qquad \nu_1(t) = \nu_0(1 - \dot{\tau})$

is basic because it shows that we must put a programmable oscillator on Quasat to eliminate the effect of the delay. Alternatively, this oscillator can be kept on earth, provided a phase-locked loop (PLL) circuit is installed on Quasat to keep the link with earth working. The latter solution is more likely to be adopted because the Quasat spacecraft has got to be kept as light as possible, in view of launch and structural requirements. In any case, the programmable oscillator appears to be the only additional hardware needed to use Quasat for SETI purposes.

3. FORM OF THE DELAY EQUATION.

The expression of the DELAY EQUATION is essential to program the Quasat oscillator correctly. Tentatively, we may write it as follows

$$\tau(t) = \tau_0 + \tau_{source}(t) + \frac{2\pi}{c}(B_x \cos\delta \cos H + B_y \cos \sin H + B_z \sin\delta)$$

$$+ \tau_{instrum}(t) + \tau_{atmos}(t)$$

where τ_0 is a delay offset, $\tau_{source}(t)$ is the delay due to source structure, the third term is the geometric term, $\tau_{instrum}(t)$ represents the variable instrumental delays in the system, $\tau_{atmos}(t)$ is the variable atmospheric term, B_x, B_y and B_z are the baseline components, δ is the declination and H the hour angle of the source. The τ_0 term is the sum of all the fixed delays in the system and can be determined most easily from calibration observations of unresolved sources ($\tau_{source}(t) = 0$). This provides a means of determining the offset between the clocks with high accuracy. The accuracy is proportional to $1/B$ (B = signal bandwidth) and for high signal/noise ratio can be of the order of nanoseconds, provided the variations in $\tau_{instrum}(t)$ are small, and provided due account is taken of the (slow) motion of the satellite.

4. CONCLUSIONS.

The designing of some hardware of Quasat for SETI applications would, obviously, require a much deeper discussion than the above suggestions. However, we shall not attempt this now because Quasat is presently just going through its Phase-A Study, and also because of the very poor interest shown by ESA for SETI. As a matter of fact, most radioastronomers in Europe do not seem to regard SETI as a "respectable" field of science. They support Quasat only for traditional topics of investigation, like black holes, peculiar galaxies, and the redetermination of the Hubble constant. This "no-SETI" European attitude is in sharp contrast to what has been going on in the United States and the Soviet Union over the last twenty years, where SETI

has received attention and funding by NASA and other organizations.
Quasat might be a good opportunity for Europe to make her voice
be heard in SETI, if only by means of some piggyback on the water
and OH frequencies. Better, one may even see things the other way
round : a substantial interest of the SETI international community
for Quasat might enhance the satellite's chances to be selected for
a Phase - B Study by ESA in 1988.

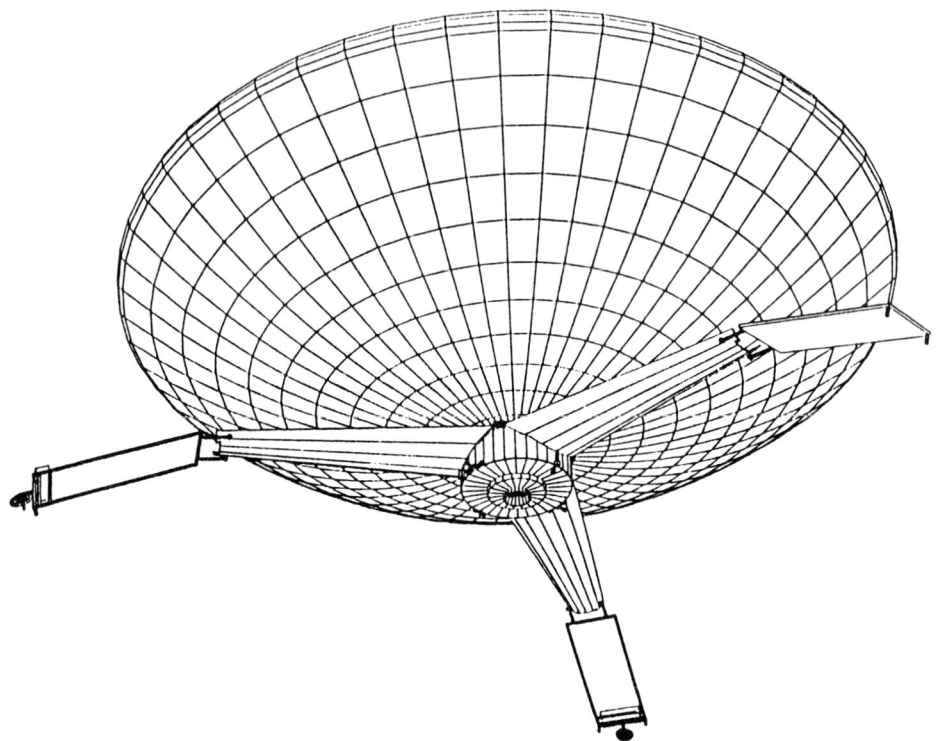

Figure 1. THE QUASAT SATELLITE IN ORBIT. This picture is taken from
Ref.[2], page 78, and, although it goes back to 1984, still reflects
the basic configuration of Quasat. In fact, the engineers at Aeritalia
Space Systems Group, who are in charge for the Quasat Phase—A Study,
have agreed that a center-fed (rather than offset) antenna is better
because easier to plan and build up. Also, the complete cyilindrical
symmetry of the whole spacecraft around the paraboloid axis appears
to be preferable, as depicted above. The only differences between the
above picture and today's actual project are in the number of solar
panels (two rather than three) and in the two small uplink/downlink
antennas (on top of two solar panels in the above picture). These
are presently envisaged to be located on independent expandable booms.
Finally the feed tower (not depicted above) is now going to carry
four coaxial feeds (rather than the three originally envisaged) for
the 92 cm wavelenght has now been added to the 1.35, 6, and 18 cm.
ones.

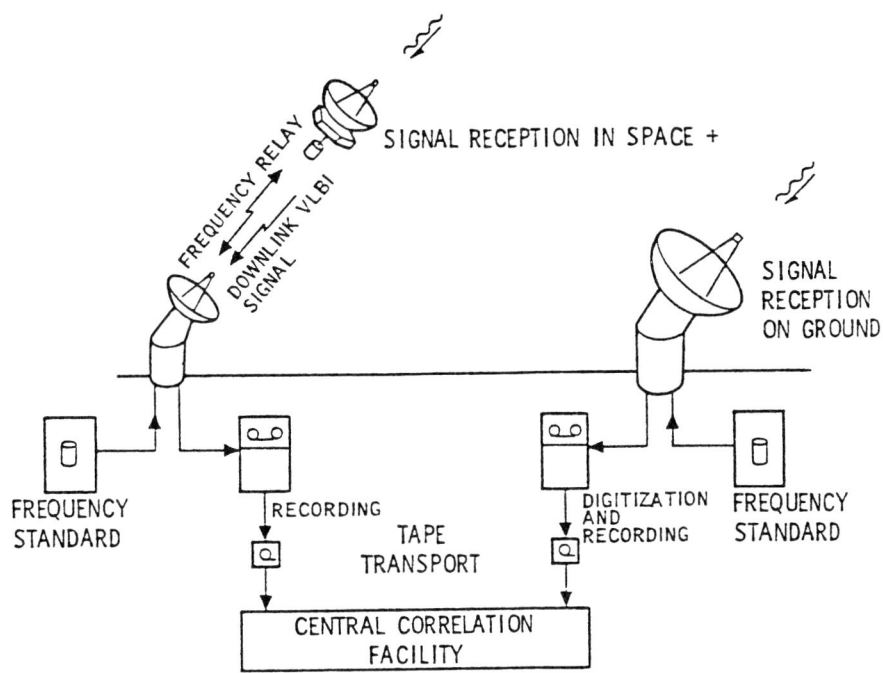

Figure 2. BLOCK DIAGRAM OF THE QUASAT SPACE VLBI SYSTEM.   Once again
this picture is taken from Ref.[2], page 27, and shows how Quasat is
is going to work in conjunction with an antenna of the existing VLBI
systems on earth. The whole system is going to provide images of a
total intensity and polarized emission of a radio source having a
higher resolution and quality than those obtainable with the ground
arrays only. For SETI, this might spell the difference from a real
ETI source and a radio galaxy happening to be on the same line of
sight.

References

[1] ESA: Quasat Assessment Study, Document SCI (85) 5, November 1985.
[2] ESA: Proceedings of the Quasat Workshop held in Gross Enzersdorf,
        Austria, 18-22 June 1984, Document SP-213.
[3] ESA: Quasat Update Report, Document SCI (86) 8, January 1987.
[4] ESA: Quasat ITT, Document AO/1-1888/86/NL/SK, June 1986.
[5] AERITALIA-Space Systems Group: Proposal for Quasat Phase-A Study,
        Document SG-PP-AI-053, September 1986.
[6] NASA: Oliver B.M. & Billingham J., eds.: Project Cyclops ,
        Document CR-114445, 1972 (Revised edition, 1973).
[7] NASA: SETI Science Working Group Report, Technical Paper 2244,
        October 1983.
[8] Papoulis A., Signal Analysis, McGraw-Hill, New York, 1977.

SETI WITH A FRENCH ACCENT

F. Biraud

Observatoire de Paris

Meudon 92190 France

ABSTRACT. A 5 day SETI program has been run at the Nançay Observatory in April, 1987, at 18 and 21 cm wavelength. No received signal can be attributed to an extraterrestrial origin. But the analysis of the receiver output shows a population of strong interfering signals, of yet unknown origin. We also discuss various methods of estimating the mean power per channel for data reduction.

1. INTRODUCTION.

Two SETI observing programs have already been run at the Nançay decimetric telescope (Biraud, 1982; Tarter, 1985). We report here the results obtained by J. Tarter and myself during a third campaign of five days duration last April. 132 more stars were observed, giving a large quantity of power spectra data for a statistical analysis of the receiver noise. The results, especially comparison with previous observations, are interesting for future work on SETI, at Nançay or elsewhere.

2. THE OBSERVATIONS.

We summarize rapidly the equipment properties and observing procedures, as they were essentially the same as during the previous observations (Tarter, 1985).

At 1.5 GHz, the Nançay telescope gives an efficiency of about 1.2 K/Jy and a system temperature of 50 K. It allows for a tracking time of at least ±35 min around transit. We used a single linear (horizontal) polarization. The spectral backend is a digital 8-level autocorrelator used here as a single bank of 1024 channels in a 50 kHz bandwidth, giving a resolution of 48.8 Hz per channel.

Each star was observed both at 18 and 21 cm wavelength. Respectively 56 and 16 overlapping spectra were measured, covering 2.24 MHz at 18 cm, centered at 1666.398 MHz, and 640 kHz at 21 cm (135 km/s) centered at 1420.405 MHz.

No off source reference spectra were taken, a star being reobserved later if it showed a potentially interesting signal. Each spectrum was

351

G. Marx (ed.), Bioastronomy – The Next Steps, 351–355.

observed for 30.3 s, thus giving a value for B_T of 1479.5.

The target stars are of spectral type F, G and K, luminosity class V, closer than 25 pc. Most of them are single. 132 stars were observed during the 5 days of observation, producing 7567 spectra at 18 cm and 2029 at 21, and a total of more than 7.8×10^6 channels of spectral data.

The reduction of the data was done at the observatory on a DEC PDP 11 computer as soon as possible after the data were collected. The analysis consisted of estimating the expected power in any given channel, and searching for large deviations in the detected power above the expected value.

3. NOISE STATISTICS AND ESTIMATION OF THE MEAN POWER.

We will see that this is a very critical point, and that a fully satisfactory solution is still to be found.

Let y be the measured power in a given channel. Normalized to a mean value of $\langle y \rangle = 1$, the probability distribution of y can be written (Oliver and Billingham, 1973) :

$$p(y) = \frac{n\,(ny)^{n-1}\,e^{-ny}}{(n-1)!}$$

and the integrated probability :

$$P(y) = \int_0^y p(y)\,dy$$

$$= 1 - e^{-ny} \sum_{k=0}^{n-1} \frac{(ny)^k}{k!}$$

The probability that noise alone in a given channel will exceed a given threshold y_T is :

$$Q(y_T) = 1 - P(y_T)$$

y_T is chosen so that the number of false alarms, to be reobserved later, is kept to a reasonable value.

3.1 Effect of a bias on $\langle y \rangle$.

It is clear that overestimating the expected value $\langle y \rangle$ will decrease the number of alarms, and vice versa. $Q(y)$ is so steep around $y = 1$ that this effect is very important.

$$\frac{Q(y+dy)}{Q(y)} = 1 - \frac{p(y)}{Q(y)} \times dy$$

The relative decrease (increase) of the number of alarms is :

$$\frac{p(y)}{Q(y)} \simeq \sqrt{\frac{n}{2\pi}} \ \frac{e^{[(n-1)\ln(y)-n(y-1)]}}{Q(y)}$$

In the range of y of interest (1.1 to 1.2), a 1% decrease in $\langle y \rangle$ will triple the number of alarms.

3.2. Effect of noise on $\langle y \rangle$.

If $\langle y \rangle$ is alternately over- and underestimated, this effect cancels to the first order. But to the second order we have :

$$\frac{Q(y+dy)+Q(y-dy)}{2\ Q(y)} = 1 + \frac{d^2Q}{dy^2} \times \frac{(dy)^2}{2\ Q(y)}$$

The number of alarms always increases (for $y_T > 0$) by :

$$\frac{d^2Q}{dy^2} \times \frac{1}{2\ Q(y)} \simeq \sqrt{\frac{n}{2\pi}} \ \frac{(ny-n+1)\ e^{[(n-2)\ln(y)-n(y-1)]}}{2\ Q(y)}$$

A 1% noise on $\langle y \rangle$ roughly doubles the number of alarms.

3.3. Study of various estimations of $\langle y \rangle$.

For our August 1984 data, we first tried to define $\langle y \rangle$ by calculating the cumulative average power spectrum by combining many spectra taken on different stars at different frequencies. This gave very bad results, because of slight variations in the exact shape of the bandpass with frequency, telescope position, and time.

We therefore decided to estimate $\langle y \rangle$ by a running mean over 41 channels adjacent to the channel under study. This has been discussed in detail in Tarter, 1985. The main drawbacks of the method are that the curvature of the spectrum (quite noticeable at both ends) gives a bias in $\langle y \rangle$, increasing the probability of false alarms, and, second, that the rms noise on $\langle y \rangle$ gives an increase of 17% on the number of alarms.

We tried a new way of estimating $\langle y \rangle$ for the reduction of our 1987 data. We determined a baseline for the whole spectrum by an harmonic fit of order p, i.e. by a sum of p harmonic cosine function fitting best the data. This is done very easily by keeping the first p values of the correlation function and taking the Fourier transform. The mass reduction of the data was made with p=25. Once more, the results are not quite satisfactory. If we plot the number of points above 1.1$\langle y \rangle$ against the channel number, the distribution shows a strong periodic variation. This periodic component is clearly visible on the smoothed spectrum taken as the baseline. The autocorrelation function does not show any discrepant point near the corresponding delay. Therefore the phenomenon does not come from a correlator failure (as was the case in our 1984 run) but probably from the fact of truncating abruptly the autocorrelation function at a point where it is still significantly non zero. The fit could be improved by taking a larger value of p, at the risk of decreasing the sensitivity to signals spread over many channels. We have to investigate further this technique, which is probably the most promising.

We should note that a faster algorithm would be to subtract the baseline, by setting the autocorrelation function to zero from 1 to p and computing a single Fourier transform, and then dividing by an average spectrum for gain (not baseline) correction. While this may more satisfactorily remove the bias in <y>, it may increase the rms noise because two non perfect estimates are used, instead of one.

4. ANALYSIS OF THE OBSERVED "ALARMS".

4.1. Histograms.

We set the threshold defining an "alarm" to 1.14<y>, as in our 1984 observations. This gave us (see Figure 1) :

Freq.	Nb of alarms	Nb per channel
18	15	2.4×10^{-6}
21	60	3.6×10^{-5}
both	75	1.0×10^{-5}

These numbers are huge : the expected value is 1.4×10^{-7} per channel. In 1984 the histogram showed a distribution of deviations parallel to the theoretical one up to 1.13, but above it by a factor of 2 (part of it being due

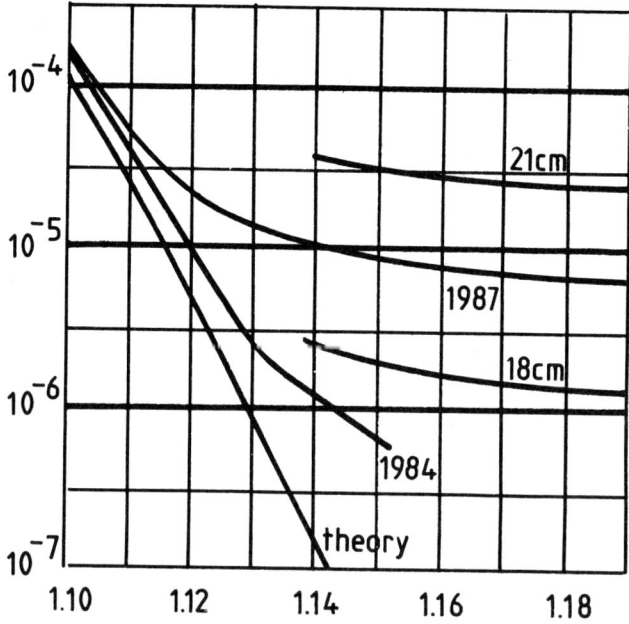

Figure 1. Probability of alarm vs. threshold.

to the noise on ⟨y⟩, see above), and a tail of rare (10^{-6}) events above 1.13. Here the tail of the distribution starts as low as 1.11 and is much more important and much flatter. The numbers at 18 and 21 cm are very different, but the slopes of the distributions are the same.

4.2. Frequency analysis.

Plotted as a function of the channel number, these 75 alarms show no significant pattern. This rules out a single interference in the receiver IF, and also a relationship with the periodic component in the baseline, as discussed above : most of these signals are at such a level that a bias in the estimation of ⟨y⟩ is not important.

It is more interesting to plot the received frequencies of these signals in RF. At both frequencies they are strongly concentrated in narrow bands.

At 18 cm, 10 out of 15 are located at 1665.9935 MHz ± 1.6 kHz. At 21 cm, all of them build up four "lines", two being at 1420.3568 and 1420.4094 MHz. This second one is especially narrow, spanning less than 2 kHz, a relative bandwidth of order 10^{-6}. Some of the scatter may come from uncertainties in the a posteriori calculation of the frequency, but the frequency is certainly not constant.

Of the 59 stars showing these "alarms" (some have more than one), 20 have been reobserved, and 10 of these detected again. The rest of them will be reobserved later this year.

We could not find any dependence on the position along the track of the focal system (which could indicate an external origin), nor on the local time. Up to now, we could not spot the origin of these interfering signals. The receiver system, and especially the various local oscillators, were the same as in 1984, when these signals did not show up. If they are not made within the receiver or the telescope equipment, they could be a serious problem for ETI searches, and the deterioration of the situation since 1984 is troublesome.

5. CONCLUSION.

Unless a real ETI signal is hidden in the not yet reobserved 29 stars showing "alarms", this search did not give any more positive results than the previous ones. But we think that the detailed analysis of the receiver noise statistics, the continuing search for a good estimate of the expected power, and the monitoring of interference, especially in the protected frequency bands, are of direct interest for the design and operation of the future multi million channels SETI machines.

REFERENCES.

Biraud, F. 1982, paper IAA-82-261 presented at Paris IAF Congress, 1982.

Oliver, B.M. and Billingham, J. 1973, "Project Cyclops", NASA CR 114445.

Tarter, J.C. 1985, paper IAA-85-473 presented at Stockholm IAF Congress, 1985.

FIELD TEST RESULTS WITH THE TARGETED SEARCH MCSA

J.C. Tarter

NASA Ames Research Center - SETI Institute
Moffett Field CA 94035 USA

ABSTRACT. In April 1985, a 74000 channel prototype of the multi-channel spectrum analyzer (MCSA) that NASA plans to use in a systematic search for extraterrestrial intelligence (SETI) was installed at DSS13, a 26 meter R&D antenna facility at the Goldstone Deep Space Network (DSN) site. Since that time the instrumentation has been used to validate the performance of signal detection algorithms using locally injected signals and the weak carriers from distant spacecraft. It has been used in an unsuccessful attempt to relocate the Pioneer 9 spacecraft, which now has been officially decommissioned. It has also been used to collect large data samples in order to study the statistical behavior of the noise in that observatory environment. The prototype has provided a limited capability to communicate with the telescope control computer and test schemes for automated operation during the Microwave Observing Project (MOP) Targeted Search mode. This paper describes results from the Goldstone Field Tests and plans to move the prototype equipment to other sites where SETI may be conducted in the future.

1. INTRODUCTION

The 74000 channel prototype MCSA was designed and constructed at Stanford University in the Engineering College laboratories directed by Professor Allen Peterson. In addition to the MCSA, a functional observing system consists of a VAX 11/750 host computer, tape and disk drives, a SUN graphics workstation, peripheral terminals, and a true power conditioner (added of necessity after brief exposure to the available power supply at the Mohave desert site). With full awareness that it would make future improvements much more difficult, it was decided to deploy this system in the field early-on to test performance in a real observational environment, at a remote site, and under conditions that approximate those that will be encountered in any future systematic automated search. Figure 1 shows the system installed in the control building of the Venus Station (DSS13) at Goldstone.

2. DISCUSSION

The objectives of the field tests were to check out the performance of the MCSA prototype itself, the signal recognition algorithms resident on the host computer, the 'VERA' supervisory software and the graphic displays used for diagnostic purposes. To date, most of these objectives have been met. In addition it was anticipated that the system could be used to characterize the radio frequency interference (RFI) environment, conduct limited SETI observations and explore opportunities for conducting innovative scientific research with a narrowband, high resolution spectrometer. For various reasons, these last three objectives have not been realized and will require additional instrumentation or an alternate site, as detailed below.

G. Marx (ed.), Bioastronomy – The Next Steps, 357–362.
© *1988 by Kluwer Academic Publishers.*

358

Figure 1. MCSA Prototype installed in DSS 13 control room along with VAX host computer.

2.1 MCSA Performance

The field test efforts have been focussed on attempting to ascertain the actual on-line sensitivity of the system, and on evaluating the stability, reliabilty and transportability of this particular architecture and implementation. By direct measurement using injected signals of known strength and by detection of the carriers from a number of orbiting spacecraft, the sensitivity of the current system has been measured to be -187 dBM. Although designed with a 10 MHz input clock and an architecture capable of producing 74000 channels of binwidth 1 Hz, the clock has been slowed down by a factor of 2. This slower speed avoids contention on the VAX input bus and results in output channels of width 0.47 Hz and an instantaneous input bandwidth of 37 kHz. This data path bottleneck has long been appreciated and levies stringent requirements on a fully automated 10 MHz system. The final system will incorporate dedicated hardware signal recognition processors implemented in a highly parallel structure along with the MCSA. During the field tests, an artifact in the output data structure was eventually traced to a failure in the downconverter unit. This failure pointed to a requirement to monitor the power level of the MCSA input to insure that it remains within tolerance. The MCSA architecture incorporates two levels of cascaded digital bandpass filters to allow observations to continue even in the presence of strong RFI signals somewhere in the input bandpass. A design goal of -80 dB isolation between adjacent channels of either bandpass filters #1 or #2 had been established. The field tests demonstrated that the prototype design afforded only -40 dB rejection in bandpass filter #1 and -60 dB rejection in bandpass filter #2. Redesign of the filters incorporating a larger number of active taps and a larger number of quantization bits in the analog/digital converter that samples the incoming data stream should provide the necessary level of rejection. In the MCSA, the most common failure mode has been the result of corroded or poorly seated board connections, with minimal failure at the level of the individual IC's. A final system implemented in part with custom VLSI may have quite different failure modes, but board connectivity is clearly a non-trivial design consideration. The stability of the system was completely determined by changes in the configuration of DSS13, beyond our control. During the intervals when the station configuration remained constant, we detected no degradation or fluctuation in the system performance; no temporal instabilities were uncovered on the time scales investigated. As one last point, the field test excercise has demonstrated that the system is relatively robust and can survive crating and transportation by amateurs! While transportable, the current prototype is hardly portable. A large increase in the packing density of the components will be required if a 10 million channel system is to prove feasible. This will be accomodated through continuing custom VLSI design efforts. As expected, location at a remote site made upgrades to the system extremely difficult, nevertheless the Stanford design team succeeded in installing and debugging a baseline processor within the MCSA to replace some of the operations originally performed by the software detection algorithms.

2.2 Signal Detection Algortithm Performance

The issues for algorithm performance during the field tests were sensitivity, efficiency, reliability, flexibility and maintainability. In contrast to many of the hardware efforts, a substantial portion of the algorithm performance evaluation was conducted in absentia using commercial phone line modem connections. The latest version of the pulse detection algorithm achieves the theoretical limiting SNR and can run in real-time, because it invokes an initial threshold to lower the data rate. The CW detection algorithm must make use of all available data to achieve the limiting theoretical SNR, and it can do this only in non-real time. Both types of algorithms achieve near optimal logical efficiency, and the current (3rd) generation algorithms are between 10 and 100 times more efficient in memory usage and computational cycles than the first algorithms installed in the host computer. Tests in the field with injected signals and spacecraft carrier waves have uncovered software bugs not evident in earlier simulation verification tests. This has led to improved reliability and more "graceful" failure modes for the algorithms. Because the algorithms are implemented as software, they are very flexible. Some of this flexibility can be maintained as the algorithm implementation moves into hardware, through the use of dedicated processors driven by microcode. The field test system located at Goldstone, has provided valuable experience in setting standards for software maintenance in a real environment of multi-user remote operations.

2.3 Performance of the VERA Supervisory Software

VERA stands for Virtual Emulation for Rendevous and Acquisition. This software has been developed under the VAX VMS operating system. It emulates the hardware configuration of the final SETI systems where a number of different dedicated hardware processors will be simultaneously operating on the data from the MCSA. VERA manages the user interface, the telescope control interface, the output reports from the signal detection processors (now software algorithms) and eventually will monitor and self check all the attached hardware. Figure 2 diagrams the current VERA configuration. During the field tests a new asynchronous mail service was installed to expedite the multi-task communications required among the virtually concurrent processors. The user interface was improved with a more consistent syntax and a user manual was prepared. The current version of VERA is quite friendly and easy to master, as judged by the learning experience of field test team members. The first elements of a scheduling assistant were installed as part of VERA. This allows an operator to efficiently select available targets for a given slice of observing time. Monitor and control elements are mostly incomplete. MCSA hardware diagnostics still run as a stand-alone process. At DSS13, real-time interactive control of the telescope is not feasible; a batch mode controller has been developed. Dynamic rescheduling capability will be required to recover from the effects of strong RFI. Future work with expert systems will be required to achieve this capability; studies of a self-organizing RFI database have begun. Eventually, all automated observations will be conducted under VERA, this will provide a common scheduling and observing environment with diagnostic display for the operator, independent of observatory site.

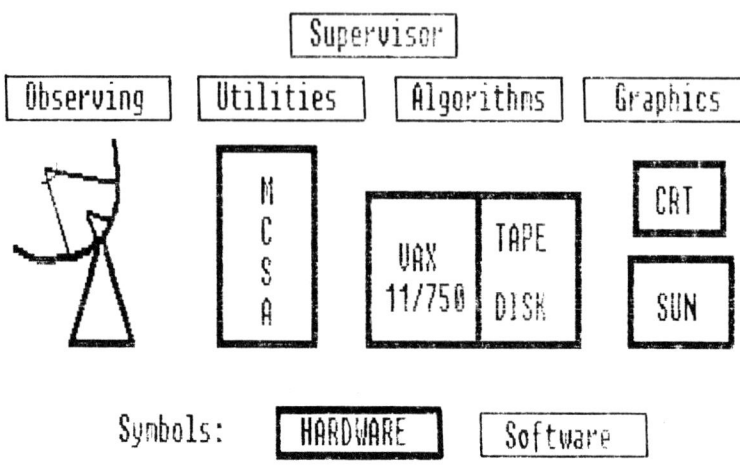

Figure 2. Schematic diagram of software and hardware components of VERA System.

2.4 Performance of the Diagnostic Display

In the final SETI system, an operator will be necessary to create an initial schedule, reschedule due to interference or do additional diagnostics on candidate signals that pass routine recognition tests. Displays will contain data about the performance of the subsystems and details of the observations. However, for the initial field tests, a diagnostic display capable of providing a picture of the signals produced by the MCSA and a time history of the observations was necessary. In the normal mode of operation, the SUN workstation provided 2 instantaneous windows on the data, the first showing all the MCSA channels from bandpass filter 2, the second displaying any one of those channels at the final DFT resolution (0.47 Hz). In addition, a subset of the high resolution output could be displayed as a temporal history, with 200 samples of data present on the screen at any moment and new samples scrolling on. The use of a mouse and a series of icon functions allows the display mode to be changed easily. Currently these icons permit selection of the intensity unit (power, amplitude, real, imaginary, logarithmic power), the scale, the threshold, the accumulation mode, the selection-of-highest-peak mode, an emergency abort and the spectral history display mode. Figure 3 shows the S band carrier from the Pioneer 10 spacecraft displayed in a grey-scale temporal history window. Figure 4 shows this same signal as a 3-dimensional dislay in the temporal history window (the orientation of the axes are selected with the mouse). Figure 5 shows the Pioneer 10 signal being accumulated along different drift rate slopes in the temporal history display window.

2.5 Future Field Test Activities

It still remains to characterise the RFI environment internal to the MCSA prototype and external to it at various observatory sites. Some internal artifacts in the MCSA have already been discovered and suppressed, but we have only been able to exclude the possibility of any non-Gaussian component to the noise at the level of 1 part in 10^7. More recorded data are required to investigate the nature of the noise at 1 part in 10^{12}. The external RFI environment

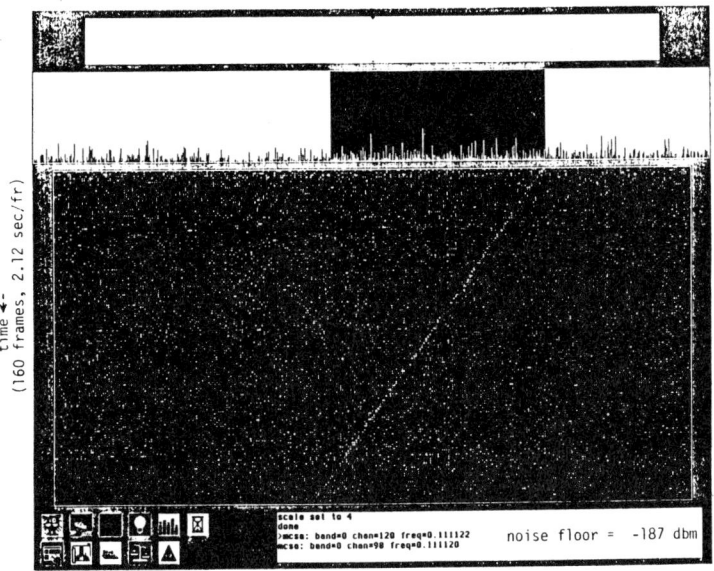

frequency --> (320 ch window, 0.47 Hz/ch)

Figure 3. Pioneer 10 Signal 6 dB above noise, drifting at 0.16 Hz/sec, grey -scale display.

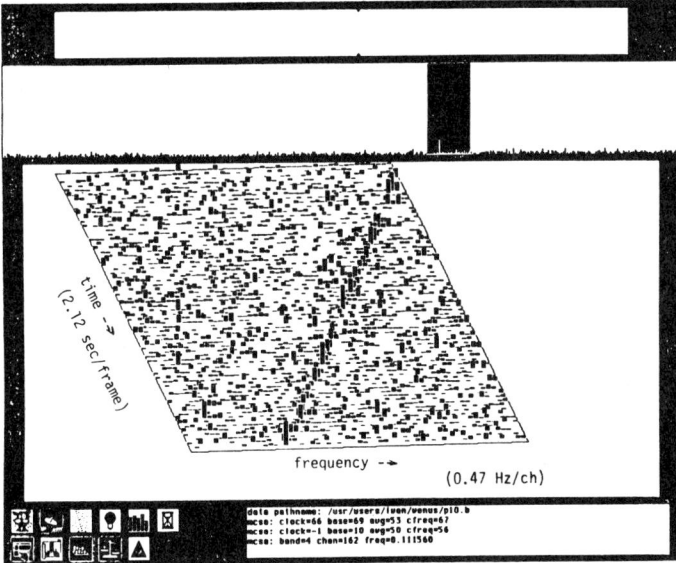

Figure 4. Pioneer 10 Signal 6 dB above noise, drifting at 0.16 Hz/sec, 3-dimensional display.

362

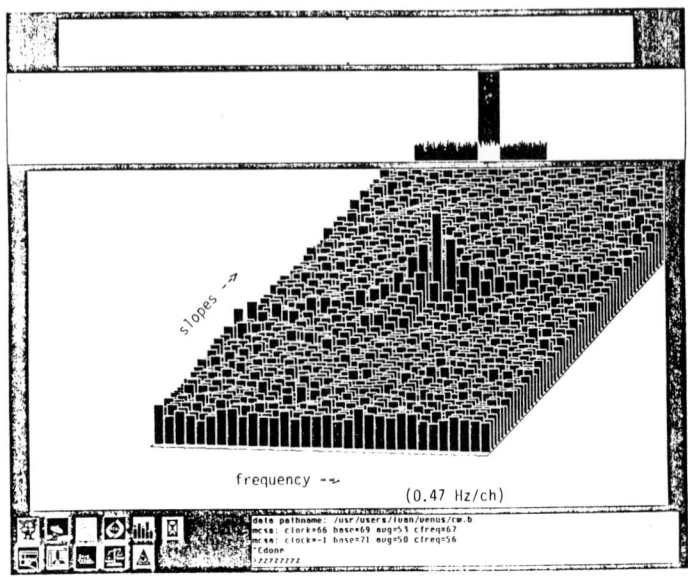

Figure 5. Pioneer Signal, CW accumulation for 100 slopes from 0 to -2 ch/frame. Accumulation peaks at slope -0.16 Hz/sec, in ch 212, with +6 dB SNR.

at any site cannot easily be studied with the current prototype since it has only 37 kHZ of bandwidth. The radio frequency surveillance system (RSSS) developed by JPL is the instrument of choice for this effort. The MCSA can be used to study the fine structure of the RFI signals that are cataloged by the RSSS. Since the universe has never been systematically explored with narrowband filters, there may well be some unique science that the final SETI systems can accomplish. As a start we plan to use the prototype MCSA for high time resolution studies of pulsars and as an accelerometer to study OH masers in circumstellar envelopes. Neither of these projects can be done at DSS13, due to a lack of sensitivity and L-band front end receivers. We look forward to attempting these observations at Arecibo Observatory in the near future.

3. ACKNOWLEDGEMENTS

Although this paper has only one author, it reports on the efforts of many individuals associated with the NASA SETI SR&T Program and with the DSS13 station at Goldstone. Ivan Linscott has supervised the upgrading and maintainence of the MCSA in the field and has participated in field test activities along with Peter Backus, Kent Cullers, Ernst Kimler, Chris Hlavka, David Harper, Jane Jordan and the author. Excellent local support and assistance have been provided at DSS13 by Juan Guarnica, Earl Jackson, George Wischmeyer and Arrol Price.

THE BERKELEY PIGGYBACK SETI PROGRAM: SERENDIP II

S. Bowyer and D. Werthimer
Space Sciences Laboratory, University of California
Berkeley CA 94720 USA
V. Lindsay
San Francisco State University
San Francisco CA 94132 USA

1. INTRODUCTION

The Berkeley SETI system, SERENDIP (an acronym: Search for Extraterrestrial Radio Emission from Nearby Developed Intelligent Populations), enables continuous observing on large radio telescopes without requiring dedicated telescope time. In this system, data being collected as part of the ongoing astronomical observing program are processed automatically in a search for narrow band signals which are highly unlikely to be of astronomical origin. Any such signals detected are noted, along with relevant observational data, for further study.

This approach obviates the need for dedicated telescope time and thus allows us to accrue a large volume of data on major telescopes. A commensal SETI program such as this is not free to choose observing frequencies and sky coordinates. However, in view of the plethora of postulated frequency regimes for interstellar communication and the large number of potential sites for civilizations which have been suggested, this is not necessarily a disadvantage.

The original SERENDIP system simultaneously searched 100 channels, each 5 KHz in width (Bowyer et al., 1983). SERENDIP II is a greatly expanded version of the original system, which has been discussed in detail by Werthimer et al. (1984). For every 65 KHz band, an array processor computes a complex Fourier transform which divides the band into 65,536 channels, each .98 Hz wide. The array processor calculates the power spectrum derived from the complex FFT and, following the algorithm suggested by Cullers et al., (1984), combines adjacent bins of the power spectrum to regain lost sensitivity when a signal falls between two bins. The resulting power spectra are scanned for peaks above a previously chosen threshold. Upon detection of a peak, SERENDIP II records civil time, telescope co-ordinates, bin number, power, intermediate frequency, and synthesizer frequency on disk for off-line analysis. SERENDIP II covers a total bandwidth of 3.5 MHz, 65 KHz at a time, by taking 50 KHz steps along the band before processing the next power spectrum, allowing a 15 KHz overlap between adjacent spectra. Because of SERENDIP's wide channel bandwidth and short integration time, we are not required to make any special assumptions about "magic" reference frames or doppler drift rates.

The SERENDIP II system is currently operating at NRAO's 300 foot telescope in Greenbank, WV. In this paper we report on the characteristics of this system in combination with this telescope, as well as elements of our off-line analysis programs which are intended to identify signals of special interest.

363

G. Marx (ed.), Bioastronomy – The Next Steps, 363–369.
© *1988 by Kluwer Academic Publishers.*

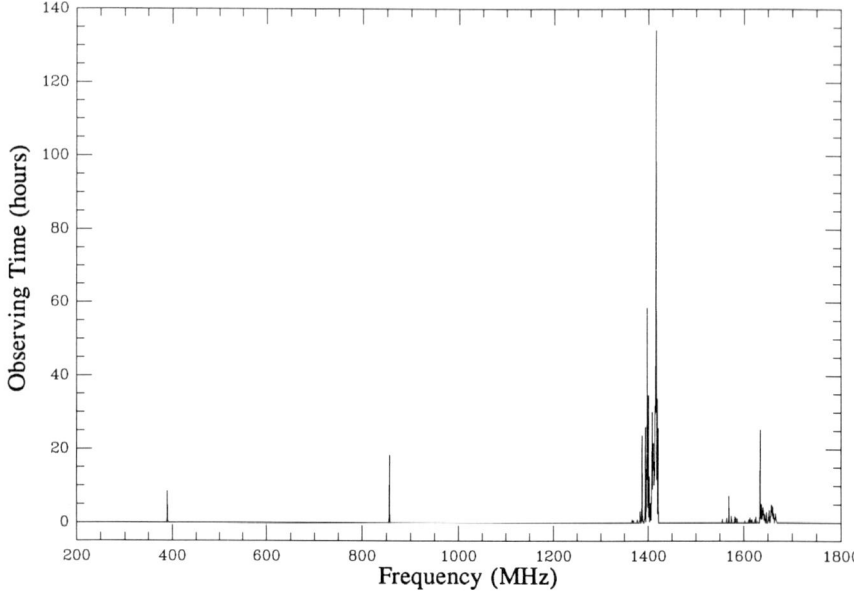

Figure 1. Amount of time SERENDIP has logged at each frequency. The peaks at 390, 856 and 1416 MHz are data taken during a search for millisecond pulsars.

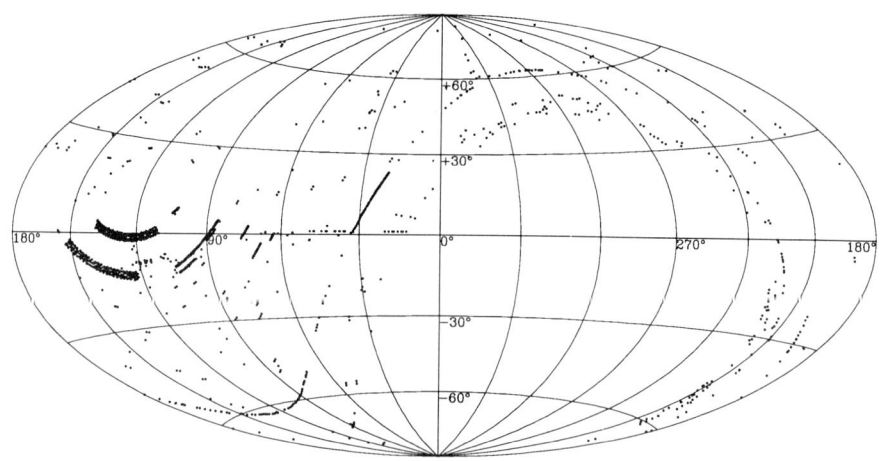

Figure 2. Galactic coordinates of SERENDIP's observations.

2. OBSERVATIONS

During the period from December 20, 1986, to March 16, 1987, SERENDIP II was in operation for a total of 885 hours or 44% of the total elapsed time. We were unable to obtain data during periods of telescope maintenance or during observing programs for which the equipment configuration did not provide an IF tap.

The distribution of these observations in frequency is shown in Figure 1. The peaks at 390, 856 and 1416 MHz are data taken during a millisecond pulsar survey (Kulkarni et al., 1984). Other frequencies which were heavily observed are due to observations of OH and H at various redshifts. In Figure 2 we show the galactic coordinates of our observations.

For each of our power spectra we determine the threshold by calculating the mean power; for any bin with power exceeding 15 times the mean power, we record the bin number and power of the event, mean power of the spectrum, telescope coordinates, local oscillator frequencies and time. In Figure 3 we show the distribution in power of events logged by the SERENDIP II system during one day's run. This distribution is composed of an exponential component expected from Gaussian noise (chi square with 2 degrees of freedom), plus a tail of higher power events. Cluster analysis reveals that the overwhelming majority of these higher power events falls into two categories: those that recur intermittently at the same bin or nearly the same bin in the baseband, and those which recur at nearly the same frequency in the RF spectrum. Although these signals may wander slightly in frequency, they typically persist at nearly the same place in the spectrum until there is some change in the configuration of the electronics. After an instrument change by the primary observer, a new pattern of interference will appear.

In order to eliminate these spurious signals we identify and reject groups of two or more events which occur at approximately the same frequency and are separated by at least two beam widths in the sky. The rejection of these spurious signals is accomplished at the cost of eliminating a part of our data which might contain a real signal. If we analyze the data in segments of two or three days of observations, we have on the order of 17,000 events in a 3.5 MHz band. These events are highly correlated, however so that portions of the band identified as containing interference occupy a relatively small fraction of the total band. In almost all the observing runs, we are able to eliminate 99% of the higher power interference while losing only 1% of the total bandwidth. In Figure 4 we show the same set of data shown in Figure 3 after removal of interference by this procedure.

The flux of these events is given by:

$$F = \frac{\sqrt{2} \, P \, T_{sys}}{P_{mean} \, G} \sqrt{b/t} = 2.5 \times 10^{-25} \, W \, m^{-2} \, \frac{P}{P_{mean}} \quad \text{at NRAO}$$

where P is the power of the events, P_{mean} is the mean power of the spectrum in which it was detected, T_{sys} is the system temperature, G is the system gain, b is the channel bandwidth and t is the integration time.

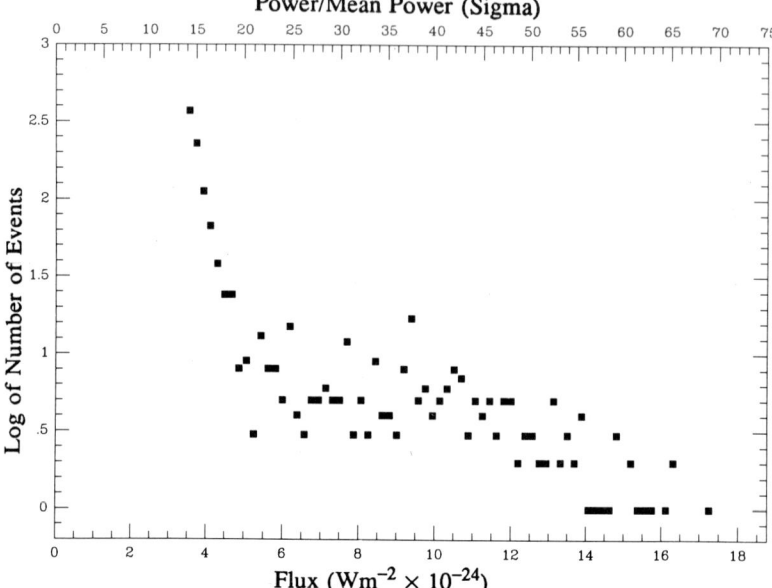

Figure 3. Distribution of events logged by SERENDIP during one day's run.

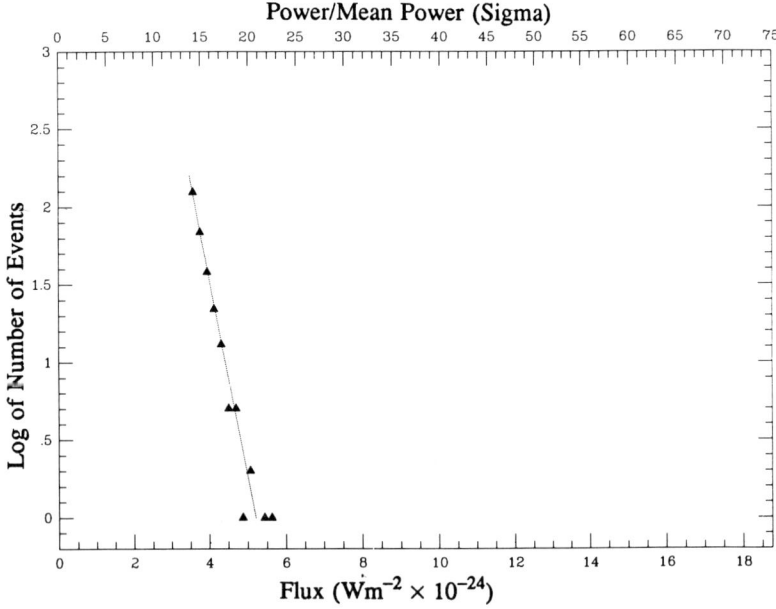

Figure 4. Same set of data shown in Figure 3 after interference algorithms have been applied.

Events with power well above the curve expected from noise are intriguing ETI candidates. During the approximately 1000 hours of operation in the current configuration, we set our threshold for the system at a power level of about 15 times the mean power. At this level some 22,000 events were recorded which survived our cleaning programs. Of these only 17 lie more than 25 times the mean power. We intend to investigate these further.

3. SENSITIVITY AND RELATIVE PROBABILITY OF ACQUISITION

SERENDIP's sensitivity, s, in W/m^2, is given by

$$s = \frac{\sqrt{2}\, n\, T_{sys}}{G} \sqrt{b/t}$$

where n is the detection threshold in terms of P_{thresh}/P_{mean}. The factor of $\sqrt{2}$ is due to the fact that we are looking at only one polarization. The system temperature of NRAO's 300 ft. telescope was measured by injecting a 2K noise source into the feed. Pointing off the galactic plane, we found system temperatures of 25K at 1400 MHz and 55K at 400 MHz. The system gain was determined using several NRAO standard sources. Comparing on-source power levels to background levels yielded a system gain of 1.4 K / Jy. With a detection threshold of 15, an integration time of one second, and a bandpass of ~1 Hz, our sensitivity to narrow band signals is ~ 4 × 10^{-24} Wm^{-2}.

As there is no reliable guide to the direction or frequency where we might find an intelligent signal, Drake (1984) proposed calculating the relative probability of acquisition under the assumptions of the uniform distribution of sources in both space and frequency and obtained:

$$p \sim cm\lambda^2 D B^{\frac{1}{4}} t^{\frac{3}{4}} T^{\frac{-3}{2}}$$

where p is the probability of detection of a signal (per unit time), c is the number of channels, m is the number of beams per unit time, λ is the wavelength, D is the telescope diameter, B is the bandwidth of a single channel, t is the integration time, and T is the system temperature. Estimated parameters for a number of existing searches are given in Table I. It should be noted that if the signal's bandwidth exceeds the channel bandwidth due to modulation or frequency drift, the relative probability scales as $B^{7/4}$ instead of $B^{1/4}$; e.g., if the signal bandwidth is 1 Hz or more, SERENDIP becomes 89 times more sensitive than Meta (Drake, 1987). The high performance of SERENDIP II is due primarily to the advantage gained by the piggyback mode in combination with a very large telescope.

Table I

	Ohio State	NASA (test MCSA)	Sentinel	Meta	Serendip II
Telescope Diameter (feet)	175	84	84	84	300
System Temperature (degrees K)	100	25	65	65	25
Total Bandwidth (KHz)	500	74	2	420	3500
Single Channel Bandwidth (Hz)	1000	.5	0.03	0.05	0.98
Channels per Beam	500	74,000	65,536	8,388,608	3,571,428
Integration Time (sec)	10	1000	30	20	1
Beams/hr	8	3.6	20	20	10
Fraction of Time in Operation (percent)	80	40	80	80	40
Relative Probability of Detection per Unit Time	0.0002	0.2	0.007	0.7	0.7

4. FUTURE WORK

We intend to continue our observations at NRAO and to analyze the resultant data. In addition to cataloging single high power events for further study, we have begun to search our data for duplicate observations of a field in which a narrow band signal recurs. In addition, we are improving our hardware system by expanding the number of channels searched simultaneously. Finally, we are improving and expanding the on-line capabilities of our software to dynamically identify and reject interference.

ACKNOWLEDGEMENTS

We wish to acknowledge the assistance of Alan Berezin, Rachael Brady and Walter Herrick for the development of the SERENDIP II software system. We thank Mike Lampton for his advice and in particular, for his comments on an earlier version of this manuscript. We appreciate the assistance of the NRAO staff and thank Intel Corporation and Mercury Computer Systems for substantial contributions to this effort. This work is supported by NASA grant NAGW-526.

REFERENCES

Bowyer, S., G.M. Zeitlin, J. Tarter, M. Lampton, and W.J. Welch (1983). 'The Berkeley Parasitic Seti Program', *Icarus, 53,* 147-155.

Cullers, D.K., B.M. Oliver, J.R. Day, and E.T. Olsen (1984). 'Signal Recognition'. In *SETI Science Working Group Report,* NASA Technical Paper No. 2244, F. Drake, J. H. Wolfe, and C.L. Seeger, (Eds.). Washington, D.C.: Scientific and Technical Information Office.

Drake, F. (1984). 'Estimates of the Relative Probability of Success of the SETI Search Program'. In *SETI Science Working Group Report,* NASA Technical Paper No. 2244, F. Drake, J. H. Wolfe, and C.L. Seeger (Eds.). Washington, D.C.: Scientific and Technical Information Office.

Drake, F. (1987). Private Communication.

Kulkarni, S., D. Backer, D. Werthimer, and C. Heiles (1984). 'Proposed U.C. Berkeley Fast Pulsar Search Machine'. In *Greenbank Workshop on Millisecond Pulsars,* S.P. Reynolds and D.R. Stinebring (Eds.). Greenbank, WV, June 1984, p. 245.

Werthimer, D., J. Tarter, and S. Bowyer (1984). 'The SERENDIP II Design'. In *The Search for Extraterrestrial Life: Recent Developments,* M.D. Papagiannis (Ed.). Dordrecht: Reidel. IAU Symposium 112.

THREE PULSE/MULTIPLE STAGE CONTINUOUS WAVE DETECTION ALGORITHMS

D.K. Cullers

SETI Institute

101 First Street #410, Los Altos CA 94022 USA

Introduction

SETI searches for structured artificial signals. In a pragmatic sense, this means signals which are concentrated in frequency, time, or both (Cullers, et al. 1985). NASA's SETI efforts are directed toward the detection of sinusoidal carriers and pulses, drifting at a rate as great as 1 channel per spectrum. Pulses are sought over a range of lengths from about 45 milliseconds to 1.5 seconds. Though single large pulses may be recognized as interesting events, they cannot, by themselves, be definitive proof of artificiality. Thus, in pulse detection, attention is focused on finding sequences of repetitive regular pulses consistent with the existence of an ongoing pulse train. For CW, it is possible, in principle, to perform coherent detection on the incoming signal in the way that a Fourier transform does for nondrifting sinusoids. However, to do this for all the drifts of interest to NASA SETI at resolutions consistent with a 1000 second observation is an enormous task. Thus, most effort has been devoted to the incoherent detection of signals, i.e. integration of the power along all possible paths for a signal with constant drift rate.

Figure 1 illustrates a typical difficulty in detecting pulses using the current Multi-Channel Spectrum Analyzer (MCSA) which performs a prime factor Digital Fourier Transform (DFT) (Cullers 1986). A pulse train, when it is synchronized with the sampling frame in frequency and time, is more visible than an incoherently detected CW signal by about 6dB in average power. On the other hand, a train containing only pulses with power split between two data frames disperses in frequency and time becoming about as visible as CW.

Pulse Detection Algorithms

A bright, regularly recurring pulse is of interest whether optical or radio. The flashes attract attention because they exceed a threshold below which most events fall. The regularity of the event confirms the fact that its cause is not random. If the pulse is a sinusoidal radio wave, narrow in frequency, it may well be an artificial signal.

A SETI beacon could be constructed by regularly sweeping the sky with a relatively narrow beam antenna. Within the beam the illumination is very bright since the energy normally lost in all directions is concentrated along only one. With a beam, one can make recurring bright pulses at relatively low average power. A lighthouse uses just such a principle in the optical.

At a radio receiver, a high threshold can be set to eliminate most random noise. Bright, low duty cycle, incoming pulses will cause the power in a few samples to exceed this threshold. Signal detection is sensitive because only a few large samples are analyzed which, primarily, contain signal. This is why pulse detection can be performed at nearly the best possible sensitivity, that of a matched filter (Cullers 1986).

Since 1981, NASA SETI has developed a set of algorithms to detect regularly spaced pulses with drifts up to ± 1 channel per spectrum. Until recently, whatever the details of the detection procedure, pulse trains were detected based on pairs of events exceeding threshold. Based as it was on pulse pairs and on the requirement that most of the pulses in a train pass threshold, detection, although sensitive, was

371

G. Marx (ed.), Bioastronomy – The Next Steps, 371–376.
© *1988 by Kluwer Academic Publishers.*

computationally intensive. This was because thresholds had to be set low in order that most pulses in a predicted train pass.

Figure 2 compares the sensitivity of another simple pulse detector which looks for three regularly spaced pulses at a high threshold with that of a theoretically possible detector which tests the total power of every possible regularly spaced pulse train. Retaining 10^{-5} of the pulse field with a computational saving of 10^{10} results in only a 3dB loss in sensitivity. The simple detector can, if data are thresholded in the MCSA, be implemented in a dedicated, commercially available CPU. This surprising result led to an examination of detection strategies and three pulse detection was tried on low threshold detectors.

The following discoveries were subsequently made. Three pulse detection could be combined with total power detection in the computationally intensive algorithms to increase their sensitivity to long pulse trains. Long trains are now sought first by searching for three commensurately spaced pulses at a high threshold. If the train has many pulses, it is almost certain that three of its members will be found even at relatively high thresholds. Then, other members with a consistent spacing are sought and tested on a total power criterion. Thresholds do not remain high in this second stage. The result is sensitivity within about .5dB of the zero threshold curve in Figure 2. For the same amount of computation, methods that extrapolate pulse pairs rather than triplets have a sensitivity about 1dB worse than this.

Incoherent CW Detection

It was suggested by B. M. Oliver in 1984 that incoherent CW detection could be performed by processing the data in stages (Cullers 1985). The first stage would be relatively short, resulting in a much reduced number of discriminable drifts at each frequency when compared with the same procedure used over the entire 1000 second observation. In subsequent, longer stages, detections could be made more certain, the paths of interest in the first stage being subdivided consistent with the better discrimination possible in the longer stage. To keep the number of accumulators to a minimum, the increase in discriminable drift paths would be compensated by the thresholding of data, keeping the memory usage constant at roughly that of the first stage.

Figure 3 demonstrates the surprisingly good multi-stage detection sensitivities that result from an optimized three stage process (Cullers and Deans, 1986). This procedure can be extended to many stages resulting in more memory and computational saving. The strategy of Figure 3 has a sensitivity within .3dB of incoherent addition over 1000 seconds. A six stage process has been implemented in the VAX11/750. The first stage is only 17 frames long with loss in sensitivity of less than 1dB and memory saving by a factor of 60.

Hanning Windows and Interspectra

Figure 4 shows the frequency response of a Hanning window (Harris 1978). The time weighted window, and the standard uniform DFT weighting, are shown in Figure 5. This window is to be used in a second generation NASA prototype MCSA because of its excellent adjacent channel rejection which significantly decreases interference from strong unwanted signals. The Hanning time window, shown in Figure 5 weights data more heavily at its center than at its ends. Using such a window to detect CW that has an equal power at all times improves detection. Overlapping data sample frames by 50% in time and applying Hanning windows to the resultant spectra improve the estimation of the power spectral averages along a CW signal path, almost as if all the windows were statistically independent (Welch, 1967). Use of overlapping windows, called interspectra, increases the probability as well that some window will closely match pulses of interest.

In comparing the MCSA, circa 1984, with present design (Duluk, et al., 1985), CW detection fairs very well. The lengthened data window, from 1 to 1.5 seconds keeps the noise power in a Hanning window of the new MCSA equal to that in the DFT window of the old one. Hanning channels respond in essentially the same way as the old DFT windows to signals drifting at a maximum rate of 1Hz/s. They have a broader main lobe compared to the DFT windows from which they are composed, but the new DFT windows are narrower. The data processing problem is slightly greater since there are 1.5×10^7 Hanning windows in the

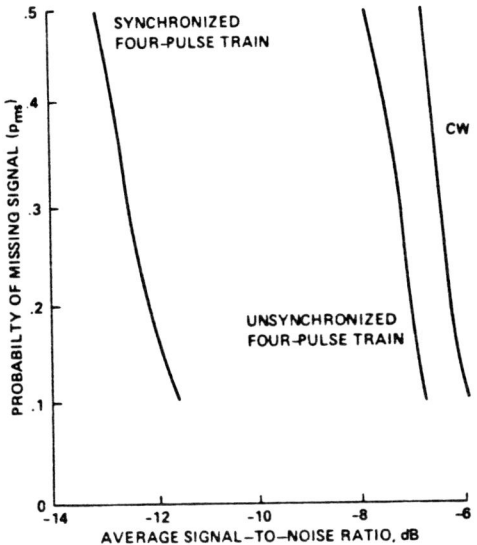

Figure 1
Comparison by Monte Carlo simulation, of the performance of a pulse and
CW detector for a 10^6 channel MCSA during a 1024 second observation
where the threshold permitted about one false alarm per observation and
drift rates greater than ± 1 channel/spectrum were excluded.

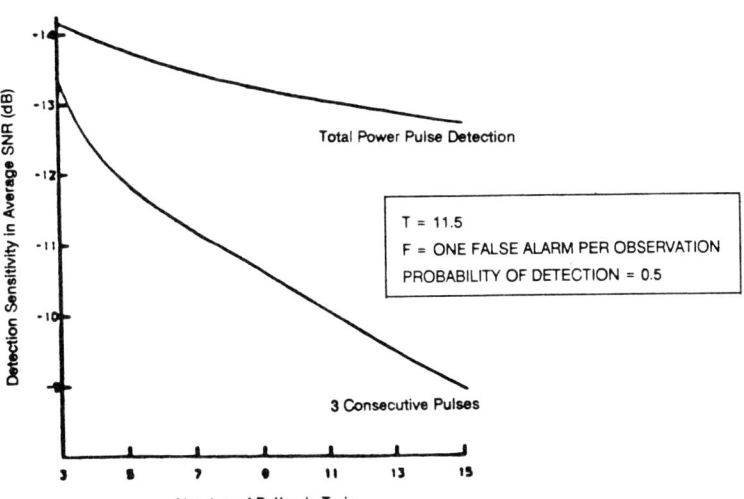

Figure 2
Comparison of sensitivity between 3-pulse detection and zero threshold total
power pulse detection.

374

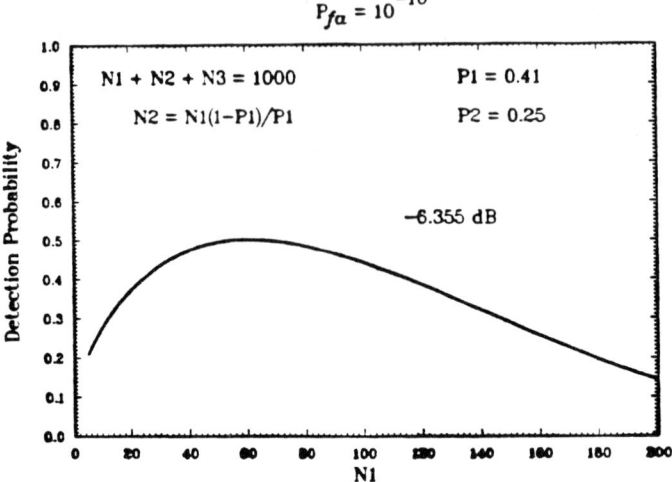

Figure 3
At the peak of its detection curve sensitivity for a 3 stage incoherent
CW detector comes within about 0.3 dB of that for full incoherent
detection. N1, N2, and N3 represent the number of spectra in each of
the three stages. P1 and P2 represent probability of false alarm in the
first two stages. The joint probability for all stages is given by P_{fa}.

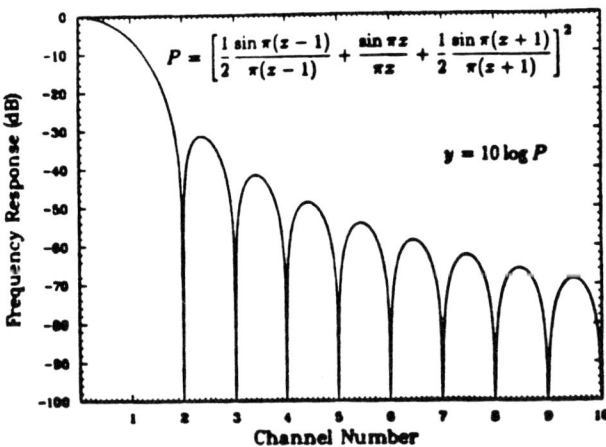

Figure 4
Frequency response of a Hanning Windowed DFT. The susceptibility
to RFI is greatly reduced.

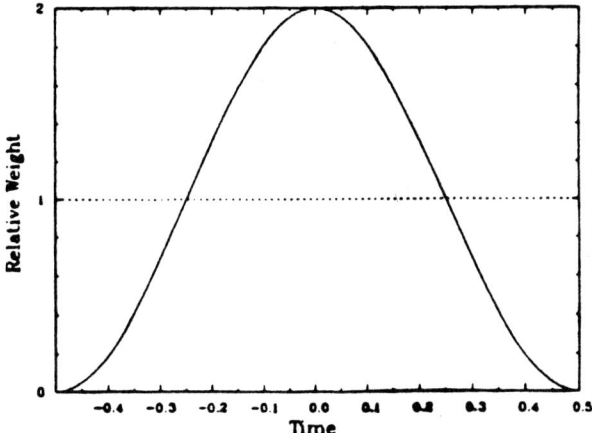

Figure 5
Weighting of a Hanning Window in time is non-uniform. Use of
overlapped spectra increases sensitivity to a CW signal.

new MCSA as opposed to 10^7 DFT windows in the old. Lastly, in average sensitivity, the new system wins hands down. The number of independent samples along a signal path is greater with the new system because of the introduction of interspectra with a net result that the average sensitivity to CW in the new system is a fraction of a dB greater than the peak sensitivity of the old.

Although any digital spectrum analyzer will provide only a discrete set of resolutions and data frames, extraterrestrial pulsed signals may have any of a continuous set of parameters. Numerical simulations of the average response of the Hanning window when bombarded with pulses of equal energy having a range of parameters, all equally likely, have shown that the average response of the new MCSA system is within less than a dB of the old system's peak response to square wave pulses.

Conclusions

Efficient algorithms now exist for detecting regular pulses. Interspectra and Hanning windows make the new MCSA sensitive to pulses and CW signals in a way that the old MCSA prototype was not. Pulse detection software can detect pulse trains with up to 7 pulses per 1000 second observation at a sensitivity within 3dB of an optimum incoherent pulse detector. Provided that thresholding is performed on board the MCSA, this can be done in a commercially available mid-size computer.

Still, problems remain unsolved. Hardware for pulse detection is now in development. Software is being streamlined to perform both sensitive pulse detection and incoherent staged CW detection on the VAX11/750. The logical complexity of pulse detection algorithms has been decreased by an order of magnitude with consequent increases in speed. Memory usage has dramatically decreased. Nonetheless, it is still impossible to do real time high sensitivity pulse detection for more than perhaps 50kHz of spectrum or incoherent CW detection for more than a few kHz on the VAX11/750. Specialized hardware is still needed for both these processes though its logic can be field tested in software.

On the horizon are possible methods of coherent or almost coherent CW detection. These involve dividing the observation into equal length stages which are analyzed coherently at all drifts and then linking them to test either total power or total numbers of events above threshold along possible signal paths. All of this, along with incoherent detection, awaits the development of specialized hardware, now in the beginning design stage.

References

Cullers, D.K., I. Linscott and B. Oliver,"Signal Processing in SETI", *Communications of the ACM*, Volume 28, Number 11, November 1985.

Cullers, D.K.,"Sensitive Detection of Narrowband Pulses", *Acta Astronautica*, Volume 13, Number 1, 1986.

Cullers, D.K., "Software Implementation of Detection for the MCSA", *The Search for Extraterrestrial Life: Recent Developments,* IAU Symposium 112, 1985.

Cullers, D.K. and S.R. Deans, "Narrowband Signal Detection in the SETI Field Test", *Proceedings of NRAO Workshop*, Green Bank, West Virginia, 1986.

Harris, Frederick J., "On the Use of Windows for Harmonic Analysis with the Discrete Fourier Transform", *Proceedings of IEEE*, Volume 66, 1978.

Welch, P.D., "The Use of Fast Fourier Transform for the Estimation of Power Spectra: A Method Based on Time Averaging Over Short, Modified Periodograms", *IEEE TRNS. Audio Electro. Acoust.* AV-15, 1967.

Duluk, J.F., I. Linscott, A.M. Peterson, J. Burr, B. Ekroot, and J. Twicken, "VLSI Processors for Signal Detection in SETI", *Proceedings of IAF Congress*, 1986.

THE CETI ELLIPSOID AND THE SUPERNOVA 1987A

W.F. Hilton

7 Stanley Road
Teddington, Middx TW11-8TP, United Kingdom

Iván Almár

Konkoly Observatory
P.O.Box 67, Budapest 1525, Hungary

Astronomers trying to receive ETI signals have the problem of where to look and in particular when to look (or listen) for CETI contact. Since ETI is intelligent, "they" will understand our difficulty and will try from their end to help establish mutual contact. It was suggested by different authors (Tang 1976, Makovetskii 1977) that the outburst of a bright nova or supernova may represent a time signal visible both to "them" and to us at various epochs, allowing for the speed of light. In 1980 Hilton put forward the concept of the "CETI ellipsoid" and suggested the Crab supernova as time signal. Stars which meet the criterion in 1987 will lie on an ellipsoid of such size that the travel time for signals from the Crab to the ETI star plus the travel time from the ETI star to Earth will just add up to 1987, and these stars should be monitored now for ETI signals. Since the distance of stars is not known accurately, the crisp and precise geometry of the CETI ellipsoid is blurred into a diffused period of observation, lasting 20 years for a star 100 l.y. distant and having a $\pm$ 10 l.y. uncertainty in its distance.

The recent supernova in LMC (1987A) is certainly an important astronomical event for all hypothetical civilizations in the near part of our Galaxy. Its outburst in Fenruary 1987 generated a new CETI ellipsoid, namely a straight line joining us to the supernova. The precise range of the ETI stars did not matter if they are transmitting signals towards us as soon as the supernova was sighted. Since the supernova is 170 000 l.y. from us, after two years the CETI ellipsoid will expand to 400 l.y. only, so all foreground stars in question will be seen within a half angle of 0.3°. In general we should examine stars lying roughly within $0.2T^{\frac{1}{2}}$ degrees of the supernova, where T is the number of years elapsed. All appropriate foreground stars of LMC should be investigated and we should transmit strong CETI signals in exactly the reciprocal direction to that of the supernova (constellation Draco). ET astronomers observing the supernova may pick up the signal as well.

References

Hilton, W.F., 1980 "The CETI Ellipsoid" paper IAF congress, Tokyo
Makovetskii, P.V., 1977, Astronomicheski Zhournal 54, 449
Tang, T.B., 1976, JBIS 29, 469

377

ENERGY SAVING INTERSTELLAR COMMUNICATION WITH MODERATE EFFORT

Jörg Pfleiderer
Institute for Astronomy
Technikerstrasse 15, Innsbruck 6020 Austria

The power consumption necessary for sending a detectable signal over galactic distances is, under optimum conditions, proportional to the bit rate of the message, to the inverse receiving area, and to the emitting beam size (G.M.Gruber & J.Pfleiderer, S.Afr.J.Phys.$\underline{8}$,43,1985). With reasonable assumptions, the power is comparable to the solar power falling on Earth if an omnidirectional signal is emitted. The body needed for thermally radiating away the waste energy has to be of planetary dimensions and, therefore, has to be put into space unless the conversion of energy into radiation is extremely efficient (thermal pollution problem). The radiation pollution problem is as serious: Any SETI receiver on the home planet will be excessively insensitive due to spill-over reception of the own emission unless the receiver is so narrow-beamed that it must also have planetary dimensions. The same would be true for the solar-cell area collecting the energy from the central star.

Power-saving measures: Both an extremely small emitting bandwidth (bit rate) and a very large planet-sized receiving antenna would place an excessively high technological burden on the receiver's side. Another possibility is an empty message (zero bit rate) with finite bandwidth. The receiver's integration time can then be made as long as needed. Examples discussed in the literature are leakage radiation ("eavesdropping") and the search for spectral characteristica indicating organic material on a planet. The possible range of such inadvertent "communication" has, however, been estimated to be only a few tens of parsecs.

The only remaining choice is a narrow emitting beam which necessitates discontinuous illumination of one target after the other with a small duty cycle. Conventional or near-conventional techniques on <u>both</u> CETI sides (Arecibo-type dishes) would suffice, and it is the only case where a true message can be transmitted without one side being excessively advanced (minimization of the combined effort). The receiver's side must then observe continuously and omnidirectionally, with many dishes (≈ 20) and many receivers per dish ($\approx 10^6$ per Arecibo dish) or search only in selected directions (B.Balázs, this volume). It remains the presently unsolved problem of the correct guess of emission parameters. G.M.Gruber & J.Pfleiderer (Sitz.Ber.Öst.Ak.Wiss.MNKL.II 191,405,1982) suggested to derive them from simple mathematical premises and gave an example.

For details see Pfleiderer (1987, in: *Proc.Symp. on Commercial Opportunities in Space*, F.Shahrokhi ed., AIAA, Progr.Ser. in Astronautics).

G. Marx (ed.), Bioastronomy – The Next Steps, 379.

OPTICAL LASERS FOR CETI

J.D.G. Rather

Kaman Aerospace Corporation

1600 Wilson Bd, Arlington VI 22209 USA

ABSTRACT.   Present research supporting SETI and CETI is entirely
focused upon microwave wavelengths and technologies. This paper demonstrates that there are compelling reasons why the search should be
broadened to optical wavelengths. New perceptions regarding the feasibility of optical laser technologies, together with reassessments of
signal-to-noise considerations, indicate both the feasibility and
desirability of interstellar transmissions at infrared, visible, and
ultraviolet wavelengths. Implicit in the rationale for SETI/CETI is
the question of ultimate motivations. It is shown that use of the
large bandwidths available at optical wavelengths greatly enhances the
utility of such endeavors. Present astronomical equipment and methods
are adequate for the initial detection, although specially constructed
receiving equipment would be required for full decoding of the signals.

I.   PHILOSOPHICAL BACKGROUND. Shortly after the invention of Light
Amplification by Stimulated Emission of Radiation in 1960, Townes[1] and
others suggested that lasers might be suitable for interstellar beacons
and/or communication. This proposition was so roundly attacked, however, by Oliver and others[2,3,4] in several technical papers in the
following decade that the superiority of microwaves for these purposes
has subsequently been accepted as "gospel truth" by most of the SETI/
CETI community. The essence of the attack was that, if it is assumed
(1) that each candidate source of emissions from a given range has the
same radiated power and the same beamwidth and (2) that maximum detectable range is the prime discriminator, then microwaves always win.
Based upon these (or equivalent) assumptions, microwaves always win
because they contain more photons per unit energy, and thus the effective noise per unit time interval is lower. In the parlance of noise
theory, the detected signal-to-noise power ratio, D_{snr}, in
photon-limited situations is given by

$$D_{snr} = \frac{P}{h\nu B},\qquad(1)$$

where P is the received power, B is the bandwidth, ν is the observed
frequency, and h is Planck's constant. What has been overlooked,
however, is that any intelligently designed communications system will
not be photon limited! Much shorter (laser) wavelengths permit reasonable optical apertures to focus the radiation on the life zone of
the target stars, thus easily overcoming star noise and minimizing
wasted energy.

381

G. Marx (ed.), Bioastronomy – The Next Steps, 381–388.

It seems very presumptuous to beg the question of optimum wavelengths for SETI on presently perceived (Earth chauvinistic) engineering limitations. Beacons, if they exist, are in all probability the artifacts of civilizations thousands or millions of years more advanced than our own because the probability of coincidental evolution is quite negligible. Indeed, such beacons, of necessity, must also be "built for the ages," since they must survive for thousands or millions of years in order to provide reasonable probabilities of detection by randomly occurring technological civilizations of questionable longevity throughout the galaxy. Any civilization that goes to the (large) trouble of building such beacons will strive to give them maximum visibility and longevity, with maximum information content and cultural value.

There is no reasonable basis for estimating the repetition rate of "The Message" since the human life span may have no universal significance. It seems most logical that there would be a hierarchical modulation scheme starting with a sort of rapidly repeating "rosetta stone" that refers the receiver of the message into ever more complex information content contained in the higher-order sidebands, so that high data content is immediately accessible in "parallel processor" fashion. Such an arrangement would not be reasonable at radio wavelengths for the large data transfer rates to be discussed below unless the message were extended in time over centuries. It is much more efficient to transmit 10^9 or 10^{10} well aimed beams simultaneously to all candidate stars, with each beam efficiently modulated to deliver maximum information in minimum time whenever and wherever it is received. Hence the importance of the highest possible frequency consistent with the galactic environment must be contemplated.

II. ASTRONOMICAL CONSTRAINTS. An obvious early optical search method might utilize moderate dispersion automated spectroscopy to examine millions of spectra for highly non-thermal emission lines. While spectrographic search methods are conceptually simple, let us consider potential problems. First, there is the question of spectrographic resolution and "star noise". Even in the depths of dark spectral lines, there is a large amount of residual star light within the resolved bandpass to contribute to the background noise figure. Hence, in order to be discovered, the laser must be bright enough to outshine the star as observed by a spectrograph. Here it is important not to confuse a simple detection of the presence of a signal with detailed demodulation of the message content. A simple detection can be achieved with a telescope of modest aperture if the signal power is greater than the background star noise in the resolved bandwidth and the observation is integrated for a long enough time -- as by a photographic plate. Having found the signal, we can then construct large and sophisticated hardware to gather enough photons to detect the message.

An obvious criticism of the use of visible wavelengths is the problem of the interstellar dust that confines our unobscured line of sight to a couple of thousand light years in the plane of our galaxy. Nevertheless, all of the glow that we call "The Milky Way" is coming from stars that lie above and below the galactic plane. Hence hundreds of billions of stars are still accessible even from our non-optimum location; and, if the posited advanced civilization resides outside the galactic plane, it won't worry about the dust. Interestingly, those out-of-plane stars, comprising perhaps 90% of the stars in our galaxy, are often older, low mass Population II stars, and hence are important candidates for the sites of old civilizations.

It should also be noted that the interstellar dust absorption and star-noise problems both disappear for wavelengths longward of a few microns in the infrared and shortward of soft X-ray wavelengths. Numerous kinds of lasers are already feasible for IR, visible, and near ultraviolet wavelengths. It is a virtual certainty that free-electron lasers, and probably other types of lasers, will also eventually reach even into the X-ray regions. So the entire electromagnetic spectrum might be considered possible territory for SETI/CETI.

III. CANDIDATE LASER SYSTEMS. There is a wide variety of presently conceivable laser devices potentially capable of performing the inter-stellar beacon/communications mission. One of the most interesting is an enormous phased array of highly efficient solid-state light emitting diodes coupled with coherent hetrodyne light detectors.

The burgeoning of very large-scale integrated circuits, automated microcircuit production methods, light emitting diodes, and many other related breakthrough technological accomplishments was not contemplated two decades ago when the philosophical foundations of SETI were initially being established. Now it is by no means frivolous to predict the feasibility of advanced methods of automated production in space that could utilize the raw materials of an asteroid, a moon, or perhaps a small planet (like Mercury) to create an enormous coherent phased array of solid-state detectors and laser light-emitting diodes. Visualize, for example, virtually the entire surface of Mercury covered with a hybrid, solar-powered, coherent detector/laser transmitter array and the necessary automated machines to build and maintain all of its components. The resulting planet-sized spherical array could then function as follows: (1) A local "fly's-eye" sub-assemblage of detectors would sense starlight impinging normal to the surface. The distributed computer controlling the array would expand and contract the required scale of the local sub-array from a few centimeters to a few tens of meters (as required for adequate signal-to-noise) to simultaneously image essentially every star in the galaxy for a substantial fraction of the planet's year. (Think of a real-time, whole-sky, full spectrum survey with the potential plate scale equivalent to the surface area of a planet! Only the portion of the sky near to the sun would be inaccessible at any given time.) Contiguous sensors in the array would sense a tremendous wavelength range, providing detailed information for automatically classifying each star and estimating its distance. (2) This information would then be transferred to adjacent coherent light emitting diodes in a synthesized array whose total power and beamwidth are specifically tailored to deliver a detectable signal to the calculated ecosphere of each star. If we use Mercury as a prototype, about 2×10^{17} watts of solar power is available for conversion to laser radiation -- about a megawatt per star in the galaxy (or much higher peak power in pulsed mode). Of course, the transmitted power would not be apportioned equally, however. The beam to each star would be tailored specifically with a power level and array aperture size required to illuminate economically only the ecosphere of each star. Such an advanced system would also contain in its memory the astronomical wisdom of the ages and added information gleaned from its own prodigious observing capabilities. It would not be troubled by pointing and tracking details such as stellar proper motions and the light-travel time across the galaxy because, over the ages of its operation, the required data would be continuously refined. Thus the overall system would be optimized to transmit the "perfect" power and beamwidth to each star, thus conserving the overall energetics of the beacon system.

IV. LASER CAPABILITIES. Because of the short wavelengths of lasers, some great advantages accrue from the ability to radiate narrow beams from modest-sized apertures. In particular, transmission of information across huge distances in the galaxy can be accomplished without the inverse-square loss so familiar in radio work. A direct consequence of this advantage is that the "star-noise," naively believed to be a principal nemesis of laser communications, becomes increasingly less important as the range increases. The possibilities become immediately apparent from some simple calculations.

The diffraction-limited transmitted beamwidth is given by

$$\Theta = 2.44 \, \lambda/D_T \quad , \tag{2}$$

where λ is the laser wavelength, D_T is the transmitter aperture, and the divergence angle Θ is specified to the first dark ring of the diffraction pattern. The logical procedure would be to estimate the distance of each target star and then adjust D_T so that the diameter of the focal spot at the target star agrees with the diameter of the "life zone", about 21 light-minutes. Figure 1 shows the required optics diameter to produce this spot diameter as a function of range. It can be seen that, at the sodium absorption line wavelength of 0.599 μm, a 1 km diameter aperture can illuminate the life zone of a star at 27,500 light years with very little power loss.

Figure 2 shows the basic considerations to obtain a signal to noise ratio adequate to simply confirm the presence of an intelligent signal against the background of star noise. The laser and its home star are shown at the orgin. The beam divergence angle Θ is adjusted to match the life zone diameter D_L at any range R. A receiving aperture of diameter D_R is shown for two ranges at A and B, where the receiving aperture is assumed to be on or near a planet within the life zone. Clearly, the receiver at Star A sees the same laser flux as the receiver at Star B. The received signal power, P_R, is simply

$$P_R = P_L \, D_R^2/D_L^2 \tag{3}$$

where P_L is the laser power. But, the receiver at Star A sees greater associated star noise power, P_N, according to the inverse square law:

$$P_N = P_{\Delta\lambda} \, D_R^2/16R^2 \quad , \quad (4)$$

where $P_{\Delta\lambda}$ is the power emitted by the star within the acceptance bandwidth $\Delta\lambda$ of the receiver. The signal to noise power ratio is, then

$$\frac{P_R}{P_N} = 16 \, \frac{P_L}{P_{\Delta\lambda}} \cdot \frac{R^2}{D_L^2} \quad , \quad (5)$$

If we assume an intended information bandwidth greater than 10 GHz for the laser transmitter, the associated $\Delta\lambda$ at $0.599\,\mu$m is 0.01 Angstrom, typical of the resolution of many existing spectrographs. For a $5800°$K star, the source noise power in this bandwidth will be $P_{\Delta\lambda} = 4.8 \times 10^{20}$ watts, as given by Planck's equation. Hence, substituting these values, we can determine the worst-case (i.e. ignoring benefits from Fraunhofer lines, etc.) required laser power for an acquisition signal-to-noise ratio of 100, which we express in terms of the number of light years, n:

$$P_L = 4.8 \times 10^{12}/n^2 \quad (6)$$

The only stipulation is that the transmitter aperture must be large enough so that n is less than the Rayleigh range. Plotting the required laser power versus distance in Figure 3, we see the dramatic advantage of lasers for the "beacon" acquisition function at long distance. One megawatt of laser power transmitted from an optical aperture 100 meters in diameter will provide a signal-to-noise power ratio of 100 at a range of 5,000 light years. Checking the photon statistics, we note that 1 watt at 0.599 m corresponds to 3.0×10^{18} photons per second. Hence, for example, 1 megawatt of signal power delivered to the life zone will yield $\sim 10^{-17}$ watts/m^2, or ~ 28 photons per second per square meter. A 4 meter telescope could therefore provide an adequate detection of the beacon, although none of the modulation could be read.

Now comes the question of demodulating the message. Referring again to equation (1), it can be shown that the detection signal to noise power ratio for the entire information content contained in the 10 GHz bandwidth is $\sim 1.5 \times 10^{-8}$ per meter squared per megawatt of laser power. Hence, the required light collector array area to receive the full information content of the signal is $\sim 10^9$ m^2 for 1 mw transmitted signal or $\sim 10^5$ m^2 for 10 GW. Thus the former array diameter would be ~ 32 Km and the latter would be ~ 320 meters, either of which should be readily feasible with human technology in the 21st century. Considering probable progress in space development, the array would likely be built in space. Its complexity is not likely to exceed that of the "Cyclops" microwave array proposed in 1971, and it may even cost less if the expected progress in electro optical systems continues at its present pace. (Several concepts for extremely large, low-weight, diffraction-limited optical apertures in space are even now under intensive study. These concepts will also lead to cost reductions per unit area of optical surface.)

V. THE CONTENT OF THE MESSAGE. A case has been made for lasers for SETI/CETI based upon several arguments: (1) They are a versatile technology for producing highly directional beams that deliver maximum power on target. (2) By targeting the power rather than broadcasting it, vastly better use is made of the energy available. (3) The availability of laser power at any chosen wavelength makes possible the choice of wavelengths that show the intelligent signal in maximum contrast to its natural environment. (4) The potentially enormous brightness (i.e. watts/steradian) of a well-designed laser system makes possible delivery of high signal-to-noise signals having tremendous data rates anywhere in the galaxy. There is no need to worry about photon noise when dealing with truly advanced civilizations.

The remaining question to be addressed is, "Why is such a bandwidth needed anyway? If a civilization could build planet-sized solid-state laser or free-electron laser transmitting arrays, why not just do it at microwave wavelengths and transmit their equivalent of the 10^{14} bits now regarded as the vehicle for most of present human culture?" There is a heirarchy of answers to this query which lead to a sobering possibility.

First, consider the exponentiation of human knowledge in the past five hundred years alone. Other than codifications of law, history, the arts, and the sciences, it is necessary to consider the implications of things that we are just beginning to glimpse. The "Arecibo Message" transmitted to the stars in 1974 indicated the nature of DNA and the number of human genetic codons, but it provided no details. Suppose it is desired to transmit the complete "recipe" for just one human being. The genetic code alone will amount to about 10^{12} bits. It is usually conveniently forgotten, however, that the cellular machine required to read this code and manufacture the prescribed protein structures, nervous system and macroorganism must first exist as a germ cell in order to make use of the code. The additional information needed to describe the associated cellular structures, functions, nutrients, maintenance requirements, etc. would require perhaps another 10^{12} bits. Then suppose that in the not too distant future a coherent X-ray laser device can be devised to provide dynamic, four-dimensional holographic imaging of the functioning brain of one of the brighter members of our species. (Every atom in every molecule of every brain cell might be quantitatively recorded to characterize the memory banks, while all electrical activity might be recorded to document the thought processes.) This information will probably constitute another 10^{14} bits. So perhaps 10^{15} bits would be required to characterize just one fully educated, functioning human being. (If it was desired to record all 10^{10} examples of humanity for posterity, about 10^{25} bits would be needed!)

Let us ponder a truly advanced intelligent species. Not only will the culture be much older, with proportionate exponential increase in the amount of raw knowledge and experience, the creatures themselves will very likely have evolved to a more complex state. It is even possible that they may have developed from organic bodies to some other state, such as self replicating electronic bodies that are much more versatile than organic matter in terms of data management capabilities and adaptation to changing environments. For the sake of argument, let us assume that 10^{16} bits might characterize one premier example of such a species. Let us also assume that this species retains the one fundamental fructifying principle of _all_ lifeforms, namely the urge to perpetuate and promulgate itself.

The example given in our previous calculation showed that bandwidths of 10 GHz are readily feasible. This would permit transmission of 10^{17} bits in less than a year. Even near the visible wavelength range the bandwidth might be expanded by at least another factor of ten or more. Hence, there are definite possibilities for achieving data rates sufficient to transmit large numbers of the alien species.

The universe is vast, and the speed of light is finite. The diaspora of a physical species through the universe would be a slow and inefficient business using classical ideas about interstellar travel if there is no way to transgress the light-speed limitations of relativity. But this limitation could be the _only_ limit if species can devise ways of transmitting themselves at the speed of light. An efficient way to undertake one way interstellar expansion might be to send out nonrelativistic probes at, say, a hundredth of the speed of light. Upon reaching a usable planet of a remote star, the probe might then serve as the receiver/synthesizer for an open-ended stream of intelligent entities from the home planet.

But the most efficient means of all would be to find an already existent civilization and cajole it into building the receiver. Maybe the purpose of interstellar beacons is first to attract attention and interest. If they exist at all at microwave frequencies, the purpose of beacons may be simply to refer the interested observer to a light wavelength with high information content. At the higher frequency, the prime function of the message may be to interest, persuade, seduce, and cajole the receiving race to build a great machine (as envisioned in many science fiction novels ranging from Hoyle's "A For Andromeda" to Sagan's "Contact"). The purpose of this machine may be to receive the full content of the message and realize it in all of its implications. The most profound implication of all would be to transplant the alien culture to the new world. Hence the machine might become the instrument for superseding the present process and direction of human evolution.

Acknowledgements. The author extends his appreciation to Joseph A. Mangano, James R. Powell, and Bobby L. Ulich for helpful discussions and criticisms. Particular thanks goes to Gerald A. Ouellette for his very insightful contributions.

388

OPTICS DIAMETER TO PRODUCE 21 LIGHT MINUTE DIAMETER SPOT

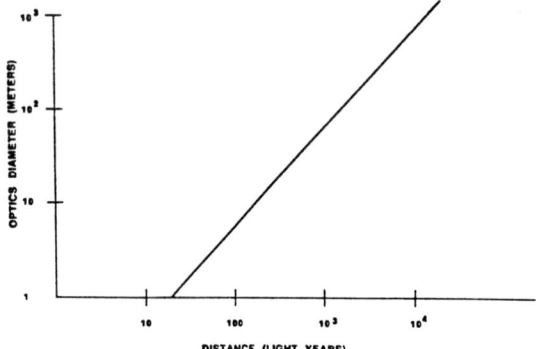

Figure 1. Diffraction-limited laser optics diameter to illuminate life-zones at different stellar ranges. Wavelength is assumed to be 0.559 μm.

Laser and Associated Star

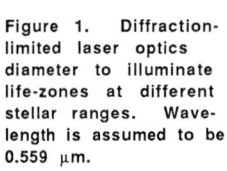

Figure 2. Parameters affecting received signal to star-noise ratio.

ADVANTAGE OF NEAR-FIELD LASER FOCUS

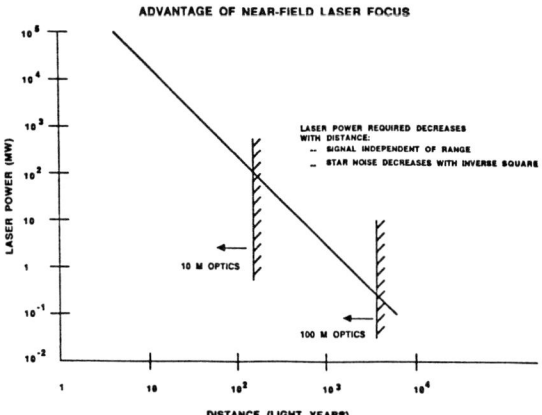

Figure 3. Required laser power versus distance to yield a signal to star-noise power ratio of 100 at the receiving aperture. Transmitter aperture diameters of 10 m and 100 m indicate diffraction-limited range for 0.559 μm wavelength.

References

(1) R.N. Schwartz and C.H. Townes, "Interstellar and Interplanetary Communication by Optical Lasers", Nature, 190,. 205, (1961).

(2) B.M. Oliver et al, "Project Cyclops", NASA Ames Research Center Report #CR114445, (1971).

(3) B.M. Oliver, "The Rationale for the Waterhole", Technical Memorandum, Hewlett Packard Corp., Palo Alto CA. (1976).

(4) B.M. Oliver, "Linear Amplification vs. Photon Detection in the Search for Interstellar Signals", Technical Memorandum, Hewlett Packard Corp., Palo Alto, CA. (1976).

SETI IN OPTICAL RANGE WITH THE 6M TELESCOPE (MANIA)

V. F. Schwartsman

Special Astrophysical Observatory
Stavropolskij Kraj, Niznij Arkhyz 357147, USSR

There exist two strategies of detecting possible artifical laser signals on the radiation background of the star, suspicious to have populated planets. The first one consists in spectroscopic detection of sharp emission lines with an unusual frequency (Schwartz R.N., Townes C.H. Nature, 190, 205, 1961). The second one is the analysis of photons coming from the object under investigation (Schwartsman V.F., Soobshchenija SAO, 19, 5, 1977). Deviations of the statistics from the poissonian characterise the magnitude variations of the object (short flares) as well as the existence of supernarrow emission features in its spectrum. The both means of signal coding seem to be possible for the connection between cosmic civilizations.

Study of the photocounts statistics of peculiar objects has been started in Special Astrophysical Observatory in 1972 with the Zeiss-600 telescope (D=60 cm). Since 1978 the observations are carried out with the 6 m telescope BTA. For the investigations of the photon statistics my colleague and I worked out and constructed an engineering and software complex MANIA (Multichanal Analysis of Nanosecond Intensity Alterations).

At present the complex MANIA allows to detect light variations of celestial objects in the range $\tau = 3, 10^{-7} - 100$ s and, correspondingly, narrow emission lines with the width $\Delta \nu \simeq 1/\tau \simeq 10^{-2} \div 3 \times 10^{6}$ Hz. Thereby, alterations of the laser line frequency within the analysed spectral range (e.g. for the additional coding of the information) doesn't decrease the possibility of the signal detection; it is only important, that the signal-to-noise ratio be not too small.

The major aim of the MANIA experiment consists in the search for black holes, the investigations of the properties of relativistic objects and flare stars. However, we also observed the objects, which are interesting for the SETI

389

G. Marx (ed.), Bioastronomy – The Next Steps, 389–390.

problem. Between them: the objects, with the purely
continuous optical spectra and considerable proper motions
(17 objects), the radiosources with the continuous optical
spectra (20 objects) and nearby (10-30 pc) solar-like stars
(8 objects).

The preliminary analysis of the observational results
didn't show any obvious artificial signal. The data (10^{10}
bit) are stored on a magnetic tape. Their treatement with
the new, more effective methods of signal detection is
planned.

The results of observations of 8 solarlike stars,
carried out in April 1986, allow to obtain an upper limit to
the power of a hypothetical arificial laser, beaming to the
Earth: $P < 10^9$ W.

Note that the total power of lasers, which are planned
to be put onto the circumterrestrial orbit in the framework
of the SDI-program is of order 10^{10} W (Bethe H.A. et al.,
Sci. Amer., 251, 4, 1984).

These are the thesis of the talk presented by died
prematurely Dr. V.F. Schwartsman to the IAU Colloquium 99.
The full text of the talk is, unfortunately, not available
to the editor.

STARS AS GRAVITATIONAL LENSES

Frank Drake

Department of Applied Science, University of California
Santa Cruz CA 95064, USA

Some fifty-one years ago, Albert Einstein (1936) noted that stars would act as lenses, creating images of distant objects with very great magnification. The focal lengths of such lenses are of the order of hundreds of astronomical units for normal stars. The detailed behavior of such stellar lenses has been described by Liebes (1964) and Eshleman (1979, 1984). Eshleman, in particular, pointed out the possible advantages of such lenses for interstellar communication.

What is the performance of such lenses? The star makes an image of a very distant point source at a minimum distance of

$$R^2/2g$$

where R is the radius of the star, and g is its "gravitational radius", defined as $g=2GM/c^2$, where G is the universal gravitational constant, M is the mass of the star, and c is the velocity of light. For the sun, this distance is 550 A.U. The star also makes an image of the point source at all distances beyond this minimum distance. Thus a detector anywhere along this "focal line" is capable of detecting the focused image created by the star.

An observer on the focal line, looking towards the star, sees a bright ring around the star, which is the light from the source which is being directed towards the observer by the focusing effect of the star. This ring becomes the effective aperture of a lens which becomes the optical equivalent to the lensing star. Its width depends on the precise circumstances, but is typically of the order of 100,000 wavelengths for a solar type star. For radio wavelengths, then, the dimensions of the lens may be of the order of several solar diameters, and the collecting area of the aperture can be enormous. The "gain" of the lens, as shown by Eshleman, is of the order of 100,000 for stars like the sun and radio wavelengths of the order of one meter. The gain is inversely proportional to the wavelength of observation and directly proportional to the mass of the star. Surprisingly, the gain is independent of the place on the focal line where the focused energy is captured.

The size of the image is proportional to the wavelength, as is to be expected, and is proportional to the square root of the distance from the star to the observer. For the sun, at the minimum distance, the image size is about ten kilometers for a one meter wavelength; at one parsec the image size would be 200 kilometers.

These image sizes result in very remarkable angular resolutions in the image. Take the case of a detector which is located 1000 A. U. from the sun, and which operates at 1-cm

391

G. Marx (ed.), Bioastronomy – The Next Steps, 391–394.

wavelength.   The size of the image is 160 meters,   which implies an angular resolution of 10^{-12} radian!   If the point source is at a distance of 10 parsecs, the spatial resolution at the source is 300 kilometers.   This is sufficient not only to detect,   say, planetary systems, but to resolve the planets!

The the use of stars as gravitational lenses not only provides enormous gain, but allows unprecedented resolutions to be achieved.

The use of the sun as a gravitational lens requires us to place detectors at least 550 A.U. from the sun, and in practice much farther, to avoid coronal effects. Thus this application is something for the distant future.

Let us consider nearby stars as gravitational lenses.   We are able to capture the images created by them without loss in gain, at least in the simplest picture.   If we use a wavelength of 1-cm,   and the star is a solar type star,   the gain will be about ten million.   If we use a 10-meter antenna to collect the focused energy,   we will achieve an overall gain of about 10^{14} . This is equivalent to the gain which would be provided by a filled antenna some 30 kilometers in diameter.   This is equivalent to 10000 Arecibo telescopes!

In this same example, let us assume that we are 50 parsecs from the focusing star, and that the source is 50 parsecs on the other side of the star.   Then the image size of a point source is 16 kilometers,   implying that the resolution at the source is 16 kilometers.   Since these values change only as the square root of the distances involved, this quality of image and resolution will be achieved for a large range of stellar and source distances. Again,   it is possible not only to detect planets, but to resolve them.   Of course,   the system allows the detection of very faint intelligent radio signals.

It is possible,   although we know not how plausible,   that civilizations will use stellar lenses to amplify signal transmissions.   In this way,   they could project very strong signals on other planets at great distances in the galaxy. However,   the alignments required are incredible.   To circumvent this problem,   in practice they may purposely defocus the beam created by the lens by placing the transmitter closer to the star than the minimum focal length this will broaden the "beam" created by the lens,   allowing planets not in "perfect" alignment to detect the signal.   This would both increase the number of planets on which the signals would fall,   and would increase the time duration over which the signal was detectable at these planets.

However, it is important to note that, whether transmissions in our direction are intentional or unintentional, the time interval during which we receive a signal may be very short. With image sizes of point sources of the order of 20 kilometers, and with stellar random motions and planetary orbital motions of the order of 20 kilometers per second,   it could well be that any signal received will persist for a time interval only of the order of one second. However, during that interval the signal may be very intense. This suggests that in our searches we should use data recording procedures and analysis procedures which are sensitive to the presence of signals which persist only a fraction of second or so.

The lensing action of white dwarf stars is of special interest. For the most massive such stars, with a mass of about 1.4 solar masses and a radius of about 1500 kilometers, the minimum focal distance is only 0.34 solar radii. For a low mass white dwarf, say of 0.4 solar masses and radius about 10000 kilometers, the minimum focal distance is about 60 solar radii. These small values of minimum focal distance are a result of the very small impact parameter for the rays which just graze the surface of the star. In any case, these examples show that white dwarf stars are capable of imaging their own planetary systems, if the distances of the planets from the star are even a fraction of the values found in the solar system. This in turn suggests that it might be fruitful to · observe white dwarf stars photometrically in the hope that we might chance upon one whose ecliptic plane passes close to the earth. In that case, as a planet passes on the far side of the white dwarf, we might detect photometric effects due to the lensed radiation from the planet.

A white dwarf in a suitably wide eclipsing binary system would image the surface of its companion star. There are two eclipsing binaries known with white dwarf components (BD+16°516 and PG1413+01), but in both cases the separation of the components is less than the minimum focal distance of the white dwarf. Thus no substantial lensing action should be expected.

Another interesting case is lensing by neutron stars. With the expected neutron star parameters of a radius of about 10 kilometers and a mass of about 2 solar masses, the minimum focal distance is only about 8 kilometers. The image size at distances of about 100 kilometers from the star is only about one meter at radio wavelengths, if all other optical effects are inconsequential. Thus the neutron star creates images of essentially all its magnetosphere. Every observer should see at all times a gravitationally lensed image of the portion of the magnetosphere which is just "opposite" to the observer, as seen from the center of the neutron star. The imaged portion of the magnetosphere is all of the "far-side" magnetosphere along a line from the observer through the center of the neutron star, and extending from the minimum focal distance to infinity.

Since the neutron star rotates, the imaged portion of the magnetosphere will be continuously changing. With one meter resolution, and typical velocities in neutron star magnetospheres, we would expect changes in the imaged radiation on a time scale of the order of a microsecond. It is tantalizing, of course, that major variations in pulsar radiation on this time scale have been observed. Perhaps there is a connection. In any case, it would appear that models of the pulsar radiation phenomena should take into account the lensing action of the star. It could well be a substantial or dominant effect in the radiation phenomenology.

Neutron stars in eclipsing x-ray binaries will image their companion stars. However, this imaging phenomenon may be compromised by the obscuration and refractive effects of the accretion disk of the companion.

Other objects in which stellar gravitational lensing may provide useful information include at least quasar nuclei, Herbig-Haro objects, old novae, and objects similar to SS 433.

References:

Einstein, A. SCIENCE, Vol. 84, 506, 1936.
Liebes, S. PHYS. REV. Vol. 133, B835, 1964.
Eshleman, V. R. SCIENCE Vol. 205, 1133, 1979.
Eshleman, V. R. SKY AND TELESCOPE, June, 1984, 493.

THE PRO'S AND CON'S OF GRAVITATIONAL LENSES IN CETI

N. Cohen
Metropolitan College, Boston University
Boston MA 02115 USA

ABSTRACT. Gravitational lenses are an intriguing means of generating intense, achromatic gain for cosmic communication (Eshleman,
1979). Yet for their vast advantage in gain, they provide some
severe constraints in modulation modes and bit rates. These limitations arise from the transience of the gain, caused by breaking
of the rigid alignment required between observer, lensing star, and
the transmitter. Except in cases where the positions of the observer and transmitter are known and corrected for with great accuracy (requiring its own large scale tracking system), significant
gain (greater than 30 dB) may only last for a fraction of a second
during the proper alignment. This would suggest a very fast bit
rate and concurrent wide bandwidth for any message transmitted with
a non-targetted strategy. Diffraction and relativistic time delay
effects are expected to add a phase problem, which will appear to
an observer measuring the total radiation of the lensed message as
a chromatic signal, if the transmitter is itself coherent. Rapid
gain variations and phase effects warrant the use of phase and
frequency modulation in this mode, and a digital encoding may prove
the most efficient. If the observer is unprepared to decode the
transient, lensed transmission, it may be misconstrued as being
spurious, terrestrial, or natural in origin.

Eshleman,V.,1979,Science,205,1133.

G. Marx (ed.), Bioastronomy – The Next Steps, 395.

THE NEXT STEPS: 20 POSSIBILITIES

A. Tough

Ontario Institute for Studies of Education, University of Toronto
252 Bloor Street West, Toronto ON M5S 1V6 Canada

ABSTRACT.   In the field of bioastronomy, at least 20 search
strategies and other next steps are possible at present.   Which of
these are most appropriate at this stage?   Twenty possible strategies
and projects are listed.   They are arranged in five clusters: (1)
develop the field of bioastronomy and its ideas; (2) search beyond
the Solar System; (3) search inside the Solar System; (4) search our
planet; (5) take action to make the contact beneficial. Three
assessments are provided for each strategy: (1) the likelihood of
success if adequate effort and funding; (2) the magnitude of benefits
to humanity if successful; and (3) the likely payoff from greatly
increased effort and resources.   Nine strategies are particularly
high priority but all nine are neglected or at least underfunded at
present.

1. INTRODUCTION

As one contemplates the vast sweep of humanity's future, it seems
likely that contact or interaction with advanced extraterrestrials
will play an incredibly significant role.   This contact could occur
next week, next year, or several hundred years from now.

What can we do now to increase the chances of early contact?
What should we do now to make that future contact as positive and
beneficial as possible for humanity?

At least 20 different strategies and projects are possible
answers to these questions.   They are presented in Table I.
Some of the strategies listed in this table have been frequently
discussed in the bioastronomy and SETI literature; others have been
suggested by only one or two writers.   It is useful at this stage in
the bioastronomy field to contemplate the total array of potential
strategies.   Better choices and priorities may result.

The 20 strategies and projects in the table are arranged in five
clusters: (1) develop the field of bioastronomy and its ideas; (2)
search beyond the Solar System; (3) search inside the Solar System;
(4) search our planet; (5) take action to make the contact
beneficial.   During the next few years, what strategies will most
likely prove highly beneficial?   Which of them are most appropriate

G. Marx (ed.), Bioastronomy – The Next Steps, 397–404.
© *1988 by Kluwer Academic Publishers.*

TABLE I.   TWENTY POSSIBLE STRATEGIES

Twenty possible strategies and projects	Likelihood of success if adequate effort and funding	If successful, how beneficial to humanity	Likely payoff from greatly increased effort
DEVELOP THE FIELD OF BIOASTRONOMY AND ITS IDEAS			
1. Strengthen the infrastructure that will enable the scientific field of bioastronomy to progress	***	***	***
2. Study how biological evolution might occur elsewhere and obtain life-relevant data about planets and their moons in the solar system and beyond	**	*	*
3. For the potential bioastronomy benefits, continue doing good astronomy	**	*	*
4. Study the likely capacities, aims, help, and methods of extraterrestrials	***	***	***
5. Study dolphins and nonhuman primates as ETI analogs	*	*	
6. Develop arrangements for obtaining potentially useful data from military and security agencies	*	**	*
7. Provide opportunities for students and the general public to learn about bioastronomy	***	**	**
SEARCH BEYOND THE SOLAR SYSTEM			
8. Search for detailed messages sent from afar by radio, particle beams, or laser beams	**	**	**
9. Try to detect astroengineering projects, high energy consumption or discharge, unusual coherence, byproducts, or other distant evidence of technological civilizations	**	*	

SEARCH INSIDE THE SOLAR SYSTEM

10. Search for a parked automated
probe that can be triggered to   **   **   **
release a message to us, or that
is (or was) sending data to its
home base
11. Watch for and communicate with
a fast flyby probe that is unable   **
to stop
12. Search for intelligent visitors
and current astroengineering   **   ***   **
projects in the Solar System
13. Search for traces of an earlier
visit to the Solar System   *
14. Try to intercept a virus that
has been encoded with a message,   **
or a spore that could be germinated

SEARCH OUR PLANET

15. Study claims of experiences
with extraterrestrial visitors,   **   **   **
spacecraft, and messages since
1940
16. Seek other evidence of current
(live or automated) surveillance   **   ***   **
or help
17. Study possible evidence of
visits to Earth before 1940   *

TAKE ACTION TO MAKE THE CONTACT BENEFICIAL

18. Send radio messages or other-
wise encourage extraterrestrials   *   ***   *
interact with us
19. Prepare for positive contact
or interaction that is successful   **   ***   **
and beneficial
20. Prepare to handle negative
possibilities (alien bandits,   *   ***   *
hostile warriors, deadly probes)

at this stage? When wrestling with these questions, it is useful to make three assessments for each strategy, as shown in Table I. Although these assessments are necessarily tentative and subjective, they are based on an extensive effort to integrate the literature about likely extraterrestrial reality (Tough, 1986).

The first column shows the likelihood of success if a reasonably adequate level of effort and funding is devoted to the given strategy. Those that are most likely to be successful are marked with three asterisks and the least likely have no asterisks at all. Intermediate levels are marked with one or two asterisks. (I have chosen asterisks in order to make the chart's 60 cells easier to grasp; words such as MEDIUM and HIGH make it hard to gain an overall picture from the chart and numbers are too precise at our present stage.)

If the given strategy is successful, how great will the benefits be for humanity's future? The second column estimates these benefits. Again the highest items are marked with three asterisks and the lowest with none. Rapid two-way communication with advanced extraterrestrials would probably have an enormous and positive impact on our civilization (***). A detailed one-way message that we decipher would also have a huge impact (**). A very brief message that simply reveals the existence and location of one distant civilization would probably have only a minor impact (*) because millions of people (50% of respondents in a recent Gallup poll) already believe that extraterrestrials exist.

Some of the 20 strategies and projects have already been tried, but only briefly and on a fairly small scale, and some have never been tried at all. One or two others have succeeded in obtaining a higher level of support and are making excellent progress.

Unfortunately, human society today is unlikely to support all 20 strategies adequately. Consequently, it is important to compare the priorities of the various possibilities, as shown in the right-hand column. Over the next few years, how large would the global payoff or benefits likely be from greatly increased attention, effort, and resources devoted to each particular strategy? Although we might wish that all 20 could receive adequate funding, the hard reality is that difficult choices will have to be made. The strategies for which additional funding and effort are likely to pay off particularly well are marked with two or three asterisks. The priority in the right-hand column is based on the other two columns and on a forced-choice assumption: if major additional support is available for only some of the 20, which of them would produce the highest payoffs for the field of bioastronomy?

Let me be explicit about one important point: I do not believe any strategy should receive less attention and funding than it now receives. Zero or one asterisk in this column simply means that major additional support will probably not pay off as well for bioastronomy as will additional support for a higher-rated strategy.

Let us now turn to each of the high-priority strategies (two or three asterisks in the right-hand column) in turn. Additional effort and resources for these nine strategies could lead to epic benefits for human civilization.

2. THE NINE TOP PRIORITIES

The first high-priority strategy in the table is to strengthen the infrastructure that will enable the scientific field of bioastronomy to develop, expand, and flourish. Although this field has made great progress in the past 10-15 years, it still lacks its own multi-discipline association and newsletter, its own journal, its own annual meeting, and a single index and abstract source. Ben Finney (1986) has suggested that social scientists and humanists might join together to lay the groundwork for the comparative study of cosmic civilizations and to form the interdisciplinary field of astrosociology. One can foresee the day when several broad-gauge Ph.D. programs and several interdisciplinary research centers will focus on advanced life in the Galaxy.

The next high-priority strategy (#4) is the study of the likely capacities, aims, principles, projects, help, and methods of advanced extraterrestrials. Each search strategy is based on major assumptions about extraterrestrial technology and psychology; enhancing our insight into their motivations and capacities can produce better choices of search strategies. We have to figure out what they are doing before we can figure out how to detect them. A sophisticated and empathic picture of extraterrestrial behavior could be developed through disciplined inquiry. Psychologists, anthropologists, sociologists, futurists, cultural evolution experts, astronomers, and biologists could all play useful roles in the inquiry.

Another high-priority strategy (#7): provide courses, conferences, books, TV programs, perhaps a semi-popular periodical, and other opportunities for students of various ages and for the general public to learn about bioastronomy. Not only will they benefit from the increased knowledge and enlarged perspective, but also they may become more supportive of government and university funding for this field.

Richly detailed messages from other places in the Galaxy may be reaching Earth right now. Detecting such a message is certainly a high-priority strategy (#8). Certain civilizations, at least at one stage of their development, may have broadcast advice, knowledge, techniques, values, ethics, principles of social and political organization, religious beliefs, even instructions for building something. Consequently, we should continue to flexibly check various possibilities within the electromagnetic spectrum. We might also check for messages sent by neutral molecular beams, neutron beams, neutrinos, or other particle beams (Bracewell, 1981) or by laser beams. We should also make arrangements now for sharing and decoding every potential beacon, signal, and message.

It is also quite possible that an intelligent civilization has chosen to send some sort of automated probe to the Solar System. The simplest type of probe or sonde (beyond a fast flyby that is unable to stop) would park or cruise in the Solar System and send data to its home civilization. In addition, a more complex probe might be programmed to release a significant detailed message to any beings who trigger it by approaching it, by directing certain radio waves or laser beams at it, or by achieving an advanced state of technology.

A probe might also be self-replicating (a von Neumann or "Santa Claus" machine) in order to send its progeny to explore other stars. Creative efforts to detect any type of probe is a high-priority strategy (#10).

It is also important to search the Solar System for signs of intelligent beings and their activities (strategy #12). The intelligent life could be extraterrestrials who have come here in a spacecraft, maybe with a propulsion system that we have not yet discovered. Two other possibilities are pointed out by Barrow and Tipler (1986). One possibility is that the intelligent "being" may be some type of supercomputer or twentieth-generation computer combined with robots. "An advanced von Neumann probe would be an intelligent being in its own right, only made of metal rather than flesh and blood" (p. 595). The other possibility is a von Neumann probe that was programmed to synthesize fertilized egg-cells of the living species that sent it and then to raise them to adulthood (p. 580). It could also build a self-sustaining space colony for them to inhabit. We could search the asteroid belt and other parts of the Solar System for signs of a space station, space colony, parked spacecraft, mining operation, materials processing plant, or some large-scale ongoing astro-engineering project. Waste heat could be a sign, for instance, as could evidence that asteroids or comets are disappearing at an artifically high rate (Tarter, 1985). Infrared data, too, could yield valuable clues.

Since the 1940's, a great many people have claimed that they have seen an extraterrestrial spacecraft and even its occupants. Most of these claims, when investigated, turn out to be the result of misperception, an inability to distinguish between fantasy experiences and reality, or even a hoax. Because there is a chance, though, that one or two of the reports are valid, some additional effort should be made to study any promising cases or avenues (strategy #15).

As we saw in the discussion of strategy #12, it is possible that some intelligent beings (or supercomputers) are visiting the Solar System or have been synthesized here. If so, there is a good possibility that they are making some sort of close-up but inconspicuous effort to observe us or even to foster our progress. They may, for instance, have some sort of inconspicuous arrangement in place that enables them to provide instant protection or some other sort of help behind the scenes (Tough, 1986). Strategy #16 suggests that we should think creatively about various possible methods of inconspicuous surveillance and help, and then figure out ways of detecting these methods.

Finally, we should increase our preparations for successful contact (#19). At least six particular steps could turn out to be useful. (a) List various ways in which contact and interaction might occur, and then think through or simulate possible scenarios for each of these. (b) Study the possible consequences of contact, particularly anything potentially negative that could somehow be avoided. (c) Develop and implement arrangements now that will ensure the wide dissemination of information about any beacon, signal, or message within the scientific community, thus making it difficult for the military or the national government of the recipient nation to impound the data (Tough, 1987). (d) Continue formulating international protocol for activities following the detection of a signal. (e) In preparation for replying to a message or for some other sort of interaction with extraterrestrials, clarify globally our goals, priorities, requests, metalaw principles, and strategies. (f) Establish an international team that has global authority to interact and negotiate with extraterrestrials. All six steps are high priority now and should not be postponed until after contact occurs.

3. DISCUSSION

We have discussed the nine strategies and projects that seem highest priority for additional effort and funding. Most of the other eleven strategies in the table are also worth pursuing if resources are available.

Today the field of bioastronomy is at an early stage of development and much remains unknown. Despite enormous strides in this field in recent years, we must still operate with a high level of uncertainty about the number, characteristics, capacities, and motives of advanced beings in our Galaxy. Consequently, at this stage, it would be premature to limit our efforts to one or two strategies while neglecting the other 18. Instead, we clearly need a multi-path approach that emphasizes a variety of strategies. At this stage in its development, the field of bioastronomy needs as much diversity as it can get: the breakthrough may come from some unexpected source. As Thomas McDonough (1987, p. 232) points out, "There is no way to be certain which strategy is best, so the mere diversity of approaches being used by so many people for so many purposes increases the chance that one of them will be the right one." Ronald Bracewell (1981, p. 350), too, has urged us to be open-minded about strategies: "All unorthodox suggestions warrant consideration, even though one might choose not to devote personal effort to following them up, because the future undoubtedly holds discoveries that are not obviously implied by the present state of knowledge, and we should be on the alert for any phenomena that might contribute to the detection of advanced civilizations elsewhere."

The potential gains for human civilization are enormous. We could benefit from knowing some basic facts about intelligent life in the Galaxy, from the enlarged perspective that this new knowledge would bring, and from any practical information that we receive. Compared to the potential payoff for humanity, all nine of the high-priority strategies are underfunded; some are grossly neglected. Effort and resources for all nine should be increased dramatically. As one contemplates the long-term future of human civilization, it is clear that few other investments today have a better chance of providing such significant benefits for future generations.

4. REFERENCES

Barrow, John D., and Tipler, Frank J. (1986). The anthropic cosmological principle. New York: Oxford University Press (Clarendon press).

Bracewell, Ronald N. (1981). 'Manifestations of advanced civilizations.' In John Billingham (Ed.), Life in the universe. Cambridge, Mass.: MIT Press. Pp. 343-350.

Finney, Ben. (1986, October) 'The impact of contact.' Paper IAA-86-471 presented at the Congress of the International Astronautical Federation, Innsbruck.

McDonough, Thomas R. (1987). The search for extraterrestrial intelligence: Listening for life in the cosmos. New York: Wiley.

Tarter, Jill C. (1985). 'Planned observational strategy for NASA's first systematic search for extraterrestrial intelligence (SETI).' In Ben R. Finney and E.M. Jones, Interstellar migration and the human experience. Berkeley: University of California Press. Pp. 314-330.

Tough, Allen. (1986). 'What role will extraterrestrials play in humanity's future?' Journal of the British Interplanetary Society: Interstellar Studies, 39, 492-498.

Tough, Allen. (1987, October). ' A critical examination of factors that might encourage secrecy.' Paper IAA-87-586 presented at the Congress of the International Astronautical Federation, Brighton.

Session 7

WHAT IF WE SUCCEED?

We have been conducting SETI work since 1960 with a steadily accelerating pace. We have already accumulated more than 150 000 hours of SETI, and with the far more sophisticated NASA SETI Program that will be starting in the early 1990's, we are reaching the point where the first contact may occur any time. Obviously this would be a historic occurrence with potentially profound effects on our civilization. But how are we going to react? Would the information be shared with the whole scientific community and the public, or would it be kept as top secret? Are we going to respond to the message received? What are we going to say? Should there be an international authority to handle all these problems?

I. S. Licevitsh *reviewed the problems that different societies have had when they established contacts with older and more advanced civilizations, as has happened in the Pacific, in Africa, in Central and South America. He said that contacts always had selfish motives, therefore he was rather pessimistic about the results of a possible contact.* Andrew Fraknoi *described a very interesting telephone survey of 18 science journalists. He asked: did they think that the first contact is likely to occur in their lifetimes? How did thay think they would hear about it? How will they try to verify it? What criteria will their editor use to decide whether to print it or not? etc. They all agreed that if the media would take it seriously, the public would take it seriously too.* Jean Heidmann *recounted the story a G star with a strong radio emission. Finally work with the VLA showed the radio source being separated from the G star by just 18 arc seconds.* Donald Goldsmith *emphasized that the scientific community must prepare for such an occurrence, because it may be too late to do so after the first signal would have been received. If there were no general accord, there may be different replies. He proposed that the IAU, acting through Commission 51, establish a committee to draft a short reply message.*

Following these presentations, a very lively Panel Discussion was organized by Frank Drake and Jill Tarter. *As we are advancing to more elaborate SETI work, implications from a possible success are begginning to attract the interest of the scientific community as well as that of the public.*

Michael D. Papagiannis

THE TALE OF A CANDIDATE SIGNAL

J.Heidman

Observatoire de Paris

Meudon 92195 France

ABSTRACT. I report on the story of a candidate Seti signal which lasted (and rested) for six years in a corner of my specific scientific work on galaxies. I shall sketch briefly the slow acquisition of observational data, with its ups and downs, and show the influence of chance personal contacts with colleagues and the effect of advances in other research fields. The shift of my interest towards Bioastronomy, triggered by the Patras Assembly, and the sudden appearance of a (not so bright!) idea had their roles also, together with probabilist changes of mood and the realization that I ought not to keep this unsettled candidate for myself. Alas, in the end it was only a nice extraterrestrial adventure...

PROLOGUE

It was March 18, early. A cold , windy, bone-deep wet sad morning was dawning painfully on the gloomy bavarian suburb. We were all waiting in front of the Tourotel, gathered from around the world, as a disparate flock of randomly selected human beings. Some had a confortable thick furcoat, some wear a jacket, some were even just in shirt sleeves, shivering on the freezing sidewalk:
     -"You know, I got here faster than my bag!"
Somehow I managed to sneak by Geoffrey:
     -"Hi! Did you get my letter, about a week ago?"
     -"No, I was already on my way..."
Whoof, here is the bus, splashing dirty drops and puffing nasty smoke. At least we shall get warm, and in time anyway to hear the latest about the cosmic background. Gently rocked on the soft ride, Geoffrey went on:
     -"What was your letter about, Jean?"
Closing on nearer I whispered confidentially in his ear:
     -"Well, I got an anomalous radio emission coming from the direction of a solar type star..."
I had no time to say more; instantly Geoffrey turned his head toward me, his piercing eyes deeply searching for mine and he sputtered eagerly, with an exploding joyful expectation:
     -"DID YOU GET CONTACT ?"

     Is not that a tremendous instant? This kind of excitement

407

G. Marx (ed.), Bioastronomy – The Next Steps, 407–413.

408

already happened to some of us. Was it not so for CTA 102, or when the pulsars were discovered, or when Mars appeared to some to have a system of canals?

THE BACKGROUND

When things started, I was not especially Seti-minded; with the advent in 1965 of the Nançay large decimetric radiotelescope, I entered a ten years work on the investigation of galaxies with the neutral hydrogen 21-cm line, aiming at the determination of their distances in order to arrive at a reliable value of the Hubble constant for the expansion of the universe. On this I wrote a textbook on Relativistic Cosmology and from it a popular one on the Extragalactic Adventure in which there was just a skinny chapter on ET's.
        Then these radio observations lead me and my Italian colleague Catherina Casini to the discovery of "giant clumpy irregular galaxies", galaxies record-high in the production of recently born stars. For this new subject, instead of working mainly on the spot in Meudon with my local group, I shifted to individual international collaborations with top astronomers in charge of new sophisticated instruments: the "clumpies" were far away and faint and I needed the best at all wavelengths.

A QSO?

It is thus that in 1980 I was observing the pair of clumpy galaxies Markarian 7 and 8 with the Westerbork synthesis radiotelescope in collaboration with Hugo van Woerden and Seith Shostak. The two galaxies are separated by 13 arcmin and they fit well in the 50 arcmin field of the radiotelescope. We obtained a map in the continuum radiation just outside the neutral hydrogen line with a 20 arcsec resolution and on it I noticed that a serendip faint 5 mJy source falls on a red stellar image about 15th magnitude of the Palomar Sky Survey (figure 1).
        For a year I did not bother; it could just be a chance coincidence or another quasar. However I had to write down our results for the two galaxies and thus I had to look more carefully at the "QSO". It then appeared that if it was one it would have been one of the brightest known and further a very red one, which was then unheard of. Thus in 1982, as I was working in Japan with Sin'Ichi Tamura, I asked him whether he could take an optical spectrum with the Okayama 1.8-m telescope; his result was clear enough: it was a solar type star spectrum, with no visible redshift.

THE TOP EXCITEMENT...

In view of that I of course got very excited; if the object is a solar type star its apparent magnitude should put it at least at some hundred parsec distance and make it unlikely to be the origin of such an intrinsically strong radio emission: why not an emission from its planetary neighbourhood?

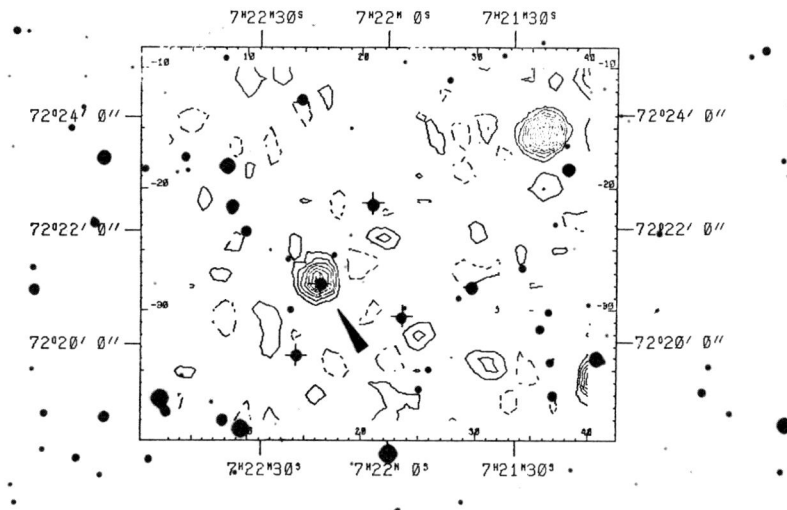

Figure 1: The object region of the Westerbork radio map superposed upon the Palomar Survey red photograph.

I was eager to check this possibility and fortunately the year before, in 1981, I had Markarian 8 observed at the VLA with Dave Heeschen and Qifeng Yin. In particular we had a continuum run made at 20 cm wavelength; however the VLA fields, being much more detailed than the Westerbork ones, we had our map made only for the immediate vicinity of the galaxy, not including the object position. So in 1983 I asked Dave whether he could still retrieve from the VLA tape bank a map of the "star" to see whether it was still on or not at this later time.

AND A BIG FALL

He did and there it was, with the same flux density. Furthermore, as the object was quite on the edge of the field, it was subject to achromatic distorsion; indeed the image was elongated in the radial direction. And this was the end of it for me, at least if one expects an ET radio signal to be quite monochromatic. First the wavelength difference between the Westerbork and the VLA observations was one centimeter. This was not too much of an obstacle because one could admit that in two years time a monochromatic radio signal could have been shifted by 5% in a kind of intentional wavelength scanning. But the fact that the VLA observation, which lasted one hour, filled completely the band width of the receiver (about 3%) meant that the emission at least did not keep in a narrow band during the exposure.

For all practical purpose this ruined my hopes for an ET origin. I even did not have the punch to ask Dave whether he could split the run in two or three separate integrations to get further insight.

BIOASTRONOMY COMES IN

By that time, at the Patras 1982 General Assembly of the International Astronomical Union was created Commission 51 on Search for extraterrestrial life, with announcement for membership. My reason to be in Patras was of course clumpy irregular galaxies for which I organized a Joint Commission meeting on Giant HII complexes outside our galaxy. There I worked out contacts with Japan astronomers in order to get access to their new giant millimetric radiotelescope in Nobeyama to observe my galaxies, which made me also meet Sin'Ichi who later got the spectrum of the "QSO".

However, after that, the sole next sophisticated instrument I aimed at was the Hubble Space Telescope for its high resolution. Somehow I started to feel that the quick spearhead investigation of clumpies was getting to its end. So I eagerly enroled in the new Bioastronomy commission in Patras. Of course this was somewhat of a problem for me: what could I bring in this field so new for me? Nothing! As a joke I told myself I could write a paper on "The catastrophic extinction of galactic life in clumpy irregular galaxies"! As a matter of fact the tremendous amount of UV, X radiation arising from their hyperactivity in star formation and evolution could very well whip out such a life in the process; the main point is to determine the frequency of such a cataclysimic phase in the life of galaxies at large, which is not yet clear.

Anyway commission 51 member I became; then I went to the first Biosatronomy symposium in Boston, then I learned from Michael Papagiannis that asteroids could be nice playgrounds for ET's...

TOO MUCH FOR A G STAR

The next most interesting thing the object could be was a particularly radio active G star. But, not being a star worker I forgot about it for years. However while working in Japan at Kiso observatory in 1985 I was very impressed by its Schmidt plate collection and asked my colleague Hideo Maehara to see whether they had records for the field of the star; maybe optical variations might show...or simply I just wished to glimpse again at my old friend! There were six plates; the star was there, steady, from 1979 to 1982, 30 years after the images from the Palomar Survey.

Thus, no variations. Just for safety I checked about possible proper motion; there was none, showing for sure that the object was not nearby. Taking account of the galactic absorption at large, its true color can be evaluated and indicated a M5 to G0 star 40 to 250 pc away. Adopting 100 pc then seemed reasonable.

If the Sun, in its quiet normal radio state, were at such a distance, its observed flux density at 20 cm would be a million times weaker than the one we had at Westerbork and VLA; i.e. the "star" is a million times more radio luminous than the quiet sun!

This was quite something, so I looked in the litterature. Solar bursts cannot help by more than a factor 100, and usually not for more than half an hour. Some stars were known to have radio flares reaching the observed level, but only sporadically. Certain binaries can have flares for as long as half a day or more but their H and K lines are in emission, which was not evidenced by Sin'Ichi's spectrum.

At this stage of poor observational level I could not go further. And, after all, the case could just be an uninteresting perspective effect, for instance between a far away radiogalaxy and a plain ordinary galactic star. In a paper published in 1981 Jill Tarter and Frank Israel reported that they searched 49 Westerbork fields for Seti or radio stars and that they found no case among 542 stellar candidates. However, in the words of G.A.Dulk in a 1985 review: "Spectacular discoveries have recently been made of quite unexpected radio emission from stars;(...) numerous other stellar radio phenomena have yet to be discovered or properly explored".

AN ET REFLECTOR?

It is in this uncertain but interesting context that I got the (not so) bright idea that the normal quiet radio continuum emission of a normal G star might be amplified by focusing it with some arificial device. In this way the fact that the emission was observed in wide bands could be explained. What was needed is an amplification by a million times; for this the star radio waves emitted in say a steradian had to be focussed in a 3 arcmin beam; this could be done by a mesh of interplanetary size in orbit around the star. For decimetric waves a metallic net with centimeter-size cells would do and its mass was estimated to be in the asteroidal range. And as we do not know of a physical law preventing the transformation of asteroids (such nice playgrounds for ET's, remember Michael) into such a circumstellar device, why not to check further in this direction?

Though I found my idea somewhat foolish I was getting convinced that at this stage I ought not to keep this affair for myself. Using great words: "What if by my carelessness humankind would loose an opportunity of contact?"

HELP!

What I needed was more observations, quite easy but covering different fields. Instead of applying myself for them at different instruments, it was more expedient that a few colleagues would have a quick look in some spare time during their own observing programs. But how to alert them? By a scientific note? It was not scientifically sound enough for that. By a popular article? It would take a long time and get distorted. By personal contacts? This is what I choose; in February 1986 I wrote a letter to 22 first rank investigators whom I know personnally to be nice people interested in the search for life in the

412

universe.
        I received 12 answers within three months, among them a new
spectrum by Derek Wills which confirmed the G type. But I had no answer
for the quick look at the VLA, so important to settle the perpspective
effect case by the accurate position it would provide. It is only in
July that, while I was visiting the splendid romanesque Saint Sernin
church in Toulouse, I fell upon Bernard Burke and asked him how to get
it; "just contact Cam Wade, he is on the spot", he said. And thus in
August Robert Hjellming, who had a run at the VLA, got the accurate
position. Result: the radio source is 5 arcsec away from the star.

THE END OF IT

Thus the perspective effect, the least interesting case, is the outcome
of this extraterrestrial adventure. This is really the end of this tale
of a candidate signal... Now that it is finished, with no result, I am
somewhat reluctant to report on it at a scientific meeting. But I was
invited to, and, after all, this human experience might help to prepare
for the real instant of a Seti success in case it comes.

APPENDIX

Among the colleagues I wrote to for help in this venture there was a
respected sage who advised me to protect my work with a short
scientific note (1). Technical details and references may be found in
it. I take opportunity of the present paper to acknowledge and give
credit for the work made by the colleagues who answered my letter:

1. Derek Wills (Texas University): from a 2.7-m telescope spectrum he
took of the star, it is mid to late G, with strong H and K, G band,
nearly no H Balmer. In his Texas radio survey he has quite a few stars
within 5 arcsec from radio sources and thinks about positional
coincidences.

2. Elisabeth Bozyan (Texas University) measured the position of the
star within 0.3 arcsec on the PSS, notes that the case resembles to
three possible radio stars reported by Windhorst in his dissertation.
Then from inspection of the PSS she notes a quite rich background
cluster, very red, not visible on the O plate; also a few very blue
objects, not on E plate, and suggests the radio source is not
associated with the star but rather with a 21st magnitude blue diffuse
object which could be a blue galaxy in the cluster or a QSO.

3. Seth Shostak (Kapteyn Laboratorium) looked back at his original
Westerbork data reduction and notes a displacemement perhaps larger
than 3 arcsec.

4. François Biraud (Meudon Observatory) and Jill Tarter (Berkeley
University and Seti Institute) observed the object at the Nançay
radiotelescope with the 1 000 50-Hz channels of its Seti system at 1

420.86 MHz and did not get a signal.

5. Woodruff Sullivan (University of Washington) tells me Paul Horowitz (Harvard University) observed the object with his Seti System and did not see narrow band emission; also that Bruce Stephenson checked in the Warner and Swasey Observatory Survey and did not see huge optical emission lines but just a featureless continuum, suggesting a K star.

6. Robert Hjellming (NRAO) obtained with the VLA at 1.49 and 4.9 GHz flux densities 3.5 and 1.0 mJy and a position with accuracy 0.07 s by 1 arcsec which is 5.3 arcsec away from the star.

REFERENCE

(1): J.Heidmann, 1986, 'Emission radio anormale provenant de la direction d'une étoile', C.R. Acad. Sc. Paris, **303**, série II, 47.

THE ARGUMENT FOR ETI IN THE EDO AGE

H. Nakamura

Takaramachi 1-3-2, Katsushika-ku

Tokyo 124 Japan

Edo Age in Japan was isolated, peaceful, self-sufficient, stationary civilization. In this Age, there was scientific speculation regarding the existence of life and intelligence on other planets and bodies not only in our solar system but also most other solar systems. It was "Yume-no-Shiro" [Expense for Dreams] by Banto Yamagata (1748-1821). It was also one of the favorite books of Prime Minister Sadanobu Matsudaira who had heavy censorship and isolation policy.

Prior to the Laplace's hypothesis, there was a original hypothesis concerning the origin of our solar system in Japan. It was "Konton-Bunban-Zusetsu" by Tadao Shizuki (1760-1806). Banto Yamagata was known the heliocentric theory and others by Shizuki's books.

Shizuki was also an opinion leader of an isolation policy for peaceful and self-sufficient stationary society. And Yamagata was a person with conservative view and belonged to the Establishment in fact. There was no sense of colonizing overseas.

The Japanese had given up the gun and colonizing overseas in the second quarter of 17C. (Perrin, 1979). The peaceful and civilized society continued for 230 years. I feel that disarmed stationary society are really an ideal for intelligent species. Is it a suggestion of answer to Fermi's question? If their thought is any guide, the extraterrestrial societies will be stationary civilizations rather than colonize the Galaxy.

However, intelligent people in Edo Age were searching for foreign informations with sharp eyes and pricking up their ears.

REFERENCES

PERRIN, N., Giving Up The Gun: Japan's Reversion to the Sword, 1543-1879, David R. Godine, Publisher, Inc.. (1979).
SHIZUKI, T., Konton-Bunban-Zusetsu, (1793)(1802)., This hypothesis was republished by Naozo Ichinohe, Uchu-Sousei-Shi, Daishokaku (1921).
YAMAGATA, B., Yume-no-Shiro, (1802-1813)., This book has been facsimile reproduced by T. Arisaka, Iwanami Shoten (1973).

G. Marx (ed.), Bioastronomy – The Next Steps, 415.
© 1988 by Kluwer Academic Publishers.

WHAT IF WE SUCCEED?
(A MEDIA VIEWPOINT)

A.G. Fraknoi

Astronomical Society of the Pacific
1290 24th Avenue, San Francisco CA 94122 USA

ABSTRACT. This paper reports on the results of a telephone survey of 29 U.S.-based science journalists about the implications of a SETI success for their work. The reporters and editors were asked about their interest in SETI, what would happen if there were an unambiguous or an ambiguous success, the attitudes of their editors or producers, and the reactions of the public. Some of their opinions should give scientists working in the SETI field interesting food for thought.

1. INTRODUCTION TO THE SURVEY

I believe Jill Tarter asked me to give this paper because — while I am trained as an astronomer — my duties at the Astronomical Society of the Pacific have given me the opportunity to work both with journalists and as a journalist. I have been a source for reporters on a number of astronomical stories, I was co-author of a syndicated newspaper column on astronomy for five years, and I have served as host of a weekly two-hour interview show about science on a San Francisco radio station. Thus I have some knowledge of science journalism from both sides.

While there has been some discussion (and a nightmare or two) among SETI researchers on how to release the news of a success, to the best of my knowledge there has been very little discussion of this subject with the journalists who will bear primary responsibility for disseminating the information to the public.

2. THE JOURNALISTS WHO PARTICIPATED

I interviewed 29 journalists working on newspapers, radio and television, all of them based in the U.S. This sample is incomplete in a number of ways — it includes only English-speaking journalists based in the U.S. and, perhaps more importantly, it includes only journalists who specialize in science news. In these respects, this study may be regarded as a progress report and as a first step in opening up a dialog between

417

G. Marx (ed.), Bioastronomy – The Next Steps, 417–424.
© 1988 by Kluwer Academic Publishers.

scientists and reporters on this subject.

On the other hand, the participants did include people employed in a variety of media, with differing backgrounds and responsibilities. And the people I interviewed are very likely to be among the journalists whom their peers would look to if a major SETI story should break. It will be their articles that are syndicated and their reporting may well set the initial tone of the public discussion.

The reporters and editors I interviewed categorized themselves in the following ways:

* 18 worked for newspapers, 3 for magazines, 4 for wire services, 2 for radio, 3 for television (one person worked for two media, so the sum is 30)
* of the newspaper reporters, 3 worked for major national newspapers, 12 worked for medium to large newspapers, and 3 worked for small newspapers
* 25 said they had done SETI stories, 4 said they had not
* 23 said they followed developments in the SETI field regularly, 2 said they followed them intermittently, and 4 said they did not follow them at all

3. THE SURVEY

In this section I shall give the text of each question I asked in the survey, cite the number of journalists who answered with each choice, and, when appropriate, present a few representative or interesting comments that the question elicited.

3.1 Do you expect that there will be contact with extraterrestrial intelligence (ETI) in your lifetime?

3 said yes, 15 said no, and 11 said "I don't know".

3.2 How do you expect to hear of a first successful SETI effort?

Among the five choices presented to them, 7 journalists picked a press conference, 4.5 said rumor, 2.5 said a telephone call from a public information officer, 2 said a news release, and 2 said a telephone call from a scientist in the field. Among those who specified their own answer, 4.5 said they would receive the news through a wire service, 4 said they had no idea, 2 said they would read about it first in The New York Times, and 0.5 said they would see it on TV. (Many journalists could not decide between two choices, so each was assigned a weight of 0.5.)

A number of journalists commented that this will be a difficult story to embargo or keep quiet and that the temptation for a leak will be very strong.

3.3 <u>Do you think a confirmed SETI success would be front-page news or a lead story</u>?

28 said yes, only one said no. The comments all emphasized that the story was one of the most important (if not <u>the</u> most important) news items of our time.

3.4 <u>Suppose all we have is a non-random string of bits in a radio signal? What criteria would you use to decide if the claimed signal detection is actually that front-page story</u>?

Almost every journalist indicated that, before publishing, he or she would call other researchers in the SETI field. If other well-known researchers took the story seriously, the journalists would be satisfied. A number of journalists said that a lot would depend on what scientists (and from what institutions) made the initial announcement. They pointed out that <u>just such an announcement</u> made by a reputable scientists would be front-page news, especially if the discovery had been discussed with colleagues outside that person's scientific group.

Among the interesting comments made by individual journalists were (I paraphrase them to make the answers succinct):

* I would ask how they can be sure this scenario is different from the discovery of pulsars — how we can be sure it is not a natural phenomenon.
* I would be sure to start the story with "...scientists claim that..." this puts the onus on them rather than on me.
* A lot depends on the competitive situation — i.e. on what other journalists are doing with the story

3.5 <u>How would the criteria used by your editor (or producer) be different from yours? In particular, would they be more skeptical than you or more eager for a sensational story</u>?

13 said their editor or producer would be more eager for a sensational story, 9 said they'd be the same as the reporter, 4 said the editor/producer would be more skeptical, and 3 indicated it would depend on the circumstances.

There was a wide range of comments concerning this question, clearly related to the kind of institution each journalist works for. There is likely to be much less divergence between the reporter and the editor on national science magazines or the large newspapers which can afford science editors, than at TV news programs or small stations where the editor is likely to have no background in science. One reporter for a large (but not national) newspaper summed up the situation many science journalists face very well by saying, "My editor would want the story to be as sensational as ethics permit."

3.6 <u>If you felt the story was pretty solid, whom would you call for com-</u>

ment?

The participants gave two kinds of answers to this question — categories of people and well-known individuals. I imposed no limit on the number of answers each respondent could give.

Among the types of people, 13 mentioned other SETI researchers, 9 mentioned theologians, 8 mentioned political leaders (5 specifically mentioned calling the White House), 4 suggested philosophers, 3 mentioned military leaders, 3 said astronomers in general, and 2 suggested poets.

When it came to specific individuals, over two dozen candidates were mentioned. Those cited most often include Carl Sagan (10 times), Frank Drake (9), Paul Horowitz (4), Bernard Oliver (3), Jill Tarter (3), Isaac Asimov (3), Phillip Morrison (3), and Andrew Fraknoi (3). Several other science fiction writers and some SETI skeptics were mentioned by one person each.

Although many journalists mentioned Sagan, Drake, and other SETI leaders, several cautioned that it would most likely be impossible for reporters to reach them by phone unless special arrangements were made to make them available to the media.

3.7 What sidebars or bounce pieces would be done to go with your story?

A sidebar in a newspaper is a box or secondary story meant to shed further light on a main article. A bounce piece in television is a follow-up story presented right after the main coverage. Here again respondents were allowed to give as many answers as they wished and were not given any suggestions to begin with.

In order of the number of replies received, here are the answers that were suggested by more than one journalist:

* The history of the scientific S.E.T.I. (14)
* Discussion of the equipment used and the technology involved (9)
* "Man in the street" interviews to get reactions (9)
* What the discovery means for humanity (8)
* Profiles of the discoverers (3)
* Earlier predictions for contact in science fiction (3)
* Interviews with science fiction writers on their reaction (3)
* Comments from UFO enthusiasts (3)
* The history of contact between different civilizations on Earth (2)
* How we decode such a message (2)
* What do we do next (2)
* Profiles of the home star of the message senders (2)

I also asked each participant if his or her institution had anything already prepared for this eventuality (the way pieces are sometimes done in advance for obituaries or for major news stories editors know are coming up and predictable). No one had anything prepared or knew of anyone who did.

3.8 <u>Suppose you heard from a reliable scientist source that research group x had found an ETI signal. You call group x and they say they are unwilling to discuss it — they don't deny it, they just won't comment. Your source refuses to let his name be used, since colleagues would be very upset at a leak. What criteria would you use to decide whether to go with the story?</u>

This was a complex, ambiguous question and drew long, thoughtful responses from most of the participants. I can summarize their answers with the following categories:

* It's hard to say; what I would do would depend on the circumstances, the source, group x, the form of their comments, etc. (13)
* I could not go with the story having only one unnamed source (10)
* If I find two sources, both off the record, and group x definitely does not deny the story, I'd go with it (6)
* I would run the story — but very carefully worded — if I thought my (one) source were reliable (6)
* I would run the story if another news organization broke it (2)

In their replies, the journalists stressed that while they might not go with the story right away, they would certainly use the information from the source to put pressure on group x. One reporter said, "I'd use every guerilla tactic I know..." while a TV journalists told me, "I would immediately have a camera crew stake out group x and just wait on their doorstep."

One reporter, working for a major newspaper, indicated that if rumors about such a discovery had been circulating among scientists and journalists, he might "on a protective basis" run a short piece about the rumors, but say that group x has declined to comment. His hope would be that such a story might "flush them out".

3.9 <u>Would you be able and willing to follow this sort of story if it developed slowly — over months and years</u>?

12 of the journalists indicated that they would try to keep up with a developing SETI sucess story in detail, 6 said they would cover major developments only, 4 said they would keep using wire service copy or perhaps interviews with local scientists, and the rest indicated that what they could cover would depend on many circumstances.

The television journalists emphasized that, if such a story developed slowly, they would need lots of interesting tape footage for ongoing coverage. They advised that scientific groups who might anticipate a success put together videotape of their equipment and techniques (very much the way NASA does in the U.S. before a planetary flyby.)

3.10 <u>Some scientists have suggested that, independent of its content,</u>

<u>just the existence of a message will be one of those stories that
changes mankind's image of itself. As a journalist, do you agree</u>?

22 of the respondents agreed, 2 disagreed, and 5 gave indecisive answers. One journalist suggested that such a discovery would be similar
in effect to the famous Apollo program photograph of the Earth from
space. Another thought that the effect of the discovery would be subtle,
"...continuing the historical trend to make us more cosmopolitan in our
thinking." One participant pointed out that there are still many people
who don't believe that humans landed on the Moon, and that there might
therefore be groups who simply don't believe a SETI success no matter
what journalists or scientists say.

4. GENERAL COMMENTS FROM THE JOURNALISTS

Although science journalists in general tend to be comparatively busy
and harried, almost all the participants in the survey found the topic
sufficiently interesting to want to make additional comments during or
after the questioning. I asked them informally if there was anything
they particularly wanted to let SETI scientists know about the topics
of the survey. Many had strong feelings or ideas and a number gave good
advice. Here is an outline of the gist of these comments:

A. This will be a hard story to keep quiet or "manage". "The news will
be out the moment you tell your spouses."

B. Be prepared for a media circus. Help reporters not to panic. Information should be distributed quickly and fairly to all media. (Don't
let the main SETI scientists talk only to the networks and the <u>New
York Times</u>.) Try to make people available to the media through the
sorts of electronic links that the NASA centers provide for important
press conferences.

C. There will be a lot of skepticism in the media and among the public.
The onus will be on the scientists to have reason to be confident
when they make their public announcement. How journalists write the
story will depend on the words and tone the scientists use.

D. Once the story is reported, there may well be hysteria in every direction. "All the nuts will come out of the woodwork." Scientist should
give some thought now to preparing for how their lives will be dis-
rupted. "You will not have a lot of time for reflection." Astronomical societies should set up phone lines and hot-lines to deal with
all the public and media questions and concerns.

E. The public will what to know what the aliens look like, what they
eat, and generally much more about them than we are likely to know
at the beginning. After the initial excitement of the discovery,
be prepared for an "is that all there is" reaction to set in.

5. CONCLUSIONS

No matter how a SETI success occurs, serious thought must be given to the method and timing of the announcement and information flow, to avoid a logistical nightmare. How the public perceives the story will be very significantly colored by how journalists understand and present it. So thinking ahead of time about how to deal fairly and accessibly with the journalists will benefit all SETI scientist in the end.

As worried (and perhaps worrisome) as these journalists sound, they are the cream of the crop in their profession. Journalists with no training or background in science, with no experience in covering scientific topics, will be much more difficult to deal with (and are likely to make up the majority of the reporters covering the story.) The more scientists are prepared with background material, visuals, and concrete plans for giving reporters equal access to principal SETI researchers, the more likely that the story will be presented reasonably in the media.

With an announcement of this magnitude, there will be great temptation on both sides for jumping the gun, leaking information, producing the wrong kind of story. Journalists who have thought about SETI are concerned that scientists will leak selectively to their favorite reporters or will hold back the story unnecessarily. Scientists are, in turn, worried that premature leaks will harm their ability to corroborate and study a message without a media "circus" disrupting their lives. It may be useful for both groups to share their concerns with each other before a message is found.

The message of this survey for scientists is that they must begin to plan for how to deal with the public aspects of a SETI success -- how to organize the necessary discussion with colleagues without leaks to the media, how to release the news of the discovery fairly and broadly, and how to assist in making sure that the flow of information to the public is as accurate and responsible as possible.

****

APPENDIX: THE JOURNALISTS WHO PARTICIPATED:

Peter Aleshire, Science/Medical Writer, Oakland Tribune
Ian Anderson, Correspondent, New Scientist and the Australian Broadcasting
    Corporation
David Ansley, Science and Medicine Editor, San Jose Mercury-News
Bailey Barash, Producer, Science News, Cable News Network
Howard Benedict, Aerospace Editor, Associated Press
Robert Cooke, Science Writer, Newsday (Long Island, New York)
Robert Cowen, Science Editor, Christian Science Monitor
Keay Davidsen, Science Writer, San Francisco Examiner
Victor Dricks, Science Writer, Phoenix Gazette

Lee Dye, Science Writer, <u>Los Angeles Times</u>
Jonathan Eberhart, Space Sciences Editor, <u>Science News</u>
Anne Gibbons, Science Writer, Peninsula Times-Tribune (near San Francisco)
Gayle Golden, Science Writer, <u>Dallas Morning News</u>
Richard Harris, Science Reporter, National Public Radio
Carle Hodge, Science Writer, <u>Arizona Republic</u>
Robert Hone, Producer, KQED-TV (San Francisco)
Richard Kerr, Writer, <u>Science</u> magazine
Robert Locke, Science Editor, <u>San Diego Tribune</u>
David Lore, Science Reporter, <u>Columbus Dispatch</u> (Ohio)
Charles Petit, Science Writer, <u>San Francisco Chronicle</u>
Malcolm Ritter, Science Writer, Associated Press
Richard Saltus, Science Writer, <u>Boston Globe</u>
Lee Siegel, Science Writer, Associated Press
Bill Skane, Science and Medical Producer, CBS Evening News
Byron Spice, Science Writer, <u>Albuquerque Journal</u>
Mike Toner, Science Editor, <u>Atlanta Journal & Constitution</u>
John Wilford, Science News Reporter, <u>New York Times</u>
Hill Williams, Science Reporter, <u>Seattle Times</u>
Patrick Young, Chief Science & Medical Correspondent, Newhouse News Service

WHO WILL SPEAK FOR EARTH?

D. Goldsmith

Interstellar Media

2153 Russell Street, Berkeley CA 94705 USA

ABSTRACT

I propose that the scientific community, and specifically the International Astronomical Union, prepare a draft reply message to respond to a verified signal from another civilization. In my opinion, if the scientific community hopes to have its voice heard amidst the clamor that will follow detection of an extraterrestrial signal, it must reach a consensus prior to the detection rather than attempting a reasoned discussion afterwards.

INTRODUCTION

Recent advances in SETI, which have taken it from a poor step-child of science to an ongoing research effort, promise significant improvement within the next few years--improvement that may well yield the first verified detection of radio signals from another civilization. Although it is impossible to predict with certainty the point in our development of SETI techniques when we can expect such a verified detection, our knowledge of our own culture implies that this detection will provoke great confusion as to what, if anything, should be done in response. Furthermore, it seems equally evident that a reply agreed upon in advance, even if by a small segment of our society, will be more seriously considered than a reply presented for consideration only after the detection of an extraterrestrial signal. Although it is far less certain what the civilization that sent the signal may think of replies that arrive there, it seems reasonable to speculate that the best chance of provoking a two-way dialogue lies with those replies that are clear, intelligible, and interesting to the other civilization, rather than weak, garbled, and filled with evidence of confusion in the sender.

DISCUSSION

A successful SETI search at radio frequencies will yield, in addition to evidence of another civilization's radio transmission, the frequency or frequencies of the transmission and the power of the received signal. The latter quantity, together with an estimate of the distance of the sending civilization, will yield the power of the broadcast signal. Thus simply by detecting the signal, we shall have a good estimate of the appropriate frequency and power of any signal that we might send in reply

425

G. Marx (ed.), Bioastronomy – The Next Steps, 425–428.

Since any signal detected within the foreseeable future is likely to arise within the Milky Way, from a distance of 10 to 10^4 parsecs, the signal will have been broadcast by a civilization at a time not immense in comparison to a human lifetime or to what we call a civilization's lifetime on Earth. This fact implies that we can have a reasonable hope that the civilization whose signal we detect will still be in existence (perhaps in modified form) at the time of our detection, and likewise will still exist at the time that any signal that we send could be detected. Human nature makes it evident that a large fraction of the human population, and of the scientific community, will favor entering into a dialogue with the civilization that we detect, if that is possible. (The minority view will be considered below.) If we do seek to attract the attention of the civilization whose signal we detect, it is likely that we shall be competing for their attention with other civilizations that exist in the Milky Way, perhaps at greater distances from us than the civilization in question, and with which an ongoing dialogue may already exist. In short, we may be less interesting than we imagine to a civilization whose radio signals we intercept. This fact implies that some consideration ought to be given to the preparation of a reply message, simply because a "poorly" prepared message may fail to elicit further communication but a "well" prepared message would succeed in doing so.

What will distinguish a "poorly" prepared from a "well" prepared message? One general principle to achieve such discrimination would turn on clarity: If a reply message arrives with low power and is difficult to understand (indeed difficult to discern as a message rather than natural background noise), it is less likely to attract attention, especially favorable attention, than a reply message with higher power and greater comprehensibility. In addition, a vaguer category of "interest" will be used to rate reply messages by a civilization that has received several or many of them. In view of the difficulty of assessing the psychological makeup of extraterrestrial cultures, it is not easy to determine just what would make a message "interesting." However, unless we are dealing with a civilization seeking relatively weak civilizations to conquer, it appears that a message that reveals fundamental uncertainty on the part of a civilization (e.g., whether sending a reply is a good idea) is less likely to attract favorable attention than a message that shows that a civilization has a basic interest in maintaining a dialogue (and, still more likely to attract attention, has something to offer in doing so).

The above considerations suggest that a reply message most likely to succeed will be one that is clearly detected, easily understood, and indicative of a reasonably well-ordered civilization. If we were to detect an extraterrestrial civilization through signals that were conflicting, muddled, and fraught with local interference, we would certainly be less eager to continue a dialogue than

we would with a civilization that broadcast a single signal, free from
local interference, and relatively straightforward to understand.
Even though we should be slow to generalize from our own experience,
the conclusions drawn above appear reasonable; as always, they imply
that we are more likely to establish contact with civilizations more
like ourselves than less like ourselves (at least at first).

If the concept of preparing a single, clear reply, as interesting
as we can make it, seems a good one, still even a cursory knowledge of
our own culture makes it clear that obtaining agreement on what message
fills the bill will be difficult. Verified detection of an extraterrestrial signal will provoke a world-wide uproar immediately upon
its announcement. In my opinion, such announcement will follow the
detection by less than 48 hours; to keep such a secret is simply not
possible within human society as presently structured. The uproar
will focus on the questions (1) What do the aliens look like? and
(2) Are we in danger of imminent attack? Once these questions are
satifsfactorily answered, attention will focus on the question of
whether we should send a reply and, if so, what should be the content
of that reply. The Earth-centeredness of our society will never be
so apparent as in the public debate over the question of content.

Paradoxically, from a technological standpoint the question of
content barely exists. Even if the arriving signal has such a narrow
bandwidth that it carries little information, it is a simple matter
in the reply message sent at the same frequency to use a larger bandwidth and to direct the recipient's attention to that fact. Therefore,
to send hundreds of millions of bits of information within a few
seconds becomes a straightforward and inexpensive matter. Since
this number of bits can include enough information about our society
to satisfy nearly anyone's desires to include some arcane matter not
of general interest on Earth (let alone elsewhere in the Milky Way),
the difficulty in preparing a reply that will please most people
consists of the operational problem of obtaining general agreement
and of the lust for censorship that will exist in various segments
of our culture. The first problem can be solved by obtaining a
consensus--among scientists prior to detection of an extraterrestrial
signal; in a broader context either (preferably) prior to or after
such detection. The second problem, that of attempted censorship,
is likely to take such forms as attempts on the part of organized
religions to present their theology as human truth, or on the part
of governments to engage in a similar exercise.

Leaving aside the details of our cultural response to the concept
of sending a reply message, it is clear that a reply message agreed
upon in advance, if only by a relatively small (if influential)
segment of society will fare better than suggested reply messages
drafted only after the detection of the verified signal. Although
it may appear slightly arrogant for the scientific community to
take it upon itself to draft such a reply (in a better-organized

society, this would fall to the governmental committee in charge of
SETI projects), given the present state of our culture it appears
reasonable--at least to scientists--that the initial draft of a reply
message emerge from scientific discussion. The International Astronomical Union, and our Commission 51, are well placed to produce the
first draft of a reply message, eventually to be considered by world
organizations whose stamp of approval would be remembered and would
still be significant (one may hope) within the next few years, in the
event that a verified signal is detected within that time.

My goal is simply to start that process in motion as best I can.
A debate over what to say to another civilization may be rather
sterile as to the actual message content, as noted above; nevertheless,
it will be immensely useful as a means of increasing public awareness
on the SETI issue, of identifying the chief currents of debate (for
example, those who seek to assure that no reply would be sent might
be pressured to express their views and thus, in my opinion, to expose
them as not worthy of much attention), and, most important, of moving
toward a situation in which the public not only accepts SETI as a
worthwhile scientific endeavor but also recognizes that SETI is more
than another means to amuse ourselves: It is our potential entrance
ticket into the galactic flow of ideas.

The Flag of Earth is present in each SETI
institution: in a black sky the gonden Sun
(right), the blue Earth (middle), the silver Moon (left up). Holding the flag:Frank
Drake (chairman of the IAU commission),
on his right George Marx (vice-chairman)
and Michael Papagiannis (secretary). Picture made by Kyong Chol Chou.

INDEX

432